Pest Control: Operations and Systems Analysis in Fruit Fly Management

NATO ASI Series

Advanced Science Institutes Series

A series presenting the results of activities sponsored by the NATO Science Committee, which aims at the dissemination of advanced scientific and technological knowledge, with a view to strengthening links between scientific communities.

The Series is published by an international board of publishers in conjunction with the NATO Scientific Affairs Division

A Life Sciences **B Physics**	Plenum Publishing Corporation London and New York
C Mathematical and Physical Sciences	D. Reidel Publishing Company Dordrecht, Boston and Lancaster
D Behavioural and Social Sciences **E Applied Sciences**	Martinus Nijhoff Publishers Boston, The Hague, Dordrecht and Lancaster
F Computer and Systems Sciences **G Ecological Sciences**	Springer-Verlag Berlin Heidelberg New York Tokyo

Series G: Ecological Sciences Vol. 11

Pest Control: Operations and Systems Analysis in Fruit Fly Management

Edited by

Marc Mangel

Departments of Agricultural Economics, Entomology and Mathematics
University of California, Davis, California 95616, USA

James R. Carey

Department of Entomology
University of California, Davis, California 95616, USA

Richard E. Plant

Departments of Entomology and Mathematics
University of California, Davis, California 95616, USA

Springer-Verlag Berlin Heidelberg New York Tokyo
Published in cooperation with NATO Scientific Affairs Division

Proceedings of the NATO Advanced Workshop on Pest Control: Operations and Systems Analysis in Fruit Fly Management held at Bad Windsheim, Germany, August 5–9, 1985

ISBN-13:978-3-642-70885-5 e-ISBN-13:978-3-642-70883-1
DOI: 10.1007/978-3-642-70883-1

Library of Congress Cataloging in Publication Data. NATO Advanced Workshop on Pest Control: Operations and Systems Analysis in Fruit Fly Management (1985 : Bad Windsheim, Germany) Pest control. (NATO ASI series. Series G, Ecological sciences ; vol.11) "Proceedings of the NATO Advanced Workshop on Pest Control: Operations and Systems Analysis in Fruit Fly Management held at Bad Windsheim, Germany, August 5–9, 1985" – T.p. verso. "Published in cooperation with NATO Scientific Affairs Division."
1. Fruit-flies – Control – Congresses. 2. Fruit-flies – Control – Simulation methods – Congresses. 3. System analysis – Congresses. 4. Operations research – Congresses. I. Mangel, Marc. II. Carey, James R. III. Plant, Richard E. IV. North Atlantic Treaty Organization. Scientific Affairs Division. V. Title. VI. Series: NATO ASI series. Series G, Ecological sciences ; no. 11. SB945.F8W67 1985 632'.774 86-6697
ISBN-13:978-3-642-70885-5 (U.S.)

Softcover reprint of the hardcover 1st edition 1986

2131/3140-543210

TABLE OF CONTENTS

Preface

These are the proceedings of an Advanced Research Workshop (ARW), sponsored by the NATO Science Panel, entitled "Pest Control: Operations and Systems Analysis in Fruit Fly Management". The ARW was held in Bad Windsheim, Germany during the week of 5 August 1985. The purpose of the ARW was to bring together scientists who are interested in fruit fly problems, but who usually do not have an opportunity to speak with each other, for an intense week of interdisciplinary collaboration. In particular, the group present at the ARW contained a mix of biologists, field ecologists, mathematical modellers, operational program managers, economists and social scientists. Each group has its own professional meetings at which fruit fly problems are discussed, but the point of the ARW was to learn about the problem from the perspective of other fields, which are equally important for the ultimate management of the fruit fly problems. (A list of attendees follows this preface.) It appears that the ARW successfully met its objective of bringing together a group for interdisciplinary considerations of the problems; I hope that the proceedings do as well.

The ARW was structured with formal lectures in the mornings and workshops in the afternoons. For the morning lectures, four different topics were chosen: 1) basic biology and ecology, 2) trapping and detection, 3) control and eradication, and 4) policy issues. Each morning, one lecture from each area was presented. (The following general principle was adopted by the organizing committee: since no one at the ARW would be recognized as a "big shot" by everybody -- because of the interdisciplinary nature of the workshop -- all individuals were treated equally. That meant, for example, that when more people wanted to give morning lectures than were slots available the speakers were chosen randomly. Everyone who attended the ARW was asked to contribute a paper to these proceedings, however.

The purpose of the morning lectures was to provide a formal, overview of the problems of fruit fly management from different perspectives. The afternoon workshops were dedicated to intense, small group collaboration. For the first workshop, four subgroups were formed to consider problems of biology, trapping and detection, control and eradication, and policy. These subgroups met individually for about 2-1/2 hours and then reported on their deliberations to everyone. The second workshop was dedicated to a "fruit fly war game", patterned after the war game simulations so commonly used by military systems analysts. The third workshop was dedicated to a discussion of resistance to the sterile insect technique (SIT); this topic arose somewhat spontaneously from the group and was not planned at all by the organizing committee. The fourth and fifth workshops were dedicated to an assessment of the state of the science of fruit fly management and the determination of future research needs.

These proceedings are organized as follows. First, the research recommendations are presented. The next three sections contain the formal papers in sections on policy, economics and operations, biology, and modelling. The last section contains a description of the workshop proceedings. I thank the NATO Science Panel for having taken a chance in sponsoring such a unique, interdisciplinary meeting. I believe that their gamble paid off. I thank the organizing committee members (J. Carey, R. Hilborn, G. Norton, R. Plant) and D. Chambers for their help in selecting the attendees at the ARW. Cathy Haider and Ronda Newton did super jobs of typing under deadlines. The Departments of Entomology and Mathematics at the University of California, Davis generously provided staff support and facilities. Most of all, I once again thank each attendee at the ARW for working so hard to make it a success.

Marc Mangel
Director of the ARW
University of California
Davis, California
December 1985

LIST OF ATTENDEES

Martin Aluja-Schunemann, Dept. of Entomology, Ternard Hall, Univ. of Massachusetts, Amherst, Massachusetts 01003, U. S. A.

James Carey, Dept. of Entomology, University of California, Davis, California 95616, U. S. A.

Carrol O. Calkins, Insect Attracts, Behavior and Basic Biology Research Lab., U. S. Department of Agriculture, Post Office Box 14565, Gainesville, Florida 32604, U. S. A.

Gerlad Carlson, Dept. of Economics, North Carolina State University, Raleigh, North Carolina 27695-8110, U. S. A.

Ugo Cirio, ENEA-Laboratory Technologies for the Protection of Agroecosys., C. R. E./Casaccia, CP 2400, 0100 Roma, ITALY.

Roy T. Cunningham, USDA/ARS, Post Office Box 4459, Hilo, Hawaii 96720, U. S. A.

Domiten Debouzie, Lab. Biometri, Univ. Lyon 1, 69622 Villeurbanne, Cedex, France.

L. Demetrius, Max Planck Institute, Goettingen, WEST GERMANY.

Robert V. Dowell, Ph.D., Department of Food and Agriculture, Division of Plant Industry, Analysis and Identification, 1200 "N" Street, Sacramento, California 95814, U. S. A.

A. P. Economopoulos, Entomology Unit/Seibersdorf Lab, IAEA, Wagramerstrasse 5, Post Office Box 200, A-1400 Vienna, AUSTRIA.

Richard Farnsworth, ERS/NRED, 1301 New York Avenue, Room 424, Washington, D. C. 20005, U. S. A.

Brian Fletcher, CSIRO Division of Entomology, 55 Hastings Road, Warrawee, Sydney, NSW 2028, AUSTRALIA.

Wayne M. Getz, Division of Biological Control, University of California, Berkeley, California 94720, U. S. A.

Vicenzo Girolami, Instituto Entomologia, Universita Di Padova, Via Gradenigo 6, 35100 Padova, ITALY.

George Haniotakis, NRC "Demokritos", Dept. of Biology, Post Office Box 60228, Athens, GREECE.

Glenn Hawkes, Dept. of Applied Behavioral Sciences, University of California, Davis, California 95616, U. S. A.

Gordon Hooper, Dept. of Entomology, University of Queensland, St. Lucia, Queensland 4067, AUSTRALIA.

Philip Howse, Chemical Entomology Unit, Building 3, The University, Southampton, UNITED KINGDOM.

Marc Mangel, Dept. of Mathematics, University of California, Davis, California 95616, U. S. A.

A. Meats, School of Biological Sciences, University of Sydney, Sydney, NSW 2006, AUSTRALIA.

Harald Michaelis, LS Tieroekologie, University of Bayreuth, 8580 Bayreuth, WEST GERMANY.

Geoff Norton, Silwood Centre for Pest Management, Imperial College, Silwood Park, Ascot SL5 7PY, UNITED KINGDOM.

Richard E. Plant, Department of Mathematics, University of California, Davis, California, 95616, U. S. A.

Alan S. Robinson, Res. Inst. ITAL, Post Office Box 48, 6700 AA Wageningen, THE NETHERLANDS.

Yoram Rossler, Israel Cohen Institute for Biological Control, Citrus Marketing Board, 27 Keren Kayemet Street, Rehovot 76345, ISRAEL.

Jerry Scribner, 6048 Kenneth Avenue, Carmichael, California 95608, U. S. A.

Masakazu Shiga, Fruit Flies Laboratory, Okinawa Prefectural Agricultural Experiment Station, 4-222, Sakiyama-cho, Naha 903, JAPAN.

Tadashi Teruya, Fruit Flies Laboratory, Okinawa Prefectural Agricultural Experiment Station, 4-222, Sakiyama-Cho, Naha 903, JAPAN.

Shripad Tuljapurkar, Graduate Group in Demography, 2234 Piedmont Avenue, University of California, Berkeley, California 94720, U. S. A.

Pauline van den Driessche, Dept. of Mathematics, University of Victoria, Victoria, British Colombia, CANADA, V8W 2Y2.

Roger J. Wood, Dept. of Zoology, Manchester Univeristy, Manchester M13 9PL, United Kingdom.

I. RESEARCH RECOMMENDATIONS

RESEARCH RECOMMENDATIONS BY SCIENTISTS AT THE NATO ADVANCED RESEARCH WORKSHOP "PEST CONTROL: OPERATIONS AND SYSTEMS ANALYSIS IN FRUIT FLY MANAGEMENT" (5-9 August 1985; Bad Windsheim, Germany)

Fruit flies are an international pest complex, causing widespread crop destruction in both developed and developing countries. Recent outbreaks of medfly in California costs aver $100 M to eradicate an ongoing fruit fly control programs in Mexico, Greece, Okinawa, Guatemala, Israel and Italy cost millions of dollars each year.

A group consisting of field laboratory biologists, mathematical modellers, economists and policy scientists, and operational program managers from around the world met in Bad Windsheim, Germany in August 1985 to discuss the fruit fly problem from the perspective of interdisciplinary research. This group identified three major problem areas which are impending further progress in fruit fly control and eradication. These are:

- Inadequate basic biological data on fruit fly behavior and ecology.

- Outmoded and/or inefficient control or eradication technologies and strategies.

- Lack of understanding of the political, social, and economic constrainings on effective control or eradication strategies.

<u>For these reasons, research is urgently needed in the following areas (not ranked according to priority):</u>

- Development of methodologies for the assessment of potential losses, effectiveness and costs/benefits of different control and quarantine policies in specific situations.

- Understanding of the public decision making process in response to pest control emergencies.

- Means for monitoring the effectiveness and environmental impact of an ongoing control or eradication program.

- Determining the efficiency and predictive qualities of traps.

- Phermones for trapping, sexing, and control of fruit flies.

- Development of alternative insecticides, improved baits, and approved application strategies.

- Quality control and field effectiveness of flies produced for SIT.

- Genetics of fruit flies with respect particularly to the development of a sexing technique.

- Relative effectiveness of male only releases in sterile insect programs.

- Ecological constraints upon the colonization of a new environment.

- Population biology of fruit flies in the field.

- Evaluation of the potential of cultural and biological control.

- Species interactions between fruit flies and the taxonomy of species complexes.

- Field testing of mathematical models.

- Nutritional studies on fruit flies.

- Integration of fruit fly control into orchard management.

The following specific proposals are made:

THAT SUCH INTERDISCIPLINARY MEETINGS AS THIS ONE BE HELD ON A REGULAR BASIS. The present meeting has provided a unique opportunity for mathematicians, biologists, economists, psychologists, demographers, operational pest control experts, and decision makers to meet together to analyze problems in fruit fly pest control. The opportunity for an exceptionally wide exchange of view on pest control has clearly indicated areas for interdisciplinary action in the future. Such action needs to be monitored and reviewed.

THAT A GENETIC STOCK CENTER FOR FRUIT FLIES BE ESTABLISHED. Genetic studies are inhibited at the present time by quarantine restrictions. A stock center is required and would best be located in a temperate country at a center already active in fruit fly research. Such a center will coordinate genetic and cytogenetic studies of behavior and will facilitate the development of a genetic sexing system.

II. POLICY, ECONOMICS, AND OPERATIONS

PERSPECTIVES ON FUTURE INTEGRATED MANAGEMENT OF FRUIT FLIES IN MEXICO

M.R. Aluja
Department of Entomology
University of Massachusetts
Amherst, MA 01003/USA

P.F. Liedo
Programa Mosca Med
Apartado Postal 576
30700 Tapachula, Chiapas, Mexico

SUMMARY

We analyze the extremely complex problem of fruit fly management in Mexico suggesting the establishment of a long term, well planned and flexible country-wide orchard management program based among others on the following premises: a consideration of the entire fruit fly species complex instead of the common practice of singling out Anastrepha ludens as the unique source of problem; a multistrategy approach in which measures to control fruit flies are integrated with other pest and disease control efforts and all the other agronomic practices used in the orchard; a redirection of the control efforts putting most emphasis in the planting and orchard maintenance phase as opposed to the harvest and marketing phases; the development of novel approaches and the implementation of established control strategies based on ecologically sound principles and that are within the cultural and economic reach of the recipient (more than 70% of the fruit produced in the country comes from small scale, resource poor farmers); a biogeographical division of the fruit growing regions, with the application of management practices adapted to each particular situation; a thorough understanding of the socio-economic, socio-political, cultural and historical milieu of the farmer; the enhancement of alternative means of fruit commercialization through the creation of agroindustries and the enhancement of strong grower associations.

1. GENERAL NOTIONS ABOUT FRUIT PRODUCTION IN MEXICO

Fruit production plays an important role in Mexico's economy. Out of approximately 19,000,000 Ha devoted to agriculture, 10% or approximately 1,900,000 Ha are planted with fruit trees, producing approximately 30% of the total value of agricultural products (DGEA, 1983).

Due to the extreme diversity of the geography and environment, a great variety of fruit species are cultivated or grow in the wild.

NATO ASI Series, Vol. G11
Pest Control: Operations and Systems Analysis
in Fruit Fly Management. Edited by M. Mangel et al.

Figure 1 shows the planted surfaces of the most common fruit types commercially grown in Mexico. Note that citrus, mango, apple, peach, guava and papaya, in that order, cover the highest percentage of the planted surface.

Pests and diseases cause important losses in yield and quality of fruit, a situation that has been exacerbated by the dramatic increase of fruit growing areas in the past fourteen years. As Figure 2 indicates, there has been a six-fold increase in the production areas of mangoes, a four-fold increase in the case of peaches and apples and a two-fold increase in the case of papaya (DGEA, 1983).

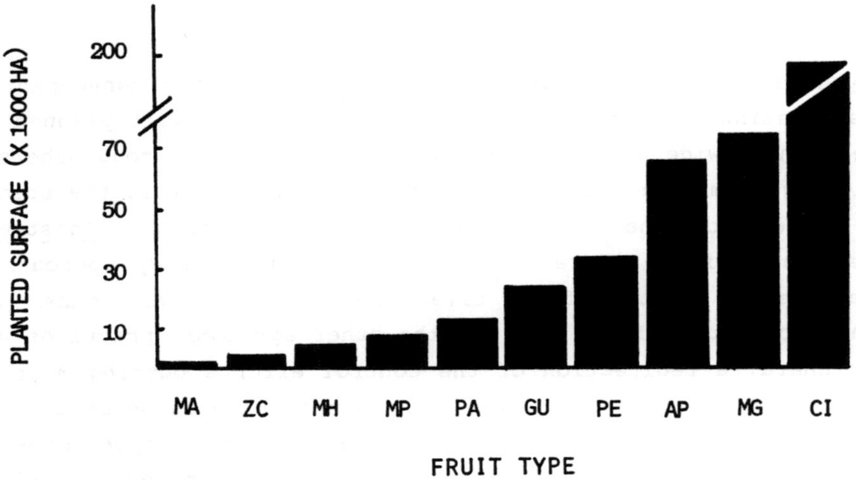

Fig. 1 - Planted surfaces of the most common fruit types commercially grown in Mexico. MA = Mamey (Calocarpum zapota); ZC = Zapote Chico (Achras zapota); MH = Tejocote (Crataegus mexicana); MP = Jobo (Spondias sp.); PA = Papaya (Carica papaya); GU = Guava (Psidium guajava); PE = Peach (Prunus persica); AP = Apple (Pyrus malus); MG = Mango (Manguifera indica); CI = Citrus (Citrus sp.).

Figures 3 and 4 show the location of the most important fruit-producing areas in Mexico. Citrus is grown in three distinct regions. The coastal desert of the state of Sonora, the subtropical and tropical regions of the states of Nuevo Leon, Tamaulipas, San Luis Potosi and Veracruz, where local mezquite grassland, tropical deciduous forest, thorn forest and tropical evergreen forests are replaced by citrus orchards. There is also a newly developed area in the Yucatan Peninsula, where tropical deciduous and evergreen forests are converted into agricultural land. (See Appendix I for a map of Mexico indicating the state location and name).

Mango is grown principally along both the Pacific and Gulf coasts in habitats where tropical subdeciduous, deciduous and evergreen forests plus thorn forests are exploited. It is important to note that Spondias sp., Jobo or Mexican Plum (Anacardiaceae), an important alternate host of Anastrepha obliqua, is also grown in the same regions on a semi-commercial basis.

Apples, peaches and tejocote or Mexican hawthorns, have been traditionally grown in the highland areas, where pine-oak forests are the dominant vegetation; in the case of apple there is a new effort to introduce drought-resistant varieties from Israel into the Sonoran desert.

Guava, although found throughout the country, is commercially grown in the central states of Aguascalientes and Zacatecas, where both mezquite-grasslands and pine-oak forests are the native vegetation.

Papaya is grown in essentially the same areas as mango. Finally, there are the various fruit species in the family Sapotaceae, which are grown in the Southeastern rain and tropical evergreen forests and the central state of Mexico in tropical deciduous forests.

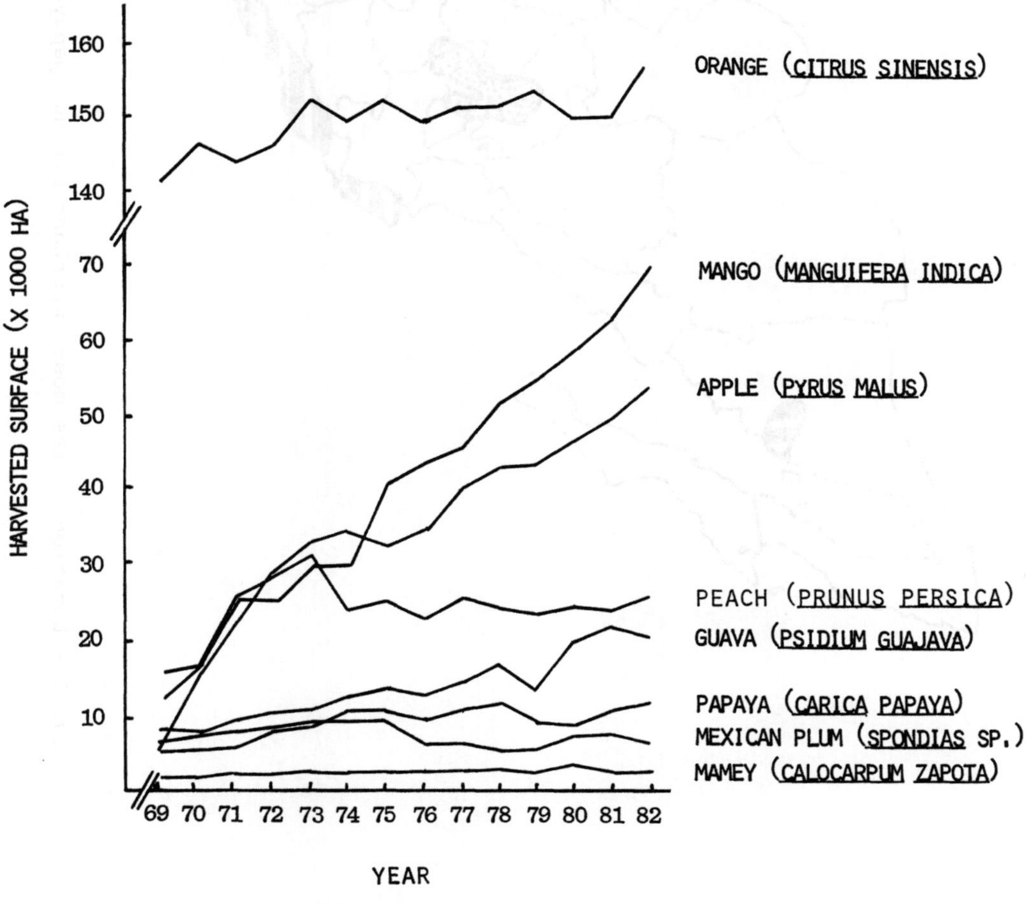

Fig. 2 - Harvested surfaces of the most common fruit types commercially grown in Mexico between the years 1969 and 1982.

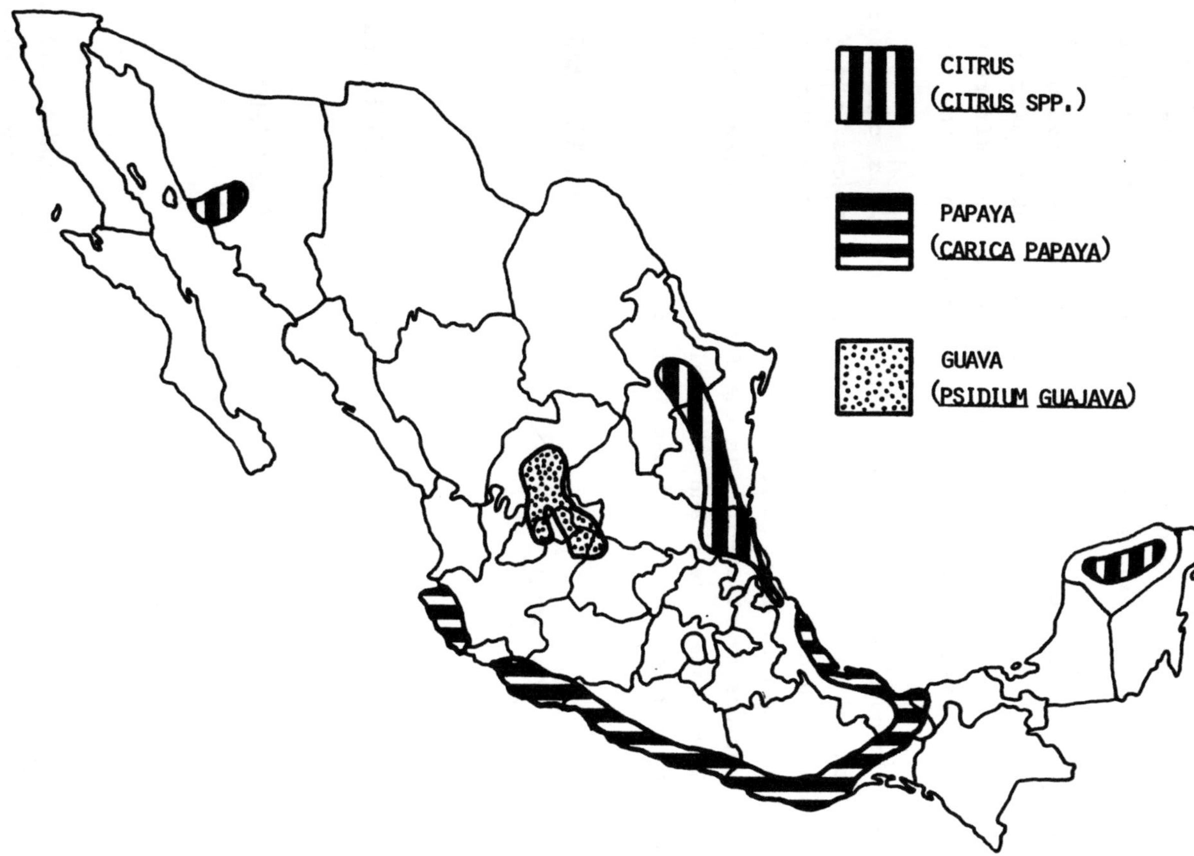

Fig. 3 - Location of the most important citrus, papaya and guava producing areas in Mexico.

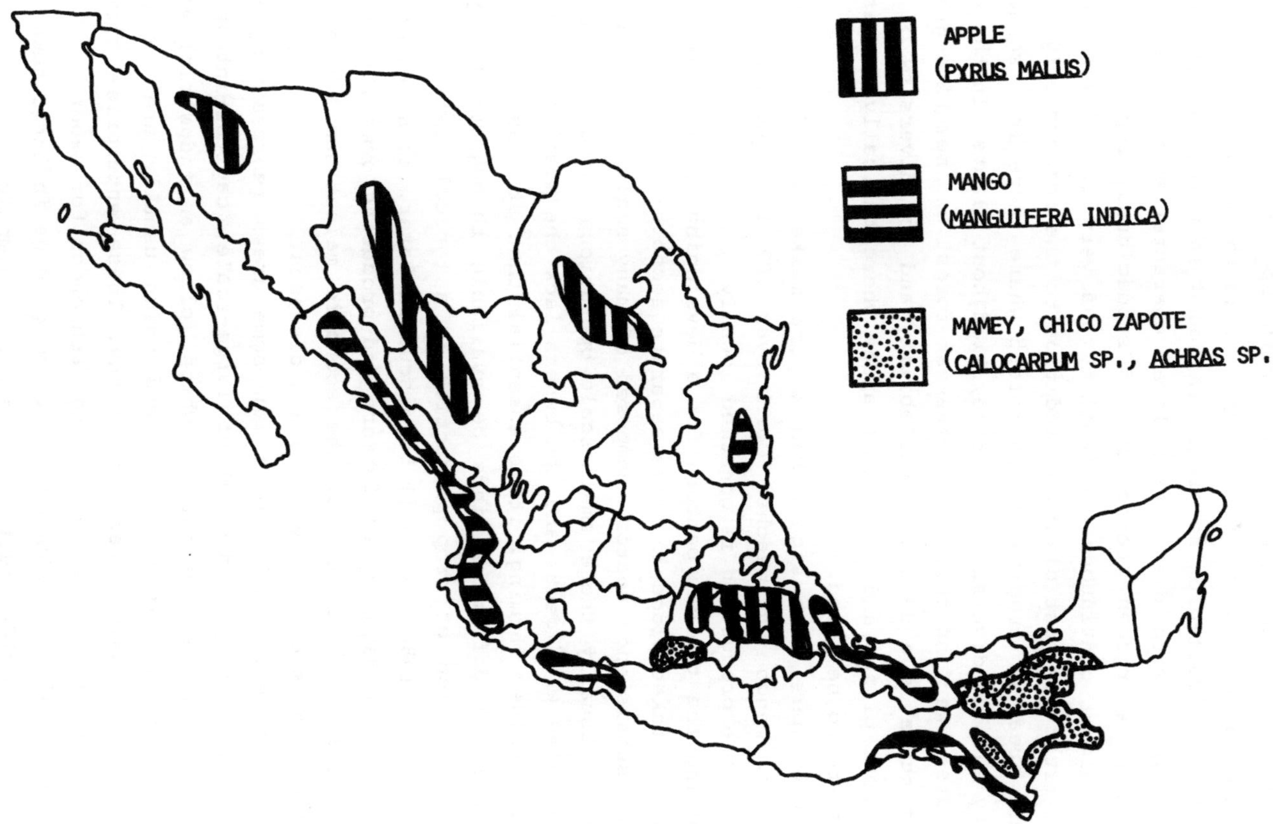

Fig. 4 - Location of the most important apple, mango, mamey and zapote chico producing areas in Mexico.

These regions are the places where these fruits are grown on a commercial basis. In addition, many fruit species are also found throughout the country in backyard gradens or in the wild.

In summary, fruit-growing areas are found in quite variable biogeographic regions with distinctive local characteristics. These include, tropical evergreen, deciduous and subdeciduous forests, where there is an almost continuous availability and very high diversity of native wild fruit fly host plants in addition to the commercially grown fruit types; areas of tropical thorn forest where there is a seasonal availability and low diversity of native, wild host plants. In the temperate climates, we find regions of pine-oak forest and mezquite-grassland, where there is a seasonal availability and low diversity of native, wild host plants and the desert areas where basically no native host plants are to be found.

Although there is a significant export market for fresh fruit, namely citrus, mango and peaches (well over 90% is exported to the USA), most of the production is consumed locally.

Following the typical pattern for a developing country, Mexico's crop production takes place in three separate sectors:

1) the subsistence sector, composed of many small, resource-poor farming units. Usually these small-scale operations cannot afford inputs, are riddled with debts and the owners lack the most basic education. In smallholder farming, it is essential that at least a certain amount of food of different kinds be available throughout the year since diversification of crop production is a typical risk-spreading device (MacArthur, 1980). As will be discussed later, this sector representing the vast majority of farmers and producing over 70% of all fruit production, is left out of the economic mainstream and has to cope with high losses caused by pests and diseases.

2) a relatively small number (in some areas representing less than 5% of the farming community) of large-scale, capital-intensive, specialized agricultural enterprises which follow an industrial model of high intervention (Levins, 1985) and which usually are the only beneficiaries of technical recommendations. These enterprise farmers tend to be profit maximizers, preferring cash crops for export.

3) small family orchards and backyard gardens in large towns and urban areas which provide fruit for personal consumption.

In general, agriculture in Mexico is a highly decapitalized activity. Although capitalization has increased, it has done so in a deformed fashion: it has been concentrated in a few hands in a few regions (Atlântida Col-Hurtado, 1979).

It is also important to note that the most important fruit growing regions, with a few exceptions, are located in areas of high concentration of rural population, with a high degree of illiteracy, a fact that will greatly influence the strategies used to control fruit pests and diseases.

2. A GENERAL FRAMEWORK FOR THE CONTROL OF FRUIT FLIES IN MEXICO.

Fruit flies are one of the key pests affecting fruit production. Currently there are four species of the genus Anastrepha (out of 19 species reported for the country), three species of the genus Rhagoletis (out of 5 species reported for the country) and one species of the genus Toxotrypana which are of economic importance (Aluja, 1984). In addition a constant threat is posed by Geratitis capitata along the Southwestern border of Mexico and Guatemala.

Fruit fly control has a long history in Mexico, with the first efforts being documented in the late 1800's (Herrera, 1904). Up to now, within field control efforts have consisted almost exclusively of insecticidal bait spray applications. Isolated actions also included the use of the Sterile Insect Technique (SIT), biological control and mechanical control. Nevertheless, growers have almost entirely relied on postharvest fumigation with Ethilene Dibromide (EDB) for the control of larvae-infested fruit. This high degree of dependency on EDB fumigations has resulted in Mexico being vulnerable to the whims of the political and economic environment of the USA and other fresh fruit importing countries.

The latter situation has changed dramatically since the U.S. Environmental Protection Agency issued a directive prohibiting the use of EDB for the fumigation of fresh citrus in September of 1984 and for mangoes in September of 1985. Unless an alternative fumigant or postharvest control method can be quickly found, the government and more importantly growers, will suddenly be forced to perform a 180 degree shift in control strategies, putting all the weight on within-field control. The shift has to be directed backwards: from a total dependency on control at the harvest and marketing stages, the grower is now forced to plan ahead and control at the planting and orchard maintenance stages.

We strongly believe that this situation offers a unique opportunity for the implementation of an Integrated Orchard Management Program, that among others includes fruit flies. Instead of searching for other fumigants or other postharvest treatments, control efforts need to be directed at the source of the problem: the field or orchard.

Figure 5 depicts what we consider to be the five essential elements of an IPM program, each carrying the same weight.

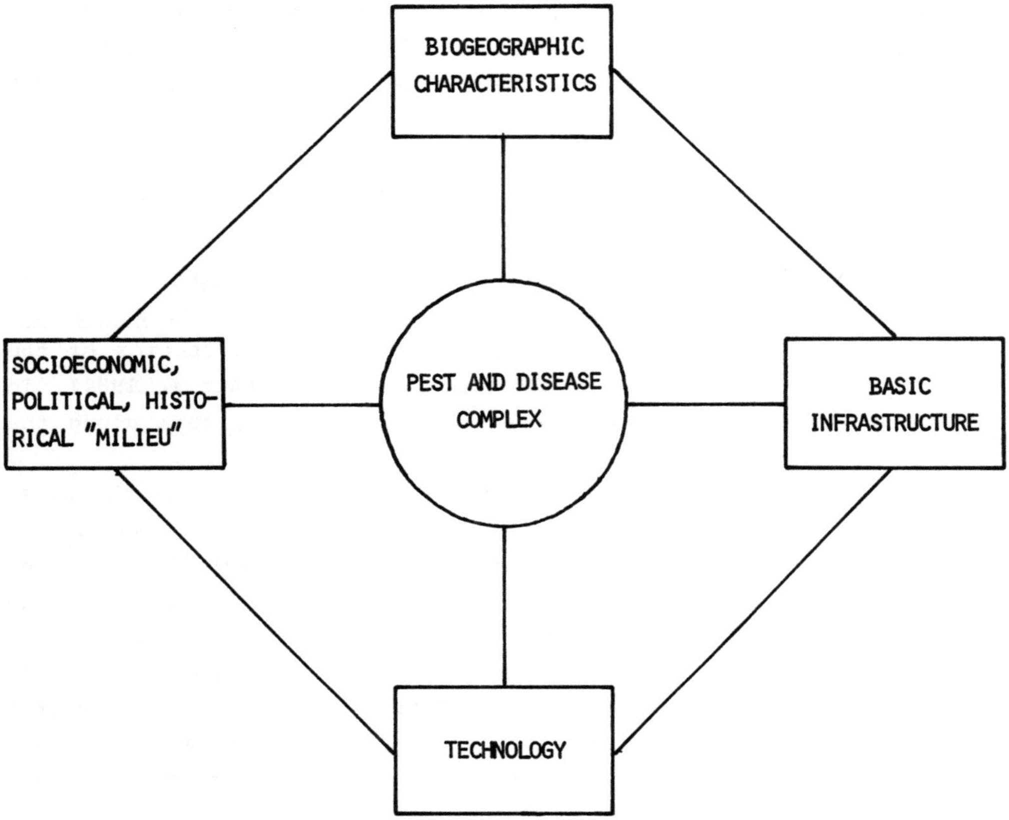

Fig. 5 - The five essential elements of an Integrated Pest Management program.

There are two fundamental questions that need to be addressed before deciding on a possible management strategy: 1) which is the target group of growers? and 2) is the goal of the program to save the fresh fruit export industry or do we want to produce clean fruit for both local and export markets? Both these questions will determine to a large extent which action paths will be followed since, as J.E. Edmunds (1981) reports, it is clear that strategies in implementation of plant protection must differ with different social situations and circumstances. As outlined in the introduction of this paper, more than 70% of the fruit in Mexico is produced by small, resource-poor farmers and yet it is only the large, resource rich farmers who presently control fruit flies and who benefit from technical assistance. Do we want to maintain the "status quo" or do we want to expand the range of our activities and include both small and large-scale growers, in which case we would introduce another dimension into an already multidimentional problem?

The same applies to the second question. If the goal is to produce clean fruit for export markets, conventional methods of control will be enough, since sufficient cash will be generated to pay the high costs of drastic interventions. Should we prefer to involve ourselves in a regional, large scale management program that involves both types of markets, a much more refined strategy will be needed. The level of pest damage and related control efforts will differ greatly for each situation. On one hand, we are dealing with a qualitative pest, where any number is unacceptable. On the other hand, the same pest would acquire a more qualitative status, where coexistence is possible, and where injuriousness is commensurate with the intensity of infestations (Geier et al., 1983). It is our contention that the problem demands a holistic approach, one that takes both production systems into account, incorporates throughout the high degree of variability involved and focuses on a long term solution to the problem.

2.1 Pest and Disease Complex

Let us start with the central object, i.e., the pest and disease complex, which not only varies in time, but which is unique for each biogeographic region. We strongly believe that for a fruit fly management program to be successful it must be part of a broader pest and disease management program. As Edens and Koening (1980) state, "expanding the system boundaries beyond the direct object of control, releases a whole new set of management and design options". As part of this philosophy, we stress that orchards can be a very important tool in reforestation efforts. Gomez Pompa et al. (1976) indicate, that by selective species enrichment with forest and fruit tree species it is possible to apply agro-silvicultural management practices. These practices are consistent with the ecological basis of secondary succession and forest regeneration processes in natural forests and can contribute significantly to the successful reforestation of large areas in the tropics. Perennial and annual crop systems with diverse mixtures of fruit trees and their associated understory herbs, shrubs and climbing vines can help to stop erosion and provide a more adequate environment for local farmer populations.

Up to now, decisions to institute a fruit fly control program were usually driven by emergency situations. Only when the export markets for fresh fruit were jeopardized by the presence of larvae in the fruit, were actions taken to control the problem. These included intensive bait spray applications, incineration of infested fruit and the quarantine of affected areas. Unfortunately these actions have only short term effects. It is also common practice to attribute all fruit fly problems to Anastrepha ludens, the Mexican fruit fly. Although this species does infest citrus and other commercial and wild fruit species in most areas of the country, it is by no means the only species of concern. Although we agree that fruit fumigation and bait spray applications can be effective against all fruit fly species, the problem demands a much more refined approach. The fruit fly complex needs to be considered as a whole and this concept incorprated into a comprehensive, countrywide orchard management program. Growers, especially poor ones, have not and will probably never accept isolated control actions. On the contrary, farmers would like to be able to control all their pests and diseases with simple, integrated actions since this is less expensive and less complicated.

2.2 Biogeographic Characteristics

The first step towards a more rational management approach should be to delimit problem areas according to their biogeographic characteristics. The strategy of implementation of plant protection should vary according to the region. Fruit flies are known to vary in their preference for particular host plants, using some hosts in certain areas and others in other areas (Aluja, 1984). Fox and Morrow (1981) suggested that "within overall phylogenetic constraints, specialization is often a flexible attribute of a species throughout its geographical range. Even in one geographical area, a herbivore may occur in more habitats than any particular food plant species and so, within any one habitat, may be more specialized than a list of all potential food plants in the area would indicate. In addition, population dynamics and phenology of the herbivore often vary among sites because of local patterns of plant availability (use of particular plant species may differ even among communities with identical lists of potential hosts)". So far, plant protection officials have recommended the same basic treatments for fruit fly control, with little regard to what particular characteristics the specific problem area had. Fruit fly species A might exploit completely different alternate host plants in area A than in Area B (as is for example the case with Anastrepha ludens), (Aluja 1984). If we want to intervene and try to interrupt the movement of flies between host and alternate host, this knowledge becomes essential.

Policy and management directives should clearly differ between the fruit growing regions in the desert state of Sonora and those of tropical states of Chiapas and Veracruz, or the highland states of Texcoco and Puebla. The key question in Sonora is to avoid the introduction of infested fruit into the isolated fruit growing regions which, due to extremely harsh climatic conditions and the absence of alternate host plants, do not harbour endemic fruit fly populations (Aluja, 1983). Since host availability and habitat suitability are essential to pest population risk assessment (Rower and Williamson, 1983), a careful examination of local conditions when establishing regulatory protocols and international marketing agreements should be made. Fruit-growing regions in Sonora should be treated as fruit fly free areas and export regulations relaxed. Meats (1981) and O'Loughlin, East and Meats (1984), have addressed the concept of bioclimatic potentials for some fruit flies in Australia, and how these could affect quarantine protocols. Efforts in the same direction are needed in Mexico.

In contrast to Sonoran orchard growers, growers in Chiapas and Veracruz should not apply funds for attempting to prevent the introduction of fruit flies. In each of these states, endemic populations are very high and environmental conditions optimal for their increase. Additional subdivisions of these regions should be made, and particular agroecosystems considered. Figure 6 compares two distinct fruit growing regions in the Soconusco, state of Chiapas, showing host availability throughout the year for two species of fruit flies: Anastrepha oblicua and A. serpentina. Although these regions are in close proximity (30 Km), they show very different local characteristics, such as different crop types and local vegetation. These conditions will determine what pest and disease management strategies can be applied. For example, the constant host availability in area B, as well as local topographical and environmental conditions, make it virtually impossible to control the A. serpentina populations. In contrast, area A offers good possibilities for the establishment of an IPM program since orchards are better defined and both topographical and environmental conditions more suited for agriculture.

In summary, there are three clearly defined biogeographical regions in Mexico where fruit trees are commercially grown: 1) isolated dry and desert areas with few or no alternate host plants, where the commercial fruit host patches are islands and the establishment of endemic fly populations highly unlikely; 2) highland regions where deciduous fruit varieties are grown and where the fruit-infesting tephritids have ususally one generation per year and their hosts are predictable in distribution and abundance from one year to the next; and 3) subtropical and tropical regions, with large numbers of cultivated and wild host plants, present throughout the year, where fruit fly populations are very high.

2.3 Socio-economic, political and historical "milieau"

When planning an IPM program, one has to determine in what socio-economic, political and historical context pest control results have

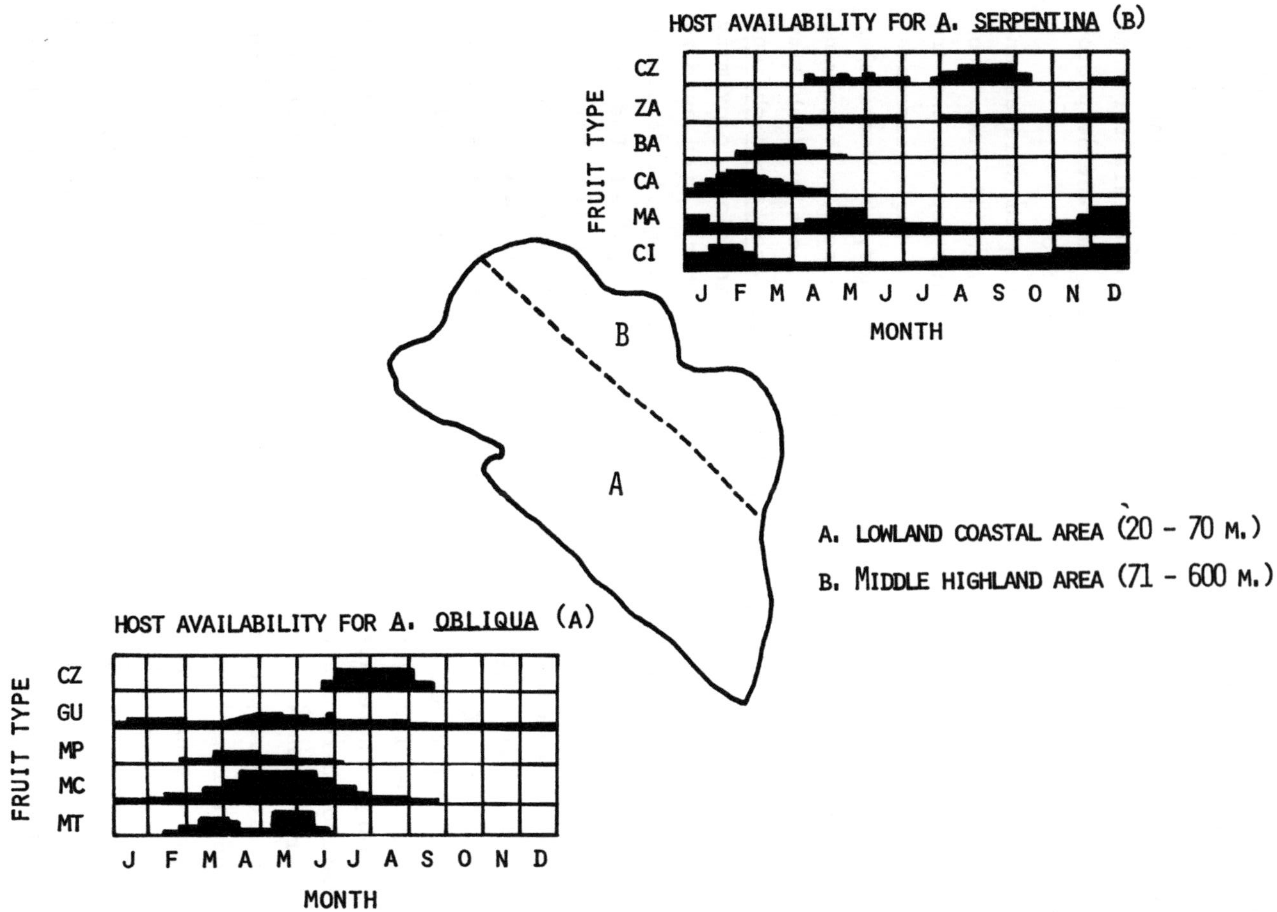

Fig. 6 - Two distinct fruit growing regions in the Soconusco area, state of Chiapas. CZ = Zapote Chico (*Achras zapota*); ZA = Zapote Amarillo (Lucuma salicifolia); BA = Baricoco (*Micropholis mexicana*); CA = Caimito (*Chrysophilum caimito*); Ma = Mamey (*Calocarpum zapota*); CI = Citrus (*Citrus* sp.); CU = Guava (*Psidium guajava*); MP = Jobo (*Spondias* sp.); MC = "Crillo Mango" (*Manguifera indica*); MT = "Ataulfo Mango" (*Manguifera indica*).

to be achieved (Brader, 1982). As already indicated, we are faced with a clearly defined situation: large, resource rich growers and small, resource poor farmers. But here too, no countrywide generalization is valid. Sonora for example, basically has only large to medium, technologically advanced growers, while in Chiapas and Veracruz small farmers who follow the traditional systems of agriculture predominate. The key element here, is the integration of these two types of production systems into one overall IPM program. Since fruit flies do not respect man-made borders, well-managed orchards will be affected by nearby abandoned or poorly-managed orchards.

Under these circumstances we should strive to secure a fair access to the knowledge and services. Even if plant protection techniques were made available and understood, the small farmers could not afford them (Sill Jr., 1982). The latter certainly applies to monitoring mechanisms and pesticides used in fruit fly control. To complicate matters even further, the production system owner (large or small) is often not directly responsible for managing pests and diseases, due to the common practice of selling the entire crop in advance as a package deal. This means that the grower completely transfers all the responsibility for orchard maintenance and harvest to a middleman who has no knowledge whatsoever of pest and disease control. The middleman usually adopts one of two alternate paths: 1) to hire a "pest killer" company or 2) to control pests and diseases himself. In both situations, large amounts of pesticides are used in excess and applied with inadequate equipment. Too often the crop system is burdened with chemical poisons, when simple cultural practices would be more efficient, economical and ecologically safe.

The political structure within the region also determines to a large degree what can be achieved in terms of a rational plant protection program. Too often, small special interest groups determine marketing procedures, export quotas, pesticides to be used and other essential elements to the fruit production systems. There are occasions where pest-induced shortages of desirable goods tend to benefit middlemen (Geier et al. 1983). In others, contractual marketing arrangements force growers to strive for zero pest levels although it seems that these standards are required by the retail buyer not the consuming public (Van den Bosh, 1976).

The history of each region also plays an important role. The citrus growing region of Xalapa in the state of Veracruz was devastated by heavy infestations of fruit flies in the sixties (Espinosa, 1983). Since this situation prevailed for several years in a row, wholesale buyers refused to buy any fruit from that region even if the grower implemented control programs. This negative image can persist for many years and greatly affect local agriculture. The situation became so desperate that some growers started to illegaly move fruit from heavily infested areas to "clean" areas and selling the fruit as if coming from the "clean" areas. This situation naturally caused tremendous damage to the "clean" areas.

Finally, there is the question of education and ethnic groups. Mexico has many different ethnic groups and a high degree of iliteracy. Many fruit-growing regions are localized in remote rural areas where extension agents are not known; even if they visited these areas, the acceptance of their ideas would be limited. It is therefore important to develop practical recommendations, which do not create technical dependency and which do not undermine local practices.

In general, we consider that the solution to all these problems is the establishment of solid grower associations and a legislative framework to back regional plant protection efforts, as will be discussed in the next section.

2.4 Basic Infrastructure

The basic infrastructure consists among others of plant protection agencies, research institutions, legislation and grower associations.

The norm is the existance of a few powerful grower associations, which only include the largest and richest growers. Small growers are poorly organized, and even when they are, their political clout is minimal. It is clear that this situation has to change, since grower associations, including pest management and marketing cooperatives, are good focal points for the implementation of a successful IPM program (G.W. Bird, 1981). There are few examples where the establishment of a local cooperative, which handles marketing of fruit in a equitable manner, generated enough capital in a time span of two years to hire private technical assistance, manage pest and disease control programs and cooperate with plant protection agents in the establishment and enforcement of quarantines and phytosanitary protocols (Zorrilla, 1983). This helped to eliminate middlemen, and in general dramatically increased the social welfare of the small farmer. Figure 7 shows a marketing scheme that we think would greatly aid in overcoming some of the current bottlenecks encountered when trying to move ahead in the control of fruit flies.

Growers associations could reduce the price of goods such as traps, bait sprays and technical assistance by buying in bulk and hiring personnel for a whole area. They could also regulate the flow of fruit out of each region, assuring that each fresh fruit shipment

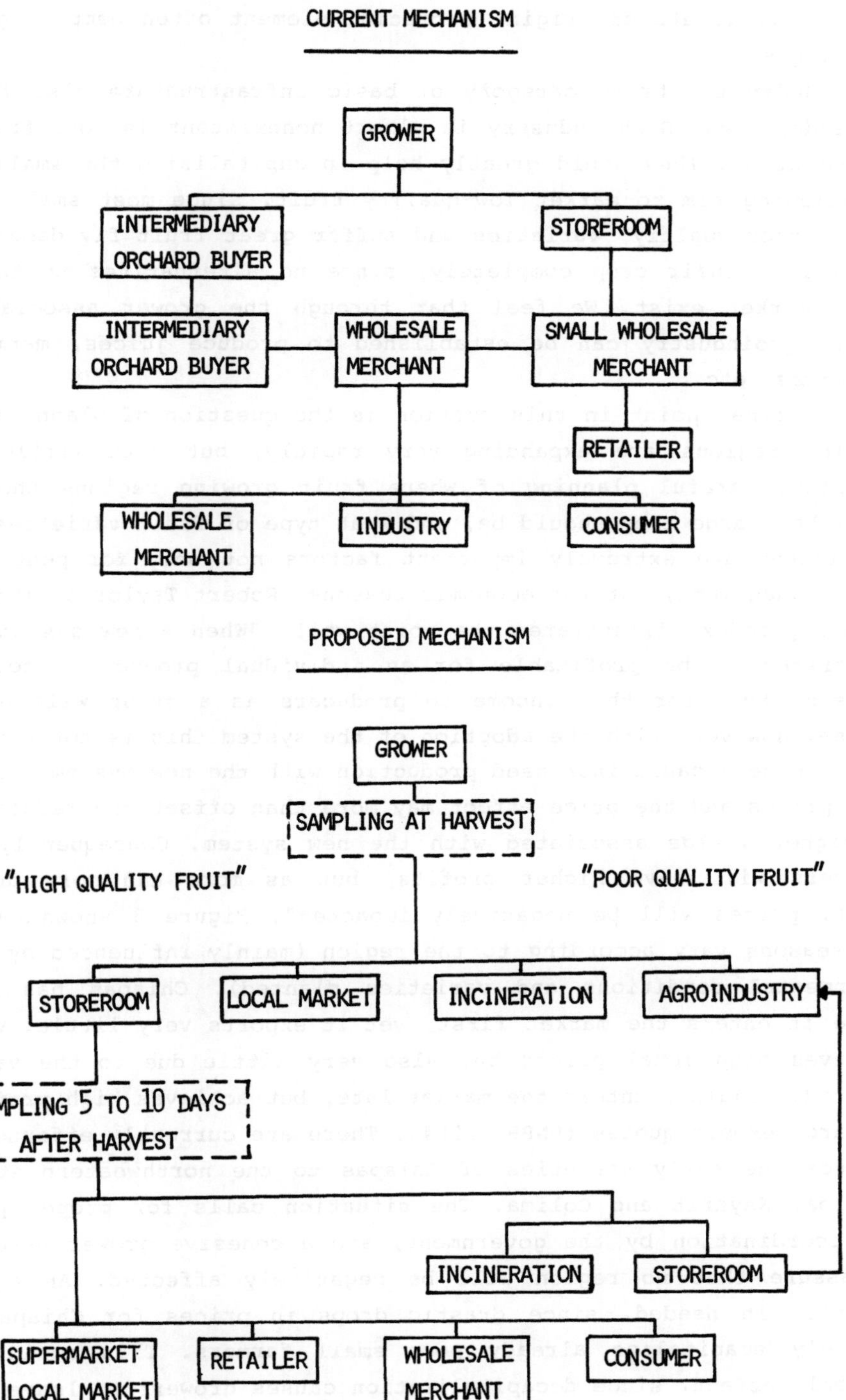

Fig. 7 - Current and proposed fruit marketing mechanisms.

has a certificate of origin, a crucial element often hampering export agreements.

Under the broad category of basic infrastructure also fall the agroindustries. Such industry is almost nonexistent in most fruit producing areas. They would greatly help in capitalizing the small grower by allowing him to market low-quality fruit. Since most small growers plant "poor-qualtiy" varieties and suffer great fruit fly damage, they often lose their crop completely, since no alternatives to the fresh fruit market exist. We feel that through the grower association, a local agroindustry can be established to produce juices, marmelades, preserves, etc.

A final point in this section is the question of planning. Fruit growing regions are expanding very rapidly, but with little global planning. Careful planning of where fruit growing regions should expand, how large they should be, and what type of fruit varieties should be planted are extremely important factors not only for pest and disease management, but for economic reasons. Robert Taylor's (1981) farm income paradox, illustrates the point well. "When a new system is demonstrated to be profitable for an individual producer, there is a tendency to infer that income to producers as a group will also increase. However, with the adoption of the system this is not necessarily the case because increased production with the new system will lower farm prices and the price effect may more than offset the reduced costs or higher yields associated with the new system. Consequently, early adopters will have higher profits, but as more and more producers adopt, prices will be negatively impacted". Figure 8 shows how fruiting seasons vary according to the region (mainly influenced by the environmental conditions and varieties planted). Chiapas has an edge since it enters the market first, yet it exports very little. Veracruz achieves high local prices but also very little due to the varieties planted. Sinaloa, enters the market late, but achieves high profits due to large export quotas (UNPH, 1984). There are currently efforts to introduce the early varieties of Chiapas to the northwestern states of Sinaloa, Nayarit and Colima. The situation calls for proper planning and coordination by the government, and a cohesive grower association to assure that no region will be negatively affected. An equitable solution in needed, since drastic drops in prices for Chiapas would severely decapitalize already poor small farmers. This could have a snowball effect, since decapitalization causes growers to leave orchard unmanaged, allowing fruit fly populations to get out of control. In Chiapas alone, the planted areas with mango have increased five fold in the past ten years as figure 9 shows. Unfortunately the most common

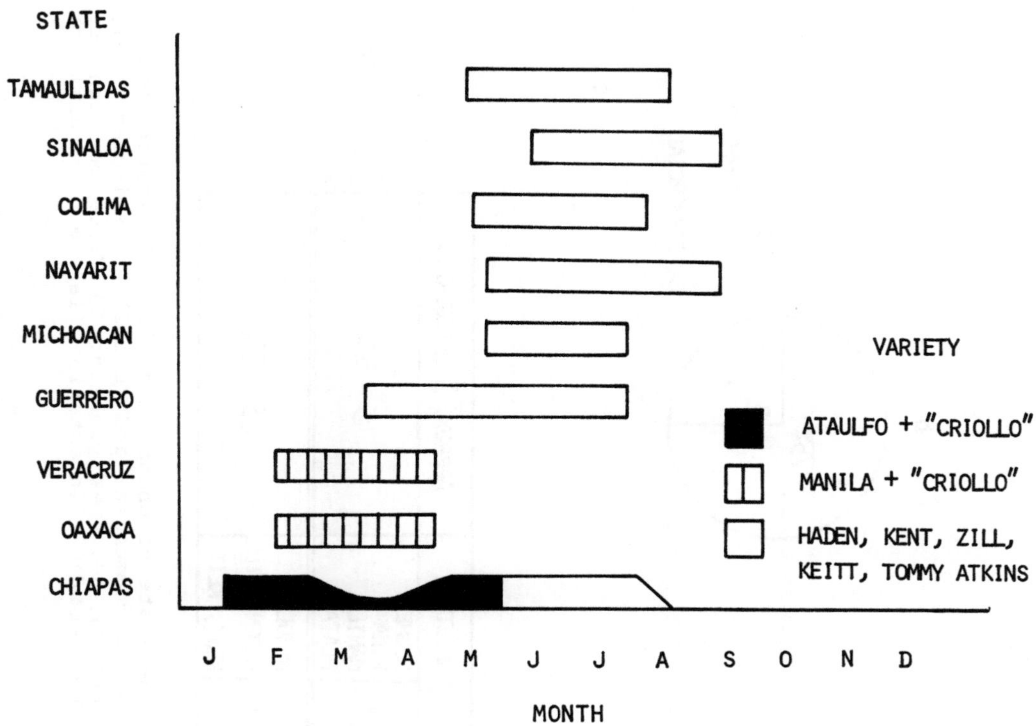

Figure 8 - Fruiting seasons of different mango varieties in the most important mango producing states in Mexico. (Source: UNPH, 1984)

varieties planted by small-scale farmers are low priced and also highly infested by fruit flies.

2.5 Technology

Included in this subject are the specific mechanisms used to manage and control fruit flies. We will omit many well known mechanisms and rather concentrate on possible future action paths. We want to underline the fact that many of the proposed management mechanisms will need to be tested in the field before a definite recommendation to the grower can be made. We nevertheless feel that it is very important to explore possible new directions since fruit fly control in Mexico has in many ways been conceptually and technically stagnated.

Since we need to consider both the needs of large-scale, resource rich and small-scale, resource poor farmers, the techincal recommendations need to be rational and easy to apply. The current practice of automatically resorting to bait spray applications has as already discussed, to give place to a more flexible approach. Although fruit flies are extremely susceptible to most insecticides on the market, thus allowing the grower to choose the cheapest and safest product, the

A

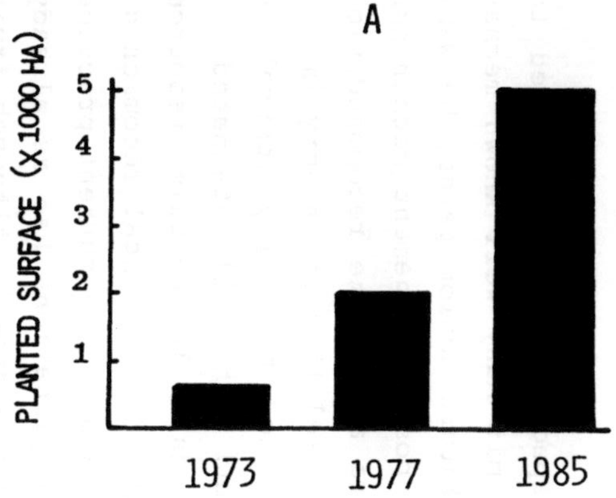

B

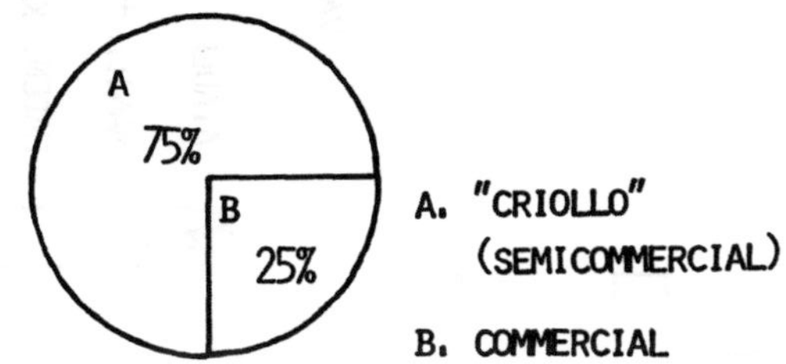

C

VARIETY		PRICE PER TON ($)
A	COCHE PINA	1500.00
	MANILA PLATANO	3000.00
B	ATAULFO	8500.00
	HADEN, ZILL KENT, KEIT	3000.00

D

VARIETY		APPROX. % INFESTATION
A	COCHE PINA MANILA PLATANO	45 - 60%
B	ATAULFO HADEN, ZILL KENT, KEIT	5 - 15%

Fig. 9 - Some relevant elements of mango production in the state of Chiapas, Mexico. A) Planted surface for the years 1973, 1977 and 1985; B) Proportion of commercial and "criollo" (semi-commercial) mango varieties in relation to the total number of hectares planted: C) Price per ton for the different mango varieties (1979) and D) Approximate % infestation (larvae in fruit) for the different mango varieties.

costs are still too high for a small production unit. Furthermore, although there are no records of insecticide resistance in fruit flies and this event seems highly unlikely due to the particular characteristics of the pest, applications should be kept at a minimum. Bait sprays are a big step in that direction since they allow for alternate row sprays. Nevertheless the effectiveness of bait sprays in attracting flies has yet to be demonstrated conclusively. Also, there is evidence indicating that bait sprays affect adversely non target organisms (Ehler and Endicott, 1984).

We have already discussed the need to consider the different biogeographic regions and the need to try to implement an area-wide management program. Once the biogeographic region has been clearly defined the next level we need to deal with is the habitat or the agroecosystem. The vegetational component of agroecosystems can be viewed as a mosaic of annual and perennial crop fields, forest patches, pasture lands, fallow fields, orchards, swamps and tree plantations (Altieri, 1983). Although it is useful to consider individual portions of the system, such as a crop island subject to colonization from several sources, a regional perspective is necessary. As Rabb (1978) suggests, pest management researchers and decision makers should consider "large unit heterogeneity" as an appropriate alternative to the more common single commodity, closed system approach. However, as Barfield and Stimac (1981) indicate, defining the boundaries of such an open system is often problematic. Insects such as fruit flies that can move over large distances (Fletcher, 1974) and can exploit a variety of habitats seasonally, present a challenging problem concerning the system boundaries: 1) how crop-insect systems are defined in terms of environment, mobility and diversity of acceptable host plants and 2) how crop systems are coupled through flows of pest insects and beneficial organisms among crop and non-crop plants. Even though the grower perceives a clear boundary around his field, the pest manager and ecologist have to be concerned with the mechanisms causing movement of pests across the boundary (Barfield and Stimac, 1981). For example, the type and distribution of non-host vegetation grown in the vicinity of hosts may have a strong influence on the density and movements of tephritid populations (Prokopy, 1977). Furthermore, crop area expansion and field aggregation in regular shapes, necessarily increases the perimeter or boundary for immigration from the surrounding habitats.

Larval stages of tephritids must occur in fruit and host availability becomes, in many situations, the major regulator of final adult population size. This factor, and the ones discussed before, have let us to believe that host and non-host vegetation manipulation within and around crop fields based on sound ecological priciples should be an integral part of any fruit fly management program. Although this concept has been extensively reviewed and applied to annual crops (Altieri, 1983; Altieri et al., 1984, Rish, 1983 and Kass, 1978), little is known about possible applications to perennial crops such as fruit trees. The difficulty largely stems from our vast ignorance of how insects perceive the environment and go about selecting hosts for feeding and oviposition. Hassel and Southwood (1978), suggest that the forager perceives the environment at three hierarchical levels: the habitat, the patch and the food item. Young (1982) identifies four critical variables in the habitat: "patch structure of host plants, habitat species densities of the host plants, the degree of distictiveness among host plant patches of a particular species in terms of features attractive or repulsive for oviposition and the degree of chemical and physical distinctiveness among patches of different host plant species cooccuring in the same habitat".

Once the herbivore has identified the habitat in which to forage it must then decide between different patches within that habitat. Unfortunately, as Denno (1983) indicates, little is known for a single species about 1) how host plant quality, competitor, predator and parasite density integrate to dictate patch quality and deterioration; 2) what effect a deteriorating patch has on the reproductive success of the herbivore and what the success might be in an alternative newly colonized patch; 3) what environmental cues are used by herbivores to measure patch deterioration and trigger migration or dispersal; 4) if indeed there are identifiable patches with spatial and temporal differences in favorableness and what the long term dynamics of the patch are and 5) how well herbivores track variable hosts by leaving poor quality patches and moving to better quality ones".

Let us review several scenarios in which one could manipulate the vegetation in the case of the fruit flies. Kareiva (1983) considers three ways in which vegetation texture in a patch can vary: 1) the distance between individuals of the same species can be altered (density); 2) the aerial or geographical extent of a stand of host plants can be altered (patch size) and 3) the frequency and identity of non host plant individuals can be altered (stand purity or diversity).

In the case of the papaya fruit fly (Toxotrypana curvicauda), Landolt and Hendrichs (1983), Landolt (1984) and Landolt (1984) have shown that females only visit host plants for oviposition during a short time period in the afternoon and that both females and males are very wary of predators, thus reducing to a minimum the amount of time spent on the host. The flies reside mostly in unidentified non-host plants surrounding the papaya plantations, from which the flies fly to the papaya trees. This results in a marked edge effect in the activities, with border rows visited most. Figure 10 shows a patch design which we consider could be effective in reducing the number of ovipositions in the commercial block. A trap crop row, separated from the main commercial block by a distance still to be determined, would concentrate the ovipositions outside of the commercial block. This design would fit well into the modular units proposed by Gliesman et al. (1981) for the lowland tropical areas of Tabasco, where among others, a polyculture of cassava, corn and papaya is recommended. Of course, one has to first determine if cassava or even corn would not be used as a refuge site by the papya fruit fly. Another possible control measure is the selective application of bait sprays. Sharp and Landolt (1984), have shown that the papaya fruit fly is "attracted" to decomposing products of sugar as opposed to proteins. Since papaya is grown where sugar cane plantations are common, the necessary baits would be readily available. Here again, it it necessary to identify refugia sites and feeding sites first, since application of bait sprays to the orchard may be ineffective due to the fact that flies enter the orchard to oviposit, not to feed. At any rate, we consider that this approach should be intensively explored.

If local growers have established a strong grower association and fruit is sold to highly demanding markets, vapor treatment as described in the U.S. Federal Register (1984) could assure 100% mortality of any larva. Since this requires expensive installations it can only be carried out through a cooperative effort by growers.

Another possibility for manipulating the vegetation is to use alternate host plants. There is an open debate as to whether it is advisable to eradicate them or rather try to use them as a control tool. We think that both are viable solutions depending on the specific situation. For example, we have already indicated that eradication is advisable in desert areas, such as the state of Sonora. Here, almost any fruit fly host is introduced and therefore not part of the local flora. No important ecological damage would be inflicted by their removal and an important reduction in the threat of an establishment of fruit flies can be achieved. Growers should by all means avoid introducing links between highly isolated orchards.

In areas where alternate host plants are abundant and part of the native vegetation, eradication would be ill-advised. Here several options should be explored. One is to take advantage of their "trap crop effect". Aluja et al. (1986) have shown that some host plants are highly "attractive" and flies concentrate in them in great numbers. Some

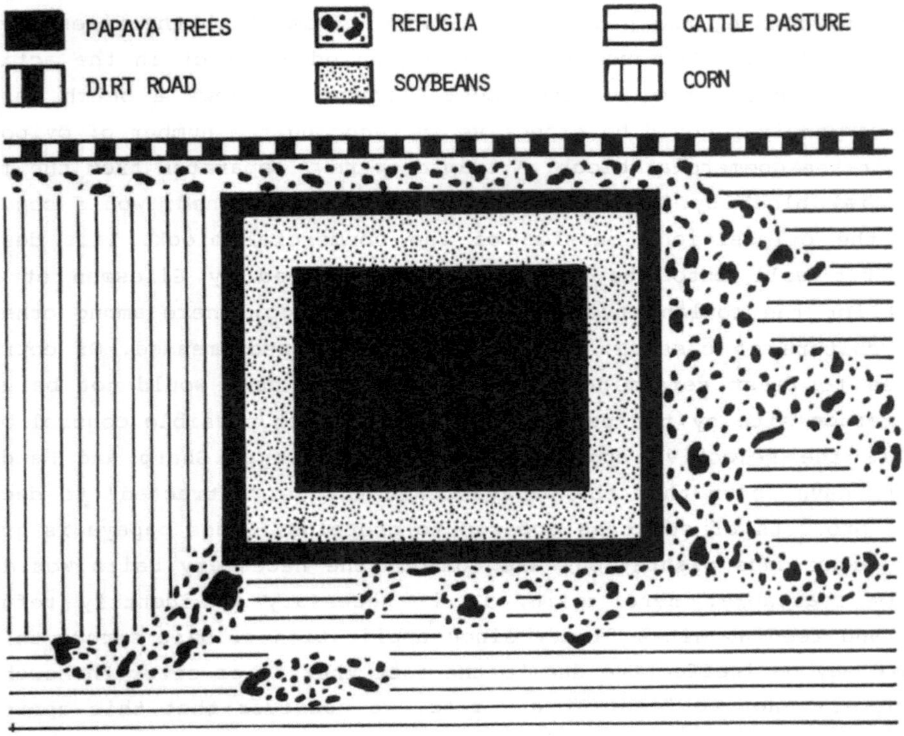

Figure 10 - Papaya orchard design for the control of the papaya fruit fly (Toxotrypana) curvicauda).

host fruits have infestation rates of up to 300 larvae/kilogram of fruit. Even if two proven host plants are grown side by side, one can observe differences in acceptability (measured by the number of larvae per fruit and the number of adults caught in McPhail traps) depending on the host species or even the variety. We have suggested (Aluja, 1984) selectively hanging traps and sampling fruit in these trees in order to obtain information about the probable number of flies dispersing into the orchard. Also, once high infestation levels are determined, concentrated and localized bait spray applications can be used to reduce the number of invading adults. One could even speculate that females might deplete their egg load in these hosts, and should they arrive in one of the less preferred hosts, which can be very close, the probability of successful oviposition reduced.

Host eradication would be also ill-advised if as Prokopy (1984) reports, host fruit stripping (or eradication in our case) stimulates long distance movements of some flies to previously uninfested areas or even can give rise to extensive attempts at egglaying in various non-hosts or less preferred host fruits. This is consistent with Fox and Morrow's (1981) suggestion that the dynamics of local specialists that have the opportunity to consume more than one type of host will be more buffered: they can switch to another potential host if their preferred one becomes scarce.

The actual planting of these highly attractive hosts with the expectation that they should successfully divert the attack by fruit flies to the commercial fruit trees, is open to debate. On the one hand it could backfire by attracting more flies than there were in the first place. On the other hand, since local people insist on having a certain number of non-commercial fruit trees around their houses or orchards, it is important to try to minimize their impact. It can well be that by proper placement within the orchard they can indeed serve a useful purpose, as we shall see later.

Let us analyze two practical examples. Figure 11 depicts orchard "Quinta Irenne", a commercial mango orchard in the Soconusco, state of Chiapas. Here six different fruit species are cultivated. Mango is the most important and there are seven different low quality ("Criollo") varieties in addition to the commercial one. Aluja et al. (1986) report that eight species of Anastrepha were caught in McPhail traps although only two actually infested mango (identified by rearing larvae from infested fruit to adult stage). Traps hung in the "criollo" varieties caught over 70% of all the flies, with some few trees being particularly "attractive". In this orchard, there is a sequence of fruiting periods, in which one or more hosts are available at all times, serving as stepping stones for the different fly populations, which are kept at high enough levels to cause severe damage to practically all fruit species.

Instead of the current design we envision one that would allow the owner to maintain all the diversity but in a more structured, block-like fashion. In orchards, temporal arrangements of trees is a fixed variable, but the spatial are not; even the temporal arrangement can be manipulated by early harvest of fruit, selective harvest, etc. Blocks would greatly facilitate the mapping of the orchard. Maps, besides providing information on tree age, size, spacing and variety, can play an important role in the management and implementation of pest, disease and nematode control and in the implementation of agronomic practices, such as irrigation, fertilization, etc. Control strategies can be planned for problem areas, blocks or varieties, rather than the entire farm. Bait sprays can be applied in border rows, infestation spots (highly preferred hosts or varieties), and the collection of all fallen fruit facilitated (instead of having to search around all the

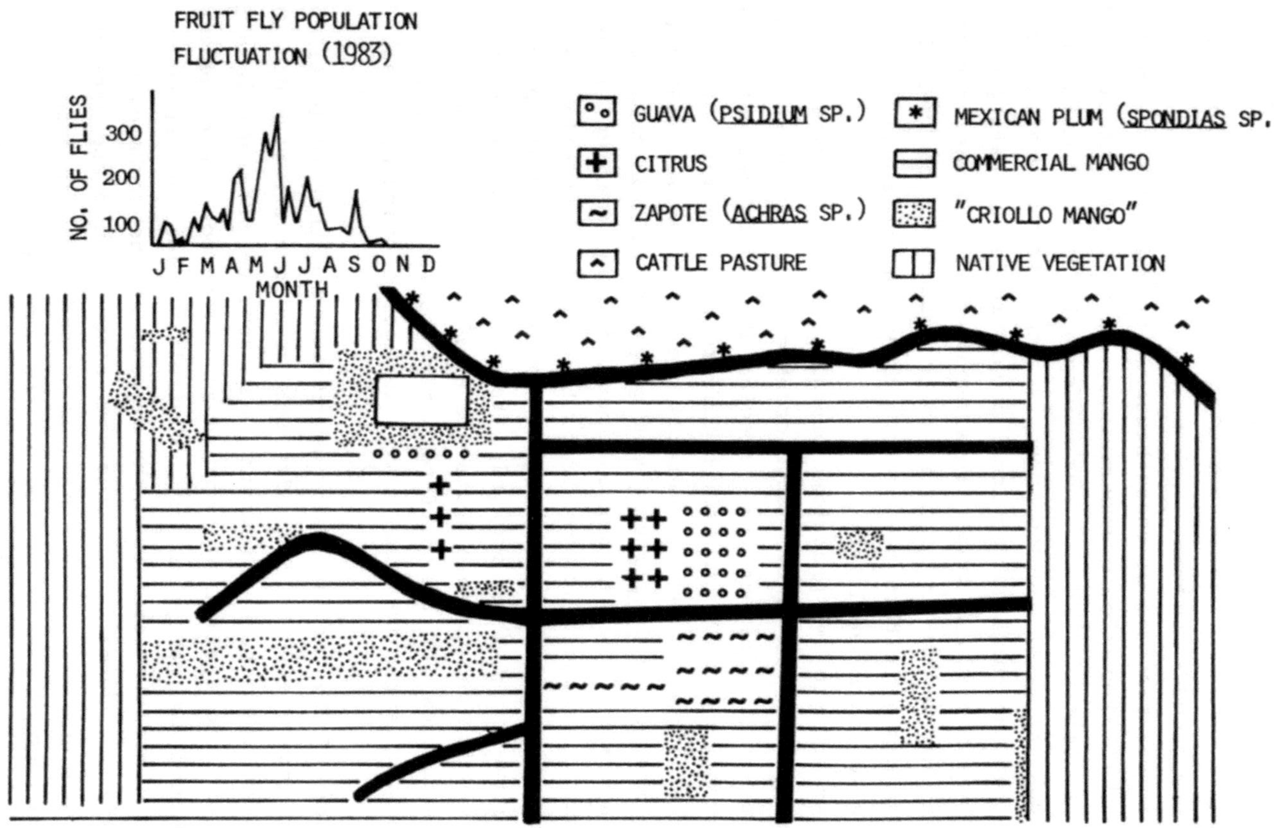

Fig. 11 - Finca "Quinta Irenne" a commercial mango orchard in the Soconusco, state of Chiapas, Mexico.

orchard, efforts can be concentrated in those areas where fruiting is at its peak). It would also facilitate the strategic placement of traps to monitor dispersal and migration into the highly valued blocks, and would allow one to maintain a predator and parasitoid pool. Blocks of highly preferred hosts could be placed on the edges, where dominant winds hit the orchard.

The fact that there is a clear difference in which varieties of mango are exploited for oviposition opens the possibility for some manipulation. It is interesting to note that the less preferred variety (actually the commercial one) was introduced to the region approximately 50 years ago. The "criollo" varieties have been used by local people for much longer and one could speculate that fruit flies have not fully adapted to it. But as Prokopy (1984) indicates, in Ceratitis capitata it appears that fruit size has a strong influence on acceptance pattern, while fruit taxonomic status has little effect. If the same pattern holds true for some Anastrepha species it is interesting to note that all the "criollo" varieties tend to be consistently small, while the commercial ones are all much larger. Also, one can speculate that commercial varieties have much higher pressure in the laticiferous channels causing the eggs to be ejected from the skin; or that the commercial varieties have properties toxic to the eggs. The two latter speculations are based on the fact that indeed these varieties release much more fruit exudates once damaged and that the fruit is much more caustic before reaching full ripeness. Finally criollo varieties are much more odorous, which could influence the host finding process.

As in the case of the alternate host plants, the following question arises: should these highly attractive varieties be left intermixed with the commercial plantation to concentrate ovipositing females and to allow for localized bait sprays? Or should they be removed from the commercial block with the expectation that flies will seek them out and avoid ovipositing in the less preferred varieties? All these factors should be investigated in the future. We are aware that once an orchard has been established, it would be very difficult to change the distribution of the trees. Also, since fruit trees are highly adapted to constantly disperse, it is difficult to protect fruit trees by introducing heterogeneity into the orchards. Nevertheless, since so many new orchards are being planted, it is very important to apply some of these concepts or at least start some experiments.

The second case, Finca "Buena Vista" (Figure 12) is more complex, since it encompasses a larger area and several non-commercial orchards, abandoned orchards and backyard gardens not owned by the large scale grower.

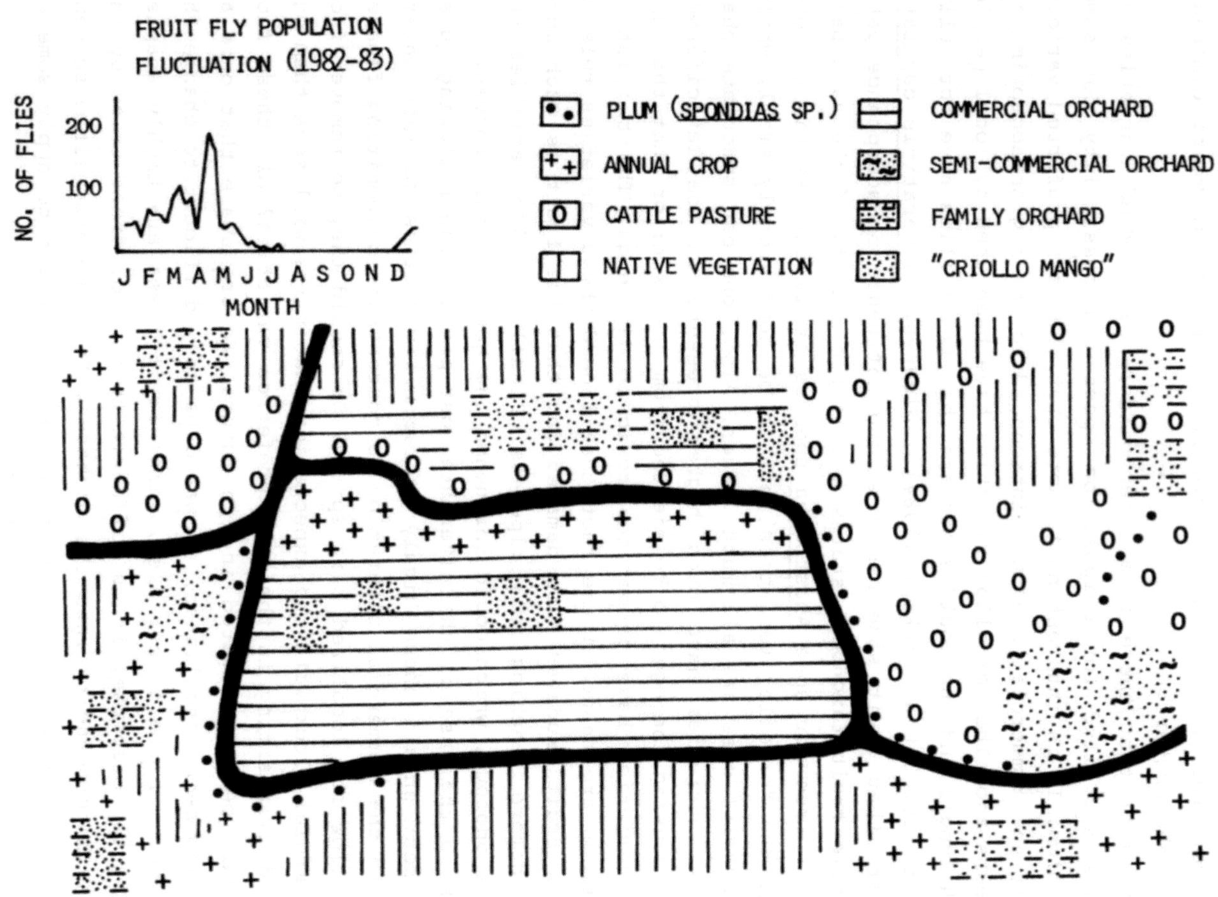

Fig. 12 - Finca "Buena Vista" a commercial mango orchard in the Soconusco, state of Chiapas, Mexico.

The boundaries of this system are vaguely defined and a area wide IPM program is required. The large commercial orchard is contantly invaded by fruit flies dispersing from surrounding patches. Between 60 to 70% of all the flies caught in McPhail traps came from border rows (Aluja et al., 1986). Since the large grower is interested in the export market and highly specialized markets in Mexico City, the economic treshold level is very close to 0 larvae per fruit. Even though it is a well kept orchard with irrigation, fertilization, pruning, etc. and little intermixing of varieties, there are nevertheless some wild host plants planted as fences or planted in backyard gardens owned by the orchard keeper, that harbor large populations of at least one *Anastrepha* species (*A. oliqua*) and serve as within-sources of infestation.

The overall management strategy calls for the control of endemic populations by applying bait sprays in alternate rows and on localized spots like the natural fences mentioned before, constant sanitation (collection and destruction of all fallen fruit), and the separation of any infested fruit at the time of harvest (if fruit are collected when the skin is still green, infested fruit will have yellow dots in areas of larval activity). Additionally bait sprays should be directed at border rows to kill invading adults and a scouting program established to determine the need for interventions. The latter should be based on inspection of McPhail traps and sampling of fruit. For a more detailed description of particular procedures see Aluja (1984). All the latter actions are affordable for the grower, whose benefits will outweigh the costs.

But what shall be done with the small growers and backyard gardens surrounding this large resource rich agricultural enterprise? No expensive inputs can be afforded, since the fruit varieties planted bring very low prices in the market (see Figure 9). Preliminary studies by Aluja et al. (1986), have shown that parasitoid populations are common. Numbers could be substantially enhanced by building "fruit beds" (see Aluja, 1984 for details) and by artificial augmentation as described by San Juana (1977). Parasitoids play a minor role in commercial orchards, but Monteigh (1971) indicates that they might be useful in wild and neglected trees. Frick et al. (1954) also report that culturally important cultivars normally bear fruit much larger than their native counterparts. Hence they provide tephritid larvae a proportionately greater degree of protection against parasitoids which cannot reach the larvae with their ovipositors. Varieties grown by small farmers and wild trees bear much smaller fruit than the commercial ones and could therefore facilitate parasitization of larvae.

Sanitation through the collection and destruction of all fallen fruit should also dramatically reduce the number of flies within these patches. This can be achieved by either manual collection or by allowing pigs to forage for fallen fruit. Sanitation can be extremely thorough in these situations, since the farmer tends his trees individually. Mechanical protection of the most valued fruit using bags can also allow for the harvest of clean fruit with pesticide applications.

A key point here is the cooperation of different growers. If fruit fly populations can be suppressed by the adoption of such cultural practices, they are well worth a communal effort or, if needed, by legislative ruling. Also, by sharing resources, storage facilities can be built to allow for early harvest. Infestation levels usually increase with ripeness and early harvest would allow the fruit to escape ovipositions; it would also allow the growers to market their product more efficiently and on more equitable terms.

Refugia or shelter sites could, in theory, also be manipulated. Broad-leaved, densely foliated trees may be the most readily identifiable plants as shelter sites (Malavasi et al., 1983). Prokopy and Roitberg (1984) reports that the type and distribution of non host vegetation growing in the vicinity of hosts has a strong influence on the density of movements of tephritid populations. It is important to explore the possibilities of using these refugia sites to concentrate fly individuals outside of the orchards and kill them by using bait-sprays. Special attention should be paid to probable effects on beneficial organisms.

There are several other specific questions that need to be addressed. One deals with monitoring tools. Avoiding to state what is widely known, let us underline the need to develop adequate monitoring devices for the different Anastrepha, Rhagoletis, and Toxotrypana species. Aluja et al. (1986), have shown that under natural conditions, 69% of the flies that approach a McPhail trap, excape being trapped. Furthermore, approximately 30% of the flies that escaped had entered the trap first. If the goal is to establish a large scale fruit fly management program in which, certain areas are to be certified for the export of fresh fruit, traps that are more efficient and easier to handle need to be developed.

There is also the question of imcompatibility of control strategies. As already mentioned before, growers not only have to deal with fruit flies but rather with a wide array of pests and diseases. Many times the recommendations of the agronomist work against those of the phytopathologist or the entomologist, or vice-versa. Improper pruning can facilitate infection by pathogens. On the other hand it has an important function in shaping the trees, allowing for a properly aereated crown, an important factor restricting outbreaks of pathogens with high moisture affinities (Palti, 1981). Bait sprays can cause considerable damage to mango if not properly applied. Also for some fruit species, bait sprays can cause phytoxicity if protein hydrolysates are used where there are residues of copper compounds from fungicidal sprays. This introduces an interesting issue, Since Opp et al. (1985), have indicated that bordeaux mixture strongly reduced oviposition attempts of R. pomonella on apple and hawthorne. Since this compound is widely used in Mexico, its use as a probable oviposition deterring substance for some of the Anastrepha species should be investigated. The effect of bait sprays on non target organisms has been previously discussed in this paper.

The use of irrigation to cause aestivating pupae to emerge during a time when there are no hosts available is an interesting possibility that calls for further research. At least in the case of Guava, irrigation can almost completely control flowering and fruiting periods. Since there is a lag time between the period of irrigation and the availablity of fruit for oviposition, emerging flies could be killed with bait sprays or forced to migrate out of the orchard. Excess of water can also cause considerable pupal mortality. In this respect it is important to note that heavy rains, which wash off honey dew and plant secretions from foliage, can cause a drastic reduction in the fertility of gravid females. Therefore, large captures of these females in traps does not necessarily translate into heavy oviposition damages, at least during the peak rainy period.

A very interesting option has been reported by Greany (1982), who applied gibberelic acid prior to color break causing the fruit to appear unripe from outside, even though the normal ripening process was taking place inside the fruit. He reports substantial reductions in the number of eggs laid in these treated fruits. This method could be very useful in the citrus growing region of Nuevo Leon and Tamaulipas where it is a common practice to leave the fruit unharvested for lengthy periods of time in order to allow prices to rise. Fruit fly attack during these periods is high if the fruit is not protected by insecticide treatments. We consider that the use of gibberelic acid, combined with proper design of orchards to have the most susceptible fruit (grapefruit is always preferred over oranges by ovipositing females, Aluja, 1984; Greany et al. 1983) protected or used to attract the flies, which would then be selectively killed by localized bait spray applications sprayed from the ground, is an alternative worth further investigation. Even though the United States Department of Agriculture successfully carried out a program of intensive aerial bait spray applications over an area of 4000 Ha in which 300,000 fruit had to be individually inspected to assure a probit 9 level of infestation (Gongora, 1984), we consider this approach economically, socially and ecologically unacceptable.

In the same vein, we consider that the probit 9 concept used so far for the certification of fresh fruit for export should be substituted with a more rational approach. Landolt et al. (1984) have proposed a maximum infestation level based on the probability of a mating pair surviving a shipment. They suggest that treatments and other quarantine methods can be tailored to fit the crop, pest and infestation rates in the field, avoiding both unacceptably high risks and situations of costly and unnecessary overkill. Webb and Landolt (1984) have proposed using an accelerometer to detect vibrations produced by chewing and moving actions of larvae when sampling fruit to detect infestations, an approach that is worth further investigation since it would be of great value in Mexico.

Finally we would like to discuss the use of the Insect Sterile Technique (SIT). This method has been successfully applied in the desert areas of Sonora and Baja California against *Anastrepha ludens*. Even though SIT should be viewed as a viable alternative under these conditions, its use in any other regions of Mexico is improbable, due to the great number of species present in a single orchard or region. The eradication of one species would allow another one to take over, and the costs of releasing sterile flies of different species would be astronomical. Even in Sonora, care should be taken to always identify invading species since the fruit that reaches that region comes from southern states, where *A. ludens* is only one of a complex of fruit fly species.

3. CONCLUSIONS

We conclude that the future perspectives for the integrated management of fruit flies in Mexico are good, especially under current circumstances in which growers will be forced to control the problem at the orchard level. We nevertheless consider that unless a much more holistic approach is followed, the current trend of control only when there is an emergency, will be largely maintained. We have discussed in this paper a series of factors that we hope will contribute to the long term solution of the problem. Even though some of the ideas are still untested in the field, they bring up important factors that are usually neglected. Not until all the dimensions of the fruit fly problem are considered, will any important progress be made.

ACKNOWLEDGEMENTS

We gratefully acknowledge the invitation of M. Mangel to participate in this workshop. We also thank NATO for providing the necessary travel funds and the Mediterranean Fly Program, DGSV-SARH for funding part of the research here reported. We thank R. Prokopy, D. Papaj, R. Voss and L. Roth for useful comments to an earlier draft of this paper and W. Carey for his constant support. We thank J.A. Aluja for providing the funds to pay for the costs of this manuscript and M.Y. Quezada and A. Schünemann for their invaluable help in gathering part of the information here presented. M Aluja was a IAEA Fellow at the time this paper was written, and wants to thank the IAEA for its important support.

REFERENCES CITED

ALTIERI, M.A. 1983. Agroecology: the scientific basis of alternative agriculture. Division of Biology Control, Univ. of California, Berkeley, USA. 162 p.

ALTIERI, M.A. 1983. Vegetational designs for insect-habitat management. Environ. Management. 7:3-7.

ALTIERI, M.A. and D.K. LETOURNEAU. 1984. Vegetation diversity and insect pest outbreaks. CRC Crit. Rev. Plant Sci. 2(2):131-169.

ALUJA, M.R. 1983. Reporte de la comision a la Cuidad de Hermosillo, Sonora con el fin de evaluar la problematica que presentan las moscas de la fruita a la produccion de frutales. DGSV - SARH. Mexico D.F. Mexico. 15 p.

ALUJA, M.R. 1984. Manejo integrado de las moscas de la fruta. Programa Mosca Med, DGSV - SARH. Mexico D.F. Mexico. 241 p.

ALUJA, M.R.; M. CABRERA; G de la ROSA; E. RIOS; H. CELEDONIO-HURTADO; J. GUILLEN and D. MOTA. 1986. Ecological studies of fruit flies in the Soconusco, Chiapas. In preparation.

ALUJA. M.R.; M CABRERA; H. CELEDONIO-HURTADO and J. GUILLEN. 1986. Behavior of some Anastrepha species in nature: the effect of McPhail traps. In preparation.

ANONYMOUS. 1984. Vapor heat treatment for papayas. U.S. Federal Register 49(169):34196-34197.

COLL-HURTADO, A. 1982. ¿Es Mexico un pais agricola? Siglo XII Editores. Mexico D.F. Mexico. 214 p.

BARFIELD, C.S. and J.L. STIMAC. 1981. Understanding the dynamics of polyphagous, highly mobile insects. in Proc. IX Int. Conr. Plant Prot. Vol I:43-46.

BIRD, G.W. 1981. Implementation of Integrated Pest Management. in Proc. IX Int. Congr. Plant Prot. Vol I:404-407.

BRADER, L. 1982. Recent trends of insect control in the tropics. Ent. exp. & appl. 31:111-120.

D.G.E.A. 1980. Anuario estadistico de la produccion agricola de los Estados Unidos Mexicanos. SARH. Mexico D.F. Mexico. 1-286 pp.

D.G.E.A. 1980. La produccion agricola en Mexico en los ultimos diez años. SARH. Vol. IV (8):7-42. Mexico D.F. Mexico.

D.G.E.A. 1983. Consumos aparentes de productos agricolas 1925-1982. SARH. Vol VII (9):20-120. Mexico D.F. Mexico.

DENNO, R.F. 1983. Tracking variable host plants in space and time. in Variable plants and herbivores in natural and managed systems. (R.F. Denno and M.S. McClure eds.) pp. 291-341. Academic Press. New York. USA

EDENS, T.C. and H.E. KOENING. 1980. Agroecosystem management in a

resource-limited World. BioScience 30: 697-701.
EDMUNS, J.E. 1981. Implementation of plant protection in developing countries. in Proc. IX Int. Congr. Plant Prot. Vol I:393-397.
EHLER, L.E. and P.C. ENDICOTT. 1984. Effect of malathion-bait sprays on control of insect pests of olive, citrus and walnut. Hilgardia. 52:47 pp.
ESPINOSA, P. 1983. Personal communication.
FLETCHER, B.S. 1974a. The ecology of a natural population of the Queensland fruit fly, Dacus tryoni. V. The dispersal of adults. Aust. J. Zool. 22:189-202.
FOX, L.R. and P.A. MORROW. 1981. Specialization: species property or local phenomenon? Science. 211:887-893.
FRICK, K.E.; H.G. SIMKOVER and H.S. TELFORD. 1954. Bionomics of the cherry fruit flies in eastern Washington. Wash, Agr. Exp. Sta. Tech. Bull. 13:66pp.
GEIER, P.W.; L.R. CLARK and D.T. BRIESE. 1983. Principles for the control of arthropod pests. 1. Elements and functions involved in pest control. Prot. Ecol. 5:1-96.
GLIESSMAN, S.R.: R. GARCIA and M. AMADOR. 1981. The ecological basis for the application of traditional agricultural technology in the management of tropical agro-ecosystems. Agro-ecosystems. 7:173-185.
GONGORA, R. 1983. Plan piloto de manejo integrado de las moscas de la fruta en la zona citricola de Nuevo Leon. DGSV-SARH. Monterrey N.L. Mexico. 19p.
GOMEZ-POMPA, A.; C. VASQUEZ-YANES and S. GUEVARA. 1972. The tropical rain forest: a nonrenewable resource. Science 177:762-765.
GREANY, P.D.; S.C. STYER: P.L. DAVIS: P.E. SHAW and D.L. CHAMBERS. 1983. Biochemical resistance of citrus to fruit flies. Demonstration and elucidation of resistance to the Caribbean fruit fly, Anasthrepha suspensa. Ent. exp. & appl. 34:40-50.
GREANY, P.D. 1982. Personal communication.
HASSELL, M.P. and T.R.E. SOUTHWOOD. 1978. Foraging strategies of insects. Annu. REv. Ecol. Syst. 9:75-98.
HERRERA, A.L. 1905. El gusano de la naranja. Boletin de la Comision de Parasitologia Agricola. Tomo II:307-415.
KAREIVA, P. 1983. Influence of vegetation texture on herbivore populations: resource concentration and herbivore movement. in Variable host plants and herbivores in natural and managed systems (R.F. Denno and M.S. McClure eds.) pp. 259-289. Academic Press. New York. USA.
KASS, D.C.L. 1978. Polyculture cropping systems: review and analysis. Cornell Univ. College Agric. Life Sci. Ithaca. N.Y. USA.69p.
LANDOLT, P.J. and J. HENDRICHS. 1983. Reproductive behavior of the papaya fruit fly, Toxotrypana curvicauda Gerstaecker (Diptera: Tephritides). Ann. Entomol. Soc. Am. 76:413-417.
LANDOLT, P.J. 1984. Behavior of the papaya fruit fly Toxotrypana curvicauda Gerstaecker (Diptera:Tephritides), in relation to its host plants, Cerica papaya L. Folia Entomol. Mex. 61:215-224.
LANDOLT, P.J. 1984. Reproductive maturation and premating period of the papaya fruit fly, Toxotrypana curvicauda (Diptera:Tephritidae). Fla. Entomol. 67:240-44.
LANDOLT, P.J.; D.L. CHAMBERS and V. CHEW. 1984. Alternative to the use of probit 9 mortality as a criterion for quarantine treatments of fruit fly (Diptera:Tephritidae)-infested fruit. J. Econ. Entomol. 77:285-87.
LEVINS, R. 1986. Perspectives in IPM: from an industrial to an ecological model of pest management. in Ecological theory and IPM practice. (M. Kogan and D.C. Herzog eds.). in press.
MacARTHUR, J.D. 1980. Some characteristics of farming in a tropical environment. in Farming Systems in the tropics. (H. Ruthenburg ed.). pp. 19-29. Clarendon Press. Oxford. G.B.
MALAVASI, A. and J.S. MORGANTE. 1981. Adult and larval population fluc-

tuation of Anastrepha fraterculus and its relationship to host availability. Environ. Entomol. 10:275-278.
MALAVASI, A.; J.S. MORGANTE and R.J. PROKOPY. 1983. Distribution and activities of Anastrepha fraterculus (Diptera:Tephritidae) flies on host and nonhost trees. Ann. Entomol. Soc. Am. 76:286-292.
MEATS, A. 1981. The bioclimatic potential of the Queensland fruit fly, Dacus tryoni, in Australia. Proc. Ecol. Soc. Aust. 11:151-161.
MONTEIGH, L.G. 1971. The status of parasites of the apple maggot, Rhagoletis pomonella in Onterio. Can. Entomol. 103:507-512.
O'LOUGHLIN, G.T.O.; R.A. East and A. Meats. 1984. Survival, development rates and generation times of the Queensland fruit fly, Dacus tryoni, in a marginally favourable climate: experiments in victoria. Aust. J. Zool. 32:353-61.
OPP, S.B.; S.L. BUTKEWITCH and R.J. PROKOPY. 1985. Influence of orchard fungicides on apple maggot fly egglaying attempts. Z. and. Entomol. 99:248-253.
PALTI, J. 1981. Cultural practices and infectuous crop diseases. Springer. New York. USA. 243p.
PROKOPY, R.J. 1977. Stimuli influencing trophic relations in Tephritidae. Colloq. Int. CNRS. 265:305-36.
PROKOPY, R.J. and B.D. ROITBERG. 1984. Foraging behavior of true fruit flies. Am. Sci. 72:41-49.
PROKOPY, R.J.; P.T. McDonald and T.T.Y. WONG. 1984. Inter-population variation among Ceratitis capitata flies in host acceptance pattern. Entomol. exp. appl. 35:65-69.
RABB, R.L. 1978. A sharp focus on insect populations and pest management from a wide-area view. Entomol. Soc. Am. 24:55.
RISH, S.J. 1983. Intercropping as cultural pest control: prospects and limitations. Environ. Management. 7:9-14.
ROHWER, G.C. and D.L. WILLIAMSON. 1983. Pest risk evaluation in regulatory entomology. Bull. Entomol. Soc. Am. 29:41-46.
SAN JUANA, A.F. 1974. Control Integrado de Moscas de la Fruta en el Cañon del Tomellín, Oaxaca. DGSV-SAG. Mexico D.F. Mexico.
SHARP, J.L. and P.J. LANDOLT. 1984. Gustatory and olfactory behavior of the papaya fruit fly, Toxotrypana curvicauda Gerstaecker, (Diptera:Tephritidae) in the laboratory with notes on longevity. J. Georgia Entomol. Soc. 19:176-182.
SILL, W.H. Jr. 1982., Plant protection: an integrated interdisciplinary approach. The Iowa State Univ. Press. Ames. Iowa. USA. 297p.
TAYLOR, C.R. 1981. Economic analysis of crop protection systems. in Proc. IX Int. Congr. Plant Prot. Vol I: 7-9.
U.N.P.H. 1984. Directorio de la comercializacion de las hortalizas y frutas mexicanas. UNPH. Culiacan. Sinaloa. Mexico. 350p.
VAN den BOSH and P.S. MESSENGER. 1973. Biological Control. Intext. New York. USA. 137p.
WEBB, J.C. and P.J. LANDOLT. 1984. Detecting insect larvae in fruit by vibrations produced. Environ. Sci. Health. A19:367-375.
YOUNG, A.M. 1982. Population biology of tropical insects. Plenum. New York. USA. 511p.
ZORRILLA, P. 1983. Personal communication.

APPENDIX I - Map of Mexico

1 - Federal District
2 - Morelos
3 - Tlaxcala
4 - Aguascalientes
5 - Colima
6 - Queretaro
7 - Tabasco
8 - Hidalgo
9 - Nayarit
10 - Baja California Norte
11 - Baja California Sur
12 - Campeche
13 - Coahuila
14 - Chiapas
15 - Chihuahua
16 - Durango
17 - Guanajuato
18 - Guerrero
19 - Jalisco
20 - Mexico
21 - Michoacan
22 - Nuevo Leon
23 - Oaxaca
24 - Puebla
25 - Quintana Roo
26 - San Luis Potosi
27 - Sinaloa
28 - Sonora
29 - Tamaulipas
30 - Veracruz
31 - Yucatan
32 - Zacatecas

PROCESS ANALYSIS AND FAILURE AVOIDANCE IN FRUIT FLY PROGRAMS

Robert V. Dowell
Primary State Entomologist
California Department of Food and Agriculture
1220 Street
Sacramento, California 95814

Linda K. Wange
1681 Pebblewood Drive
Sacramento California 95833

ABSTRACT

As with any massive program, fruit fly programs have had their difficulties. The major problems which have resulted in failure appear to arise from four major sources: 1) adoption of the "cannot fail syndrome", 2) goal displacement, 3) incomplete consequence analysis, and 4) failure to question. Each represents an organizational mindset. Failures in fruit fly progrms are organizational not individual in nature. The "cannot fail syndrome" represents the belief that if a process is done correctly it cannot fail; therefore, why bother to monitor success, merely monitor the process. Goal displacement occurs when the execution of a program becomes more important than the original goal of the program. Incomplete consequence analysis is the incomplete identification of those functions in which failure carries unacceptable consequances. Failure to question is the blind following of historical precedents either as executed in previous programs or as found "in the

NATO ASI Series, Vol. G11
Pest Control: Operations and Systems Analysis
in Fruit Fly Management. Edited by M. Mangel et al.

literature". Examples of each type of failure are given in the text.

The likelihood of failure can be reduced by external reviews of project policies and procedures and by an effective, reciprocal information flow between project workers and management at all levels -- not just along the chain of command, but across project organizational units, also.

INTRODUCTION

An unprecedented number of exotic fruit flies have invaded California over the last six years. Eight different species have been discovered at various times (Table 1).
Six of these flies established infestations which were the targets of eradication programs. These programs were successful against five of the flies; the program against the sixth is still in progress (Table 2).

Each of the eight species of fruit flies poses the threat of causing extensive crop losses in California estimated at $910,544,000 per year, and control costs are estimated at $289,113,000 annually (Table 3). While these costs are of concern, it is the threat of greatly increased pesticide use that has convinced the California Department of Food and Agriculture (CDFA) to undertake the eradication programs listed in Table 2. Dowell (1985b) estimates that 90% of the pesticides used against the fruit flies would be used by private citizens in urban areas. This represents an

TABLE 1. EXOTIC FRUIT FLIES FOUND IN CALIFORNIA 1980-1985

FLY		YEARS FOUND (1980-1985)
Apple Maggot	- *Rhagoletis pomenella*	1983-1984-1985
Caribbean Fruit Fly	- *Anastrepha suspensa*	1983-1984
Medfly	- *Ceratitis capitata*	1980-1981-1982 1984-1985
Melon Fly	- *Dacus curcurbitae*	1985
Mexican Fruit Fly	- *Anastrepha ludens*	1983-1984-1985
Oriental Fruit Fly	- *Dacus dorsalis*	All Years
Peach Fly	- *Dacus zonatus*	1984
Solenaceous Fruit Fly	- *Dacus latifrons*[1]	1983

[1]larvae found in peppers sent from Hawaii.

TABLE 2. STATUS OF FRUIT FLY ERADICATION PROGRAMS AND NUMBER OF FLIES CAUGHT

FLY	NUMBER CAUGHT[1]	YEAR ERADICATION STARTED AND ENDED	ESTIMATED COST[2]
Caribbean Fruit fly	4	1984	
Medfly	455	1980-1982-3 sites	
Mexican Fruit Fly	177	1983-1984	
Oriental Fruit Fly	155	1983-1984-3 sites	
Peach Fly	2	1984	

1As of August 1, 1985

[2]Overmiller 1982

additional 5,818,000 kg of active material per year used by the citizens of California.

TABLE 3. ESTIMATED CROP LOSSES, CONTROL COSTS, AND INCREASED PESTICIDE USE THAT A STATEWIDE INFESTATION OF EACH FLY COULD CAUSE

FLY	ESTIMATED CROP[1] LOSSES ($1,000)	ESTIMATED CONTROL[1] COSTS ($1,000)	ESTIMATED INCREASED PESTICIDE[1] (1,000 Kg)
Apple Maggot	$146	$2,273	150
Caribbean	72,700	11,628	421
Medfly	206,000	156,000	1,818
Melon[2]	74,633	26,946	430
Mexican	86,974	20,366	1,091
Oriental	205,474	22,500	1,045
Peach	191,617	22,400	1,045
Solenaceous[2]	73,000	27,000	455
TOTALS	$910,544	$289,113	6,455

[1]From Dowell 1983, 1984, 1985a, Dowell and Odoemelam 1984, Harper and Dowell 1984, and Odoemelam and Dowell 1984.

[2]Dowell unpublished data

The number and diversity of fruit fly programs in California have given us an unparalleled opportunity to observe each programs' strengths and weaknesses. We are especially concerned with failures in these and other related programs: how they were discovered, how they arose and what steps were taken to correct the situation. These observations will form the basis for this paper.

A failure represents an inability to carry out the designated function of a program, and these failures are defined differently for the various programs. For example, a failure in a detection program occurs when a fruit fly infestation is discovered only after it is too large to deal with effectively. Failure in a quarantine program represents the movement of maggot infested produce into or out of the country or state. Eradication programs fail when the target population is not eliminated.

We view failures in fruit fly programs as arising from four sources: 1) adoption of the "cannot fail syndrome", 2) goal displacement, 3) incomplete consequence analysis, and 4) failure to question. Each of these are characteristics of an organization. They represent a mind set within which the individuals in the organization function. Failures in fruit fly programs are primarily organizational not individual in nature. These categories of failure sources are not mutually exclusive or absolute: they may overlap and any situation may have a combination of several categories.

<u>Cannot Fail Syndrome</u>

The "cannot fail syndrome" is characterized by the statement "If it is done properly it cannot fail; therefore why bother to monitor it?" It represents blind faith in the process as infallible. Ethylene dibromide (EDB) fumigation and trapping during eradication programs are two examples of this syndrome.

A flow diagram (Figure 1) shows the fumigation process formerly utilized to comply with quarantines on the movement of fresh fruit from Mexican fruit fly infested areas in Mexico out to uninfested areas. The process eliminated infested lots of fruit before fumigation, made certain that fumigation was done correctly, and monitored the success of the fumigation afterwards. The finding of a single maggot resulted in the rejection of the entire lot of produce.

Recently the United States Department of Agriculture (USDA) decided to simplify the process (Figure 2). It was stated, "Since fumigation cannot fail if done correctly, all we have to do is to monitor the act of fumigation to assure maggot free produce."

The result of this new program was that California received hundreds of boxes of mangos infested with live Mexican fruit fly larvae 7-10 days <u>after</u> fumigation. The fruit was sold in stores over an 800 km area of California. This error in judgement was discovered after consumers returned the maggot infested fruit to the stores and complained. As a result of this incident, the USDA and Mexican Government reinstituted the original process.

Another example of the "cannot fail syndrome" occured during the early stages of the medfly eradication program in California in 1980-1982, when it was suggested that it was unnecessary to trap inside the treatment zones. Proponents

FIGURE 1. Flow Diagram of Quarantine Compliance Process (pre-1985) for Fruit from Mexico.

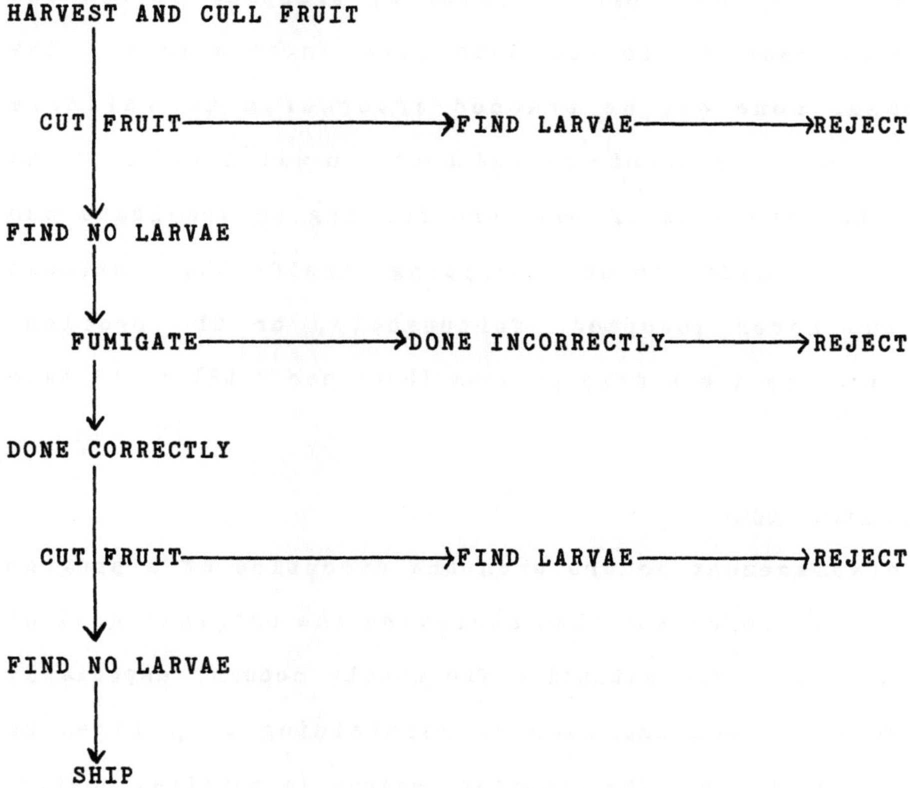

FIGURE 2. Flow Diagram of Quarantine Compliance Process in 1985 for Fruit from Mexico.

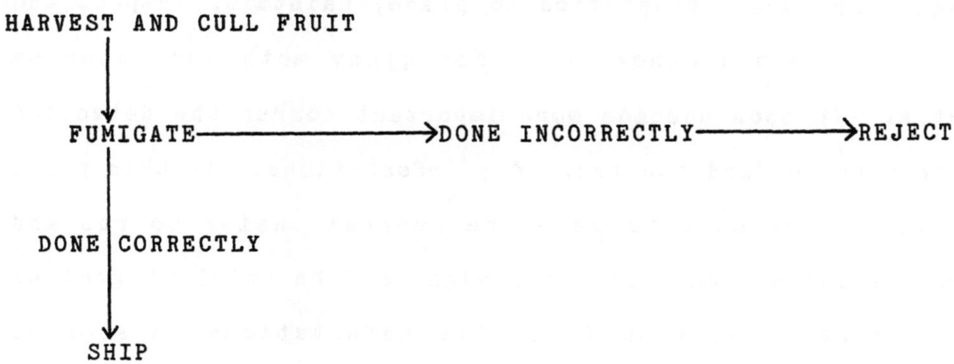

stated, "We are certain that the eradication tactics will work and that money can be saved by trapping around the treatment zone to locate satellite infestations. The treatment zone can be trapped afterwards to validate eradication." Opponents argued that you would be blind and unable to determine if you were failing to eradicate the pest. This advice to omit trapping inside the treatment zone was never accepted, fortunately, or the problems encountered by the medfly program (Scribner 1983) could have been worse.

Goal Displacement

Goal displacement occurs when the execution of a program becomes more important than achieving the original goal of that program. This situation frequently occurs, especially with routine programs such as maintaining trap lines to detect fruit flies. The trapping program is massive, costly, and time consuming, but must be done correctly to be effective. Each trapper maintains 12-17 individual fruit fly traps per 2.56 Km2 throughout their assigned area. There is frequently insufficient time to place, maintain, inspect and rotate these and other traps for gypsy moth and Japanese beetle. It soon becomes more important to run the detection traps than to find the fruit fly infestations. At this point decisions are made to make the program easier to run and these decisions sometimes lose sight of the original goal of the program: to find fruit fly infestations as soon as possible. We have three examples of goal displacement: the piggy backing of traps, the use of combolure (dorsalure), and the easing operational constraints.

The piggy backing of traps , placing more than one trap per site, is part of the official protocol (Gilbert et al. 1985) used by CDFA to run 90,000 traps (Henry pers comm.) to detect individual fruit flies and new fruit fly infestations. The goal of this detection program is to detect fruit flies; therefore, it seems reasonable to spread these 12-17 traps over the largest possible area and put out only one trap per site. However, this is an inefficient way to run a trap program. It is more economical to piggy back up to 3 traps per site (Gilbert et al. 1985) and utilize 8-10 different sites (Figure 3). This greatly simplifies the maintaining of the traps, however it also reduces the information gained by 17-53%. It may be acceptable to place 2 specific, non-interacting traps per site; but the general fruit fly traps (McPhail and Pherocon AM) should not be placed together with the specific traps (Chambers, et al. pers. comm.).

For a number of years combolure (dorsalure) has been used by the USDA and CDFA in detection programs for melon fly and oriental fruit fly. Combolure is a combination of cuelure for melon fly and methyl eugenol for oriental fruit fly. The use of combolure allowed one trap to detect both fruit flies, reduced expenses by 50%, and made program operation easier; however, traps baited with combolure catch 44-96% fewer

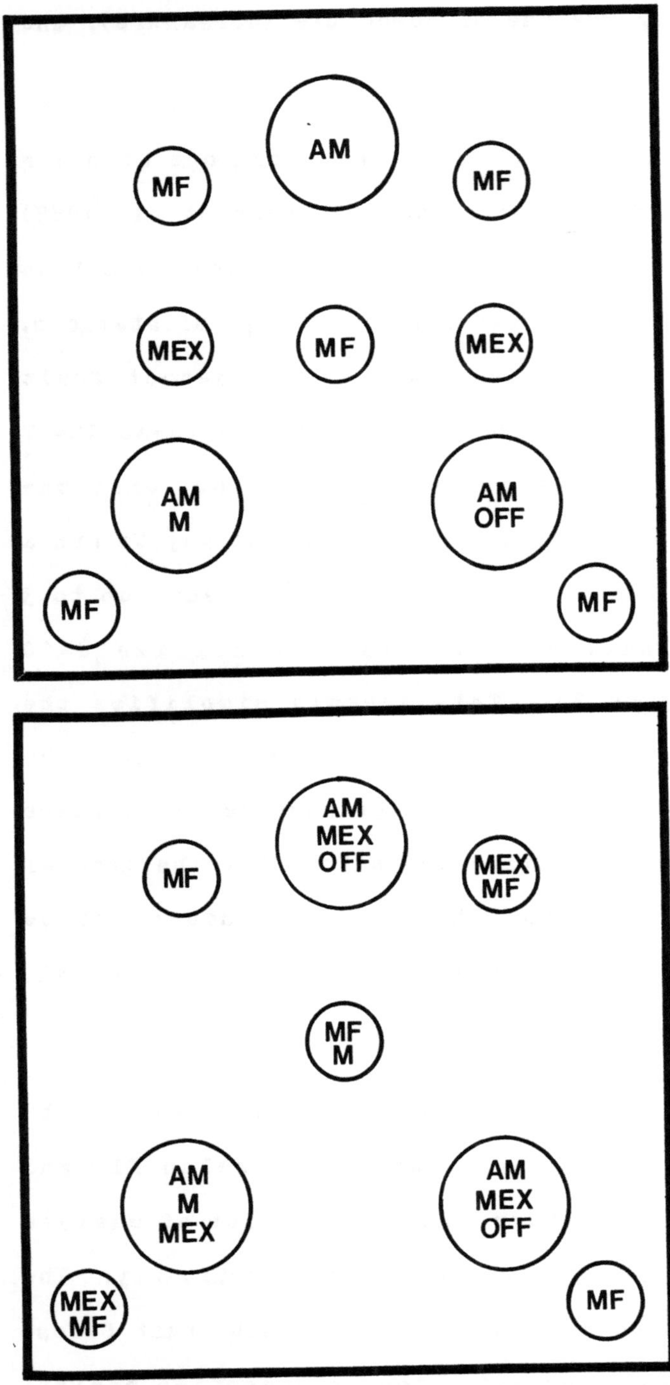

<u>KEY</u>

AM = Apple Maggot

Pherocon AM Trap

M = Melon Fly

Jackson Trap with Cuelure

MEX = Mexican Fruit Fly

McPhail Trap

MF = Medfly

Jackson Trap with Trimedlure

OFF = Oriental Fruit Fly

Jackson Trap with Methyl Eugenol

FIGURE 3. Examples of Piggy-backing Traps for Fruit Fly Detection.

oriental fruit fly than traps baited solely with methyl eugenol (Hooper 1978, and Tan 1982).

Oriental fruit fly is far more frequently trapped in California than the melon fly. Since 1956 three melon flies have been trapped compared to 827 oriental fruit flies. In addition eight infestations of oriental fruit fly have been eradicated; four having larvae present. Despite oriental fruit fly posing a greater risk of invading the U.S., it was USDA policy, until 1985, to recommend that combolure traps be used to detect oriental fruit fly (Poucher pers comm.).

Combolure traps have additional drawbacks. Melon fly and oriental fruit fly have different host ranges. Traps were hung in oriental fruit fly hosts (Gilbert et al. 1984) reducing their effectiveness for melon fly. Also, the two lures have different evaporation rates. Cuelure evaporates at a rate about 1/10th that of methyl eugenol (Cunningham pers. comm.). With servicing every 2 weeks, it is possible over a 4-6 week period for a wick to contain only cuelure. A trapper would then see that the wick has sufficient fluid and not rebait. This results in the trap having no methyl eugenol and no attraction for oriental fruit fly, the fly most likely to be present in California.

In 1985, CDFA implemented a policy of using separate traps for melon fly and oriental fruit fly (Clark pers comm.). It is interesting that in 1985 two of the three melon flies ever found in California were captured in traps baited solely with cuelure --- not combolure.

In a trapping program as large as California's, it is always desirable to ease operational constraints as much as possible. Easing of operational constraints is not always bad, but it can result in lowered trap effectiveness if all the consequences of a change in procedure are not carefully analyzed, and ultimately, the orignial goal of finding flies effectively has been displaced with managment of the program.

The steps in the development of the current trap and lure used to detect the medfly and the effect on trap catch are shown in Figure 4. Each step represents an easing of operational problems associated with running the traps: the protein lure is easier to use and more reliable than sugar; angelica oil is still easier to use and far more attractive than protein lures; the Steiner trap is plastic and dry which makes handling the trap and flies easier; Trimedlure easier to handle than angelica oil and more attractive; the Jackson trap dry, paper trap and far easier to handle than a Steiner trap. Each step makes trapping easier but also improves trap effectiveness. Overall a Jackson trap with Trimedlure is 19 - 84 times more attractive to medflies than a McPhail trap with sugar.

Unfortunately, not all changes in trapping procedures are as productive. This can be seen in the development of the trap and lure for the Mexican fruit fly (Figure 5). As with medfly, the first step replaces a sugar lure with a more attractive and consistent protein lure. The addition of borax to the protein lure stabilizes the solution, slows decay of the flies, but reduces attraction to 83% of the previous trap and lure. When the bait is made into a dry pellet trap, maintenance is greatly simplified, but this reduces trap attractiveness. At this point the trap is only 98% as attractive as a McPhail trap with sugar.

It has been suggested that plastic McPhail traps can be substituted for the glass ones currently used without loss of effectiveness (Nakahara et al. 1975). However, field tests have shown that the plastic McPhail trap is only 18% as attractive to Mexican fruit fly as a glass trap (Rhode and Sanchez 1982). While utilization of plastic traps could ease operational constraints, the resulting traps would be less than 20% as attractive as a glass McPhail trap baited with sugar. Therefore use of the plastic traps has been rejected by CDFA.

The probability of finding a fruit fly is a function of trap attraction <u>and</u> number. Although McPhail traps baited with protein tablets are less attractive to Mexican fruit fly than traps baited with liquid protein, the greater number of traps that can be run using the protein pellets as the lure offsets the loss of attraction.

<u>Incomplete Consequence Analysis</u>

This is the incomplete identification and oversight of those functions in which failure carries unacceptable consequences. In any fruit fly eradication program that uses the sterilized insect technique, the failure to identify wild flies from sterile flies would have catastrophic consequences.

During the medfly program in California (1980-1981) over 4,000,000,000 sterilized medflies were released (Churchill - Stanland et al. 1985). The flies were reared in four laboratories (Scribner 1983) and released throughout the infested zone, and an intensive trapping program was used monitor their success. Released flies were marked with florescent dye to differentiate them from undyed wild flies.

During this project the USDA and CDFA decided to determine which laboratory reared the best sterilized flies based upon the percentage recaptured in project traps (Note, we have a strong case of goal displacement here also). To do this flies from each lab were marked with a different colored dye (Scribner 1983). Unfortunately, undyed wild flies glowed under ultraviolet light in a manner very similar to sterilized flies marked with green or yellow dye (Churchill - Stanland pers comm.). In one quality control test 200 undyed, sterile medflies were placed into traps which were sent through the project identification lab. Only 10 were identified as wild; 190 were classified as sterile (Churchill - Stanland pers comm.).

In April 1981, the project thought it caught 25 wild medflies (Overmiller 1982) and that it was succeeding (Scribner 1983).

FIGURE 4. Development of Current Trap and Lure for Medfly.

TRAP/LURE	ATTRACTION COMPARED TO BASIC TRAP AND LURE
McPhail/Sugar (Baker et al. 1944)	Basic trap and lure
↓	
McPhail/Protein (Lopez and Spishakoff 1963)	1.4X Increase
↓	
McPhail/Angelica Oil (Steiner et al. 1951)	6.3-28X Increase
↓	
Steiner/Trimedlure (Beroza et al. 1961)	15.8-70X Increase
↓	
Jackson/Trimedlure (Harris et al. 1971)	19-84X Increase

FIGURE 5. Development of Current Trap and Lure for Mexican Fruit Fly.

TRAP/LURE	ATTRACTION COMPARED TO BASIC TRAP AND LURE
McPhail/Sugar (Baker et al. 1944)	Basic trap and lure
↓	
McPhail/Protein (Lopez and Spishakoff 1963)	1.4X Decrease
↓	
McPhail/Protein & Borax (Lopez and Becurri 1967)	0.8X Decrease
↓	
McPhail/Protein & Borax Pellets (Lopez et al. 1968)	0.8X Decrease
↓	
Plastic Trap (Proposed)/Protein and Borax Pellets (Nakagawa et al. 1975).	0.18X Decrease

In fact, on the basis of this quality control test these 25 wild flies caught probably represented approximately 500 wild flies: eradication was, in fact, failing. This failure was evident in June 1981, when large numbers of medfly larvae were found in fruit in residential areas (Overmiller 1982).

The decision to ascertain which lab produced the best sterile flies was made without taking into consideration that the project identification lab was already over worked and staffed with temporary employees, not scientists. There was no thought given to how this extra burden of sorting flies according to color would effect the lab. The lab was taxed beyond its abilities and disaster resulted. A critical portion of the eradication project was compromised placing the entire project in jeopardy because the lab was unable to identify sterile flies from wild flies. The resulting situation was saved only by an extensive program of aerial application of malathion and bait (Scribner 1983).

Failure to Question

It is easy for a program to rely upon past history to make decisions. Whether it is history as printed "in the literature" or represents previous practice, "it worked before", this unquestioning acceptance can lead to failure.

Perhaps the best known case of "its in the literature" concerned the ability of the medfly to survive the winter in northern California. Based upon a series of laboratory studies it was widely believed by policy makers and their advisors that the winters in northern California were too cold for the medfly to survive (Andrew et al. 1977, Flitters and Messenger 1958, Messenger and Flitters 1954). It was almost one year into the project before it was widely accepted by decision makers in California that the medfly could survive northern California winters. Adult medflies were trapped every month except February 1981, and larvae were found in fruit from November to February 1981 (Overmiller 1982). The capture of 25 medflies in April 1981, clearly showed that the fly had successfully overwintered.

During this time the area believed to be infested grew from about 51 Km^2 in June 1980, to 200 Km^2 in December 1980, to over 4,000 Km^2 by July 1981. Of note is an earlier and less widely known which study that concluded that the medfly would be able to survive as far north along the Pacific Coast as Portland, Oregon (Gjullin 1931).

At the start of any eradication program it is natural to review past successful programs and use them as a guide. The "it's worked before" viewpoint can give stability and assurance to project management. It can buffer against tinkering with a successful format; however, the same mindset can also prevent a full analysis of the current situation to ensure that these time tested tactics will indeed work in the current situation.

CDFA had eradicated an infestation of medfly in 1976 using limited ground spraying of pesticides, stripping fruit from infested properties, and the releasing of sterilized medflies (Cunningham et al. 1976). Those same tactics were utilized in 1980 against new infestations in Los Angeles and Santa Clara. They worked in Los Angeles, but failed in Santa Clara. Blind belief precluded a full analysis of the differences in the two situations. Since medfly traps were used in Northern California for the first time in 1980, the Santa Clara infestation, when found, was larger and had been in California longer than the Los Angeles infestation. From the beginning the project did not realize that they lacked a complete delimitation of the Santa Clara infestation and lacked sufficient numbers of sterile medflies to make the program work (Scribner 1983).

PREVENTING FAILURES

Problems such as we have described can arise in any organization at any time. Describing them is an easier task than preventing them from occurring. There are two major tactics that will minimize the probability of failures: 1) external reviews and 2) effective, reciporcal flow of information between workers and management across project organizatinal units.

CDFA utilizes periodic reviews of each eradication project by science advisory panels. Each panel is composed of scientists with extensive research or eradication experience with the target pest. The panels help project managment in decision making by suggesting or eliminating options in areas such as biological control, trap number, type and lure, and pesticide selection. They also act as a sounding board for project ideas. Based upon the advice from one panel, CDFA now uses liquid protein as the lure in McPhail traps used to delimit around positive trap catches of fruit flies. This policy makes the traps more difficult to maintain, but it increases trap attractiveness (Figures 5) to fruit flies.

There must be effective, reciporcal flow of information between project workers and management. This must occur through established channels, including quality control programs, and through informal channels such as personal contacts. The exchange of information should occur not just along the chain of command, but, also, across project organizational units. Information of little concern to one part of the organization may be critical to another part. Managers must constantly examine suggested changes by having them reviewed by numerous people on the project. Nothing can prevent failures of the types we have described; however, the likelihood of their occurrence can be greatly minimized.

The topic of this paper has been failures in fruit fly programs. It would be inappropriate if the reader was left with the impression that all fruit fly programs are filled with problems. The successes of California (Dowell 1985b) (Table 2) and Florida show that programs to detect and eradicate new infestations of fruit flies are regularly run in a competent, effective manner. What we wanted to show is that they are not perfect and that constant vigilance is essential to their continued success. One cannot rest on past laurels and assume that tactics used in previously successful programs are exactly what are needed for a specific, current program.

ACKNOWLEDGEMENTS

Many persons contributed to the ideas presented herein. They include Hilary Lorraine, Derrell Chambers, Glenn Hawkes, Roy Cunningham, and Donald Dilly.

Literature Cited

Andrew, C. O., Cato, J. C., and Prochaska, F. J., 1977. Potential eonomic impact of a fruit fly infestation on the U.S. Citrus Industry. Proc. Fla. State Hort. Soc. 90:29-32.

Baker, A. C., Stone, W. G., Plummer, C. C. and McPhail, M. 1944. A review of studies on the Mexican fruit fly and related species. USDA Misc. Pub. 531. 155 pp.

Beroza, M., Green, N. , Certler, S. I., Steiner, L. F. and Miyashita, D. H., 1961. New attractants for the Mediterranean fruit fly. Agric. Food & Chem. 9:361-365.

Chambers, D., Carey, J., Cunningham, R., Dowell, R., and Rice, R., 1985. Oriental Fruit Fly Program Critique. CDFA unpub. Rpt.

Churchill-Stanland, C. USDA APHIS. Guatemala Personal Communication

Churchill-Stanland, C., Stanland, R., Wong, T. T. Y, Tanaka, N., McInnis, D. O., and Dowell, R. In press. Size as a factor in the mating propensity of the Mediterranean fruit fly. <u>Ceratitis capitata</u> (Weidemann), (Diptera: Tephritidae) in the laboratory. J. Econ. Entomol.

Clark, R. A. Program Supervisor, CDFA, Sacramento, CA. Personal Communication.

Cunningham, R. T., Routhier, W., Harris, E. J., Cunningham, G., Tanaka, N., Oshinata, K., Johnston, L., Edwards, W., Rosander, R., and Vettel, W. G., 1976. Insect Eradication by Sterile Male Release: A Case Study, the Mediterranean Fruit Fly in Los Angeles. CDFA pub. 50 pp.

Dowell, R. V., 1983. Environmental assessment of the Mexican Fruit fly and its eradication in California 1983-1984 Program. CDFA pub. 95 pp.

Dowell, R. V., 1984. Environmental assessment of the apple maggot and its eradication in California. CDFA pub. 29 pp.

Dowell, R. V. 1985a. Environmental assessment of the Mediterranean Fruit Fly and its eradication in California. CDFA pub. 96 pp.

Dowell, R. V. 1985b. Surveillance and control of exotic insect pests in California. Bull. Soc. Vector Ecol. 10:52-59.

Dowell, R. V., and Odoemelam, O. O., 1984. Environmental assessment of the peach fruit fly and its eradication in California. CDFA pub. 22 pp.

Flitters, N. E., and Messenger, D. S., 1958. Effect of temperature and humidity on development and distribution of Hawaiian and Mexican fruit flies. J. Rio Grande Valley Hort. Soc. 12:7-13.

Gilbert, A. J., Bingham, R. R., Beavers, M. M., and Clark, R. A., 1985. Insect Trapping Guide. CDFA pub. 78 pp. (Revised).

Gjullin, C. M., 1931. Probable distribution of the Mediterranean fruit fly (*Ceratitis capitata* Weid) in the United States. Ecology 12:248-258.

Harper, P. A. and Dowell, R. V., 1984. Environmental assessment of the Caribbean fruit fly and its eradication in California - 1984 Program. CDFA pub. 78 pp.

Harris, E. J., Nakagawa, S., and Urago, T., 1971. Sticky traps for detection and survey of three tephritids. J. Econ. Entomol. 64:62-65.

Hooper, G. H. S., 1978. Effect of combining methyl eugenol and cuelure on the capture of male tephritid fruit flies. J. Aust. Ent. Soc. 17:189-190.

Lopez, F. and Bercerril, O. H., 1967. Sodium borate inhibits decomposition of two protein hydrolysates attractive to the Mexican fruit fly. J. Econ. Entomol. 60:137-140

Lopez, F., and Spishakoff, L. M., 1963. Finding New Lures for Mexican Fruit Fly. Calif. Citrog raph 1963:354,356-357

Lopez, F., Spishakoff,L. M., and Becerril, O. H., 1967. Pelletized lures for trapping the Mexican fruit fly. J. Econ. Entomol. 61:316-317.

Nakagawa, S., Suda, D., Urago, T., and Harris, E. J., 1975. Gallon plastic tub: a substitute for the McPhail trap. J. Econ. Entomol. 68:405-406.

Odoemelam, O. O., and Dowell, R. V., 1984. Environmental assessment of the Oriental fruit fly and its eradication in California. CDFA pub. 511 pp.

Overmiller, C., 1982. Medfly Eradication Project: Chronology of Events. CDFA pub. 130 pp.

Poucher, C. Florida Department of Agriculture and Consumer Services. Winter Haven, Fl. Personal Communication.

Rhode, R. H., and Sanchez, M., 1982. Field evaluation of McPhail and gallon plastic tub traps against Mexican fruit fly. Southwestern Entomol. 7:98-100

Scribner, J., 1983. The medfly in California: Organization of the eradication program and public policy. Hort. Science 18:47-52.

Steiner, L. F. 1957. Low-cost plastic fruit fly trap. J. Econ. Entomol. 50:508-509.

Steiner, L. F., Miyashita, D. H., and Christenson, L. D., 1957.Angelica oils as Mediterranean fruit fly lures. J. Econ. Entomol. 50:505.

Tan, K. H., 1983. Response of Dacus (Diptera: Tephritidae) to Ocimum sanctum (Linn.) extracts and different attractants in Penang, Malaysia in Fruit Flies of Economic Importance, A.A. Balkema pub. R. Cavalloro (ed).

AN ECONOMIC MODEL TO MEASURE COSTS AND BENEFITS OF ERADICATION

Richard L. Farnsworth
University of Illinois
Department of Agricultural Economics
305 Mumford Hall
Urbana, IL. 61801/USA

ABSTRACT

Fruit flies can be found in most fruit producing regions throughout the world. Countries such as the United States and Mexico use quarantines, monitoring, and eradication programs to prevent entry of fruit flies, particularly medfly. Other countries with permanent populations frequently entertain notions of large-scale eradication programs. This paper discusses the use of economic tools within a systematic decision framework to measure the costs and benefits of eradication programs. Several conclusions follow from the review and economic analysis. First, a systematic decision framework can aide researchers and policymakers define and structure the problem. Second, eradication programs can affect groups within and outside of agriculture; a comprehensive assessment of the associated costs and benefits should include all relevant impacts. Finally, a comprehensive analysis requires multidisciplinary teams to develop basic biological data, identify conflicting objectives, and measure the social, economic, and environmental consequences.

INTRODUCTION

The accidental introduction and possibly permanent establishment of a pest into new agricultural production areas can produce short-term and long-term economic, environmental, and social consequences. Immediate impacts might be increased production expenses and higher crop losses for growers located in pest infested areas. Also, additional pesticide applications or use of more toxic pesticides by growers to control the pest might aggravate local environmental and health problems. Continued dispersion of the pest to other production areas and protective actions from other states and countries, however, could substantially alter current demand and supply relationships in output and input markets, change cropping patterns, destabilize the domestic farm community, decrease a country's comparative advantage in world agricultural markets, and exacerbate national health and environmental problems.

Conversly, eradication of a permanently established pest population creates positive and negative economic, social, and environmental consequences. One could expect the following likely consequences: varying production cost impacts across grower groups, alteration of current supply conditions and output prices, a range of short-term and long-term environmental and health consequences, and changing demand in the input industries as well as consumer orientated industries.

NATO ASI Series, Vol. G11
Pest Control: Operations and Systems Analysis
in Fruit Fly Management. Edited by M. Mangel et al.

Policymakers must answer the following question: "What action, if any, should be undertaken to decrease the adverse economic, social, and environmental consequences of an accidental foreign pest introduction or a permanent pest infestation?" Unfortunately, the evaluation leading to selection of the best possible action is extremely complex as shown in the recent environmental impact statement for the proposed Hawaiian tri-fly eradication project (CICP, 1985). The Consortium for International Crop Protection, authors of the report, summarize available information and recommend a benefit-cost-risk analysis of six basic program options, including no action, at the local and national levels.

The remainder of this paper addresses two points, uncertainty and measurement of consequences, raised in the Hawaian environmental impact statement and other eradication proposals. First, a general decision process that can incorporate uncertainty and multiple objectives is quickly reviewed. Second several economic concepts and welfare theory are discussed as a means to measure economic consequences at the local and national levels. Several graphical examples using welfare analysis are included to illustrate how economics can measure benefits and costs of eradication proposals.

DECISION ANALYSIS FRAMEWORK

For a complex decision problem such as a response to a foreign pest infestation, a systematic decision framework can improve the analysis. One decision framework proposed by Keeney and Raiffa (1976) consists of five steps: preanalysis, structural analysis, uncertainty analysis, preference analysis, and optimization analysis. The five-step process can assist researchers and decisionmakers define the problem, determine feasible actions, identify individual and group consequences, associate consequences with each action, measure consequences in terms of individual and/or societal objectives, and choose the one action that optimizes a prespecified objective or objectives.

A payoff matrix (Figure 1) conveniently summarizes major components generated by the five-step decision framework: actions (A), events (E), event probabilities (P(E)), consequences (C), and preference structure (U(C)). Actions represent alternative eradication proposals. Events or states of nature (Savage, 1954: Marschack and Radner, 1972) represent factors not under a decisionmaker's control. Event probabilities, P(E), reflect the frequency in which events occur. For example, early studies assumed eradication as a proven technology. This meant eradication is achieved with complete certainty under all recommended actions. A more realistic approach might be two events, eradication feasible and not feasible with probabilities summing to one. Consequences capture the benefits and costs that result from specific action-event combinations. Consequence measurement is usually consistent with prespecified objectives. Past objectives such as cost minimization or benefit-cost analysis used monetary units. Future consequence measurements might include nonmonetary measures of health and environmental impacts. Finally, one action can be chosen in the payoff matrix that miximizes overall objectives and is consistent with the preference function U(C).

events and probabilities

$$
\begin{array}{cc|ccccc|}
 & & E_1 & E_2 & \cdots & & E_n \\
 & & P(E_1) & P(E_2) & \cdots & & P(E_n) \\
\hline
 & A_1 & U(C_{11}) & U(C_{12}) & \cdots & & U(C_{1n}) \\
\text{a} & A_2 & U(C_{21}) & U(C_{22}) & \cdots & & U(C_{2n}) \\
\text{c} & & \cdot & & & & \cdot \\
\text{t} & & \cdot & & & & \cdot \\
\text{i} & & \cdot & & & & \cdot \\
\text{o} & A_i & U(C_{i1}) & \cdots & U(C_{ij}) & \cdots & U(C_{in}) \\
\text{n} & & \cdot & & & & \cdot \\
\text{s} & & \cdot & & & & \cdot \\
 & & \cdot & & & & \cdot \\
 & A_m & U(C_{m1}) & \cdots & & & U(C_{mn}) \\
\hline
\end{array}
$$

Figure 1: Payoff Matrix

The five-step decision framework besides formalizing the decision process can direct attention to major bottlenecks. One area is measurement of consequences. The next section introduces several economic concepts and a measurement method that can be used to quantify impacts associated with eradication programs.

BASIC ECONOMIC CONCEPTS AND APPLIED WELFARE ANALYSIS

An accurate assessment of consequences for alternative eradication actions (strategies) and unknown events (states of the world) requires close cooperation between physical and social scientists. Using their training and expertise, physical scientists communicate to economists plausible output and input changes under different actions and states of the world (C_{ij}). In turn, economists assign value to the output and input changes, assess the aggregate impacts in an uncertain economic environment, and summarize the implications in terms of identified objectives.

Welfare theory provides a conceptual framework that allows economists to identify and measure consequences. After a discussion of the essential economic concepts, welfare theory is applied to a general eradication problem.

Economic Concepts

The economic concepts required to understand and use welfare theory are demand and supply curves, producer surplus, consumer surplus, and market equilibrium. First described at the firm or individual level, these concepts are expanded to a market or industry level to assess net changes in societal wellbeing.

Supply Curve: For an individual firm, the supply curve, S, represents the quantity of goods, services, or intermediate products a firm willingly supplies at different prices during a specified period of time, with all other factors held constant (Figure 2). The supply curve slopes upward to reflect a firm's willingness to supply more output at higher prices. In the short-run, the supply curve is a firm's marginal cost curve. Marginal cost is the increase in total variable cost to produce an additional unit of output.

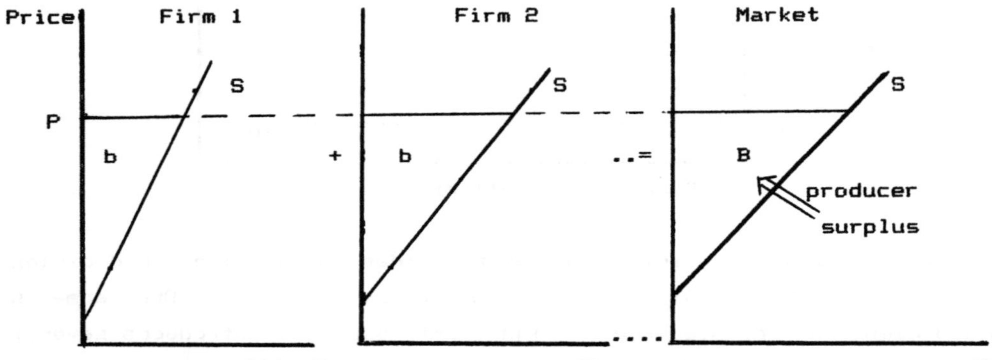

Figure 2: Individual and Market Supply Curves

The horizontal summation of individual firm supply curves produces an industry supply curve exhibited in Figure 2. In an analogous way, the industry supply curve shows the total quantities supplied by the firms in the industry at different output price levels. The industry's supply curve is the industry's short-run marginal cost curve. The supply curve shifts position if any relevant constant factors change: resource prices, substitute and complement good prices, technology, number of sellers, taxes, etc. For example, introduction of a foreign pest might decrease output and/or increase insecticide expenditures. Industry and individual supply curves would shift up and to the left to show increased costs at every output level.

Producer Surplus: Producer surplus, sometimes called quasi-rent, is defined as the area above the firm supply curve and below the market price line. This area, labeled b in Figure 2 is a measure of the difference in gross receipts and total variable costs. Producer surplus represents the total benefit of a producer remaining in business (producer surplus = profit + fixed costs). Horizontal summation of each firm's producer surplus equals aggregate producer surplus (B) at the industry level.

Demand Curve: Starting at the individual level, a demand curve, D, represents the quantity of a particular good, service, or product a consumer willingly purchases at different prices during a specified time period, with all other factors fixed at current levels. An individual's demand curve slopes downward to reflect more purchases of quantity Q at lower prices (Figure 3).

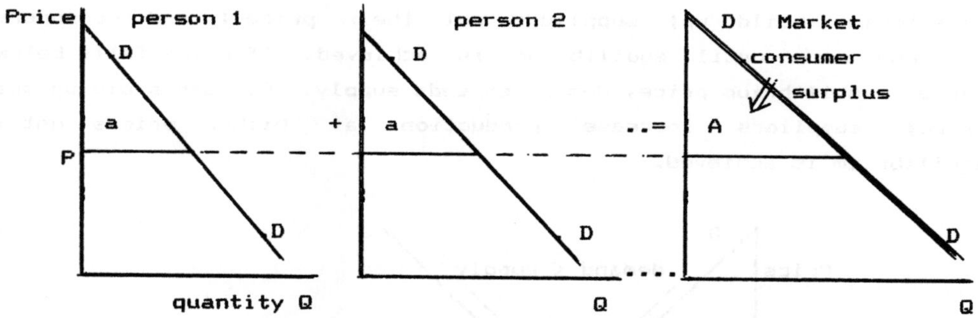

Figure 3: Individual and Market Demand Curves

Horizontal summation of individual demand curves produces a market demand curve. The market demand curve (Figure 3) shows the quantity of good Q willingly demanded by all consumers at varying prices with all other factors held constant. The demand curve shifts if relevant constant factors change: income, prices of related goods, tastes, preferences, etc. For example, if income increases, a consumer might purchase more of a commodity at every price level. An increase in income implies a shift of the demand curve out and to the right.

Consumer Surplus: A common definition of consumer surplus is the difference between the maximum amount a consumer willingly would have paid and the actual price paid for good Q. In Figure 3, consumer surplus represents the area below the demand curve and above the market price line labeled P. Again, a horizontal summation of individual consumer surpluses gives market consumer surplus.

The notion of consumer surplus is an extremely controversial subject and requires a brief digression. The controversy stems from our knowledge that satisfaction or utility a consumer receives from consumption is not observable. Hence consumer surplus as defined is simply an area without welfare significance. Fortunately under fairly broad conditions, consumer surplus attains welfare significance when it can be related to willingness to pay measures: compensating and equivalent variation. A complete discussion of the subject is beyond the scope of this paper. The interested reader, however, can find an adequate statement and resolution of the problem in Just, Hueth, and Schmitz (1982).

Market Equilibrium: Equilibrium is achieved in a market when supply equals demand. Figure 4 illustrates market equilibrium and identifies the market clearing price and equilibrium quantity. The market mechanics are quite clear. Above the market equilibrium price, suppliers produce more of good Q than consumers willingly buy. Inventories build up; suppliers cut their prices; and consumers purchase more until equilibrium is achieved. If price falls below market equilibrium price, demand exceeds supply. Consumers bid up the price; suppliers increase production at higher prices until equilibrium is achieved.

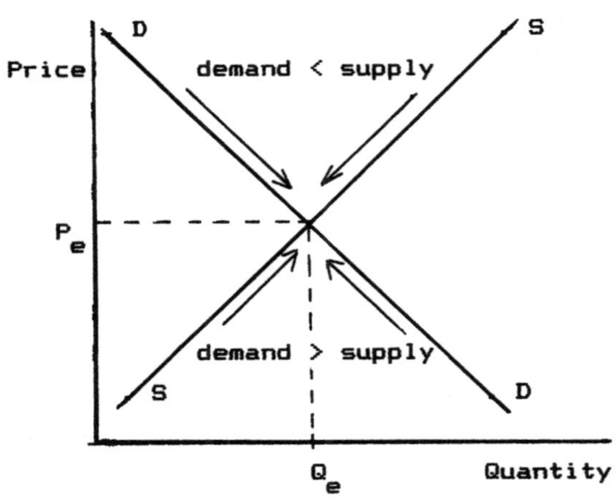

Figure 4: Market Equilibrium

<u>Applied Welfare Analysis</u>

We are now prepared to examine conceptually some of the aggregate impacts that might result from an eradication program. Three frequently mentioned situations come to mind that require investigation in a comprehensive eradication proposal assessment: successful eradication with one producer group and no changes in related markets; with different producer groups; and with adjustments in related markets. For simplicity, assume eradication can be achieved with a probability of one, consequences can be measured in monetary units, and environmental and health consequences are insignificant. The payoff matrix consists of m rows and one column. Consequences are the summation of changes in product and input markets minus program costs.

Successful Eradication and Net Social Gain: Physical scientists predict an output increase and/or input expenditure decrease given a successful eradication program. That is, the firm can now produce the same quantity at a lower cost or a larger quantity at the same cost. In addition, higher profits generally attract new firms into the industry. The aggregate impact is a shift of the supply curve down and to the right, from S to S' in Figure 5.

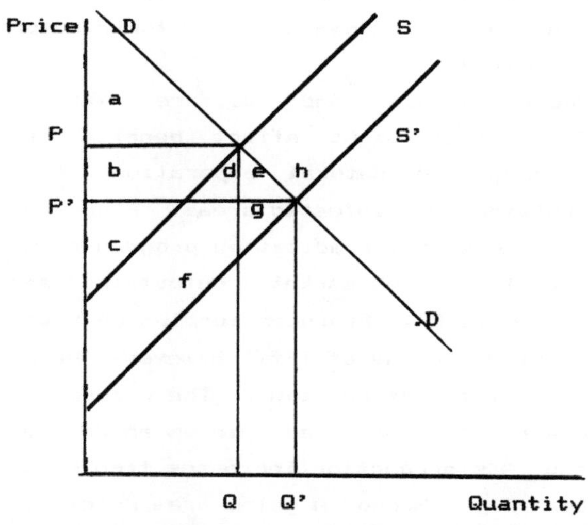

Figure 5: Successful Eradication and Welfare Gain

What can be said about a change in social wellbeing? Before an eradication program the equilibrium market price and quantity are P and Q. Consumer surplus is area a and producer surplus is area b+c. After the supply curve shifts to S', equilibrium price falls to P' and quantity sold increases to Q'. Consumer and producer surplus after a successful eradication program are areas a+b+d+e and c+f+g, respectively. The change in consumer surplus is area b+d+e (a+b+d+e−a). The change in producer surplus is area f+g−b (c+f+g−(b+c)). The net change in social welfare is d+e+f+g. In other words, society gains the amount between the two supply curves and below the demand curve assuming no changes in other markets. Eradication program costs can be subtracted from the market gains to evaluate economic feasibility.

Three additional scenarios can be examined using Figure 5. First, if one incorrectly assumes a successful eradication program does not alter equilibrium price, benefits are overstated by area h. For example, the medfly eradication proposal for Central America (1972) assumed a successful eradication program with no change in market price. If price did indeed change, the breakeven point defined as the point where cumulative benefits equal cumulative costs would have occurred later than stated. Second, eventhough society gains d+e+f+g, producers as a group are worse off if the change in producer surplus, f+g−b, is negative. The CICP (1985)tri-fly environmental impact statement mentioned this possibility. Third, the analysis can be reversed to show the impact of a pest introduction. S' becomes the original supply curve that shifts to S. Society loses area d+e+f+g. If producers face an inelastic demand for their output, producer revenue might increase (b>f+g).

Different producer groups and welfare effects: Another overlooked possibility that might affect benefit estimation is different producer groups. A natural separation is growers who produce inside and outside pest infested areas. Figure 6 illustrates a situation where a successful eradication program results in a net gain of area h+i in the output market. Output increased from Q to Q'. Price dropped from P to P'. Producer surplus changes from f+g to area g+i, a net positive change of i−f. However, the program helps only producers in the infestation zone. Their supply curve shifts from S^A to $S^{A'}$ because of increased yields and/or decreased pest control costs. Group A's production increases from Q^A to $Q^{A'}$ at the new equilibrium price P'. Group A gains area e−(c+a). At the new equilibrium price P', however, group B's production drops from Q^B to $Q^{B'}$. Producer surplus for Group B decreases from area a+b to area b, a loss of area a. Hence, even though societal net gain is area h+i minus program costs, some grower groups can experience losses which should be measured.

Welfare Effects in Related Markets: Eradication programs can significantly alter other input and output markets. In a boll weevil eradication study, Carlson and Suguiyama (1985) indicate the demand for land increased for cotton production in the eradicated zone. In the classic study by Schmitz and Seckler (1970), introduction of the tomato harvester decreased demand for migrant farm labor. Hence, the authors included unemployment cost in the aggregate analysis.

For a complete analysis, welfare impacts in related markets must be measured. A typical situation is depicted in Figure 7. Successful eradication of a pest shifts the output supply curve from S to S'. Output increases from Q to Q"; price decreases from P to P". Net welfare gain in the output market is f+h. One possible negative impact is decreased demand for insecticides shown by a shift of curve D to D' in the insecticide market. Without any further adjustments, quantity demanded drops from Q to Q'; price, from P to P'. Producers in the insecticide industry lose area a+b. Net gain for the two affected markets is measured by area f+h-a-b, not area f+h. Again, program costs must be subtracted to obtain net social gain.

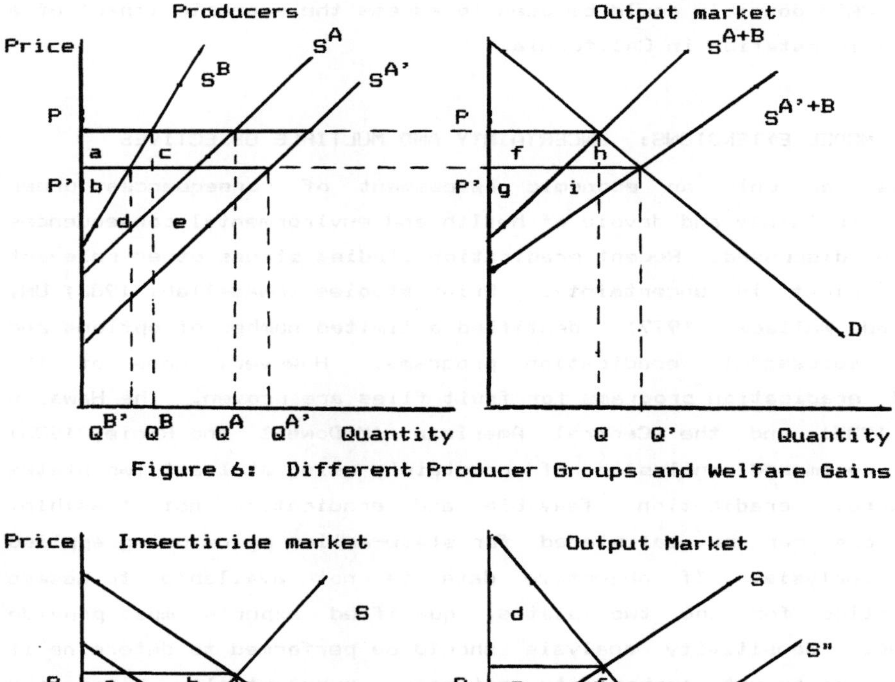

Figure 6: Different Producer Groups and Welfare Gains

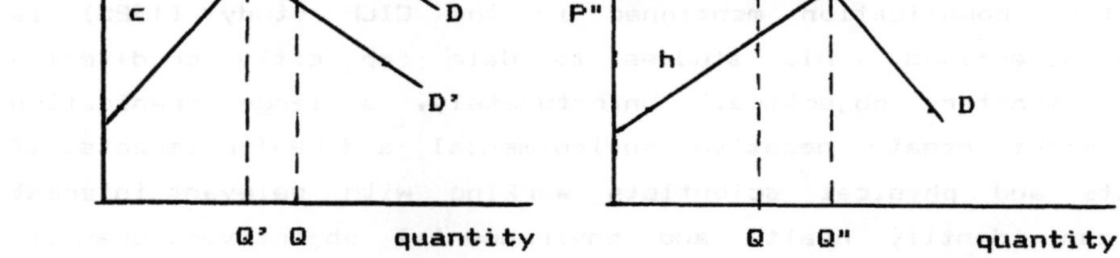

Figure 7: Welfare Gains in Output and Input Markets

For eradication projects, related markets can include agricultural and nonagricultural industries. Successful eradication programs can change crop production patterns which in turn alter demands in input markets. One very important nonagricultural industry identified in the Hawaiian tri-fly study is tourism (CICP, 1985).

Aggregate Welfare Models

The above three examples show how welfare analysis can be applied to assess benefits and costs of eradication programs. Applied welfare theory is not an abstract concept. Several aggregate models have been constructed to assess real world problems. Two current models are TECHSIM (AER, 1983) and CARM (personal communication). TECHSIM, an econometric simulation model, was used to assess the boll weevil eradication program and the introduction of an exotic pest (P. pachyrhizi) into the United States (ERS Staff Report, 1981; AER, 1984). CARM possibly could be used to assess the economic impact of a fruit fly infestation in California.

MODEL EXTENSIONS: UNCERTAINTY AND MULTIPLE OBJECTIVES

Thus far only an economic assessment of consequences under complete certainty and devoid of health and environmental consequences has been discussed. Recent eradication studies stress other relevant issues. First is uncertainty. Prior studies (Awadallah, 1983; UN, 1973; and Wallace, 1977) identified a limited number of options and assumed successful eradication programs. However, none of the proposed eradication programs for fruit flies are proven. The Hawaian (CICP, 1985) and the Central American (McDowell and Pirie, 1985) studies recommend inclusion of uncertainty using at least two states of nature: eradication feasible and eradication not feasible. Consequences can be calculated for state-action pairs using applied welfare analysis. If objective data is not available to assess probabilities for the two states, qualified experts must provide estimates. Sensitivity analysis should be performed to determine if slightly different probability estimates substantially alter the choice of actions.

Another complication mentioned in the CICP study (1985) is multiple objectives. All studies to date implicitly or directly assume a monetary objective. Unfortunately, a large eradication project might create negative environmental and health impacts. If economists and physical scientists working with relevant interest groups can identify health and environmental objectives, quantify factors to reflect the objectives, and measure the changes for state-action pairs, consequences in the payoff matrix can be vectors that reflect monetary and nonmonetary elements. The analysis could then progress through steps four and five and a preferred action identified consistent with prespecified multiple objectives.

Keeney and Raiffa (1976) discuss multiobjective decision analysis in great detail. They present several applied examples in the areas of air pollution, emergency services, and siting of airports and nuclear facilities.

Two realistic alternatives exist, however, that reduce multiple objective problems to single dimensional problems. A decision problem can be converted to one objective if health and environmental impacts are quantifiable in monetary units using willingness to pay methods similar to those found in the economics literature (Just, Hueth, and Schmitz, 1982). Another alternative is to establish aspiration levels for nonmonetary objectives (Keeney and Raiffa, 1976). Choose one action that maximizes net social gain among a set of actions that do not increase infant mortality by .001 is an example. One should note that an unlimited number of aspiration levels can be constructed. Hence, interested parties must reach consensus before this approach can be used.

CONCLUSIONS

Successful eradication programs have the potential of increasing social wellbeing. Unfortunately, real and perceived consequences can affect many groups within and outside the agricultural sector. Relevant consequences must be identified and measured. Given a systematic decision framework and welfare theory, researchers and policy makers can merge their expertise and produce a comprehensive and defensible evaluation of eradication techniques and programs.

References

Awadallah, A. et. al. 1983. "A Survey of Crop Losses Caused by the Mediterranean Fruit Fly in Egypt." Prepared for the Joint FAO/IAEA Division of Isotope and Radiation Group, 21 June.

Carlson, G. A. and L. Suguiyama. 1985. "Economic Evaluation of Area-Wide Cotton Insect Management: Boll Weevils in the Southeastern United States. North Carolina Agricultural Research Service Bulletin 473, June.

Collins, G. S. and C. R. Taylor. 1983. "TECHSIM: A Regional Field Crop and National Livestock Econometric Simulation Model." Agricultural Economics Research 35:2 (April), pp. 1-18.

Consortium for International Crop Protection. 1985. "Eradication of the Tri-Fly Complex From The State of Hawaii: Draft Environmental Impact Statement. Prepared for the Plant Protection and Quarantine , Animal and Plant Health Inspection Service, United States Department of Agriculture.

Just, R. E., Hueth, D. L. and Schmitz, A. 1982. Applied Welfare Economics and Public Policy. Englewood Cliffs, N.J.: Prentice-Hall.

Kenney, R. L. and Raiffa, H. 1976. Decisions with Multiple Objectives: Preferences and Value Tradeoffs. New York: John Wiley & Sons.

MacCrimmon, K. R. 1973. "An Overview of Multiple Objective Decision Making." Multiple Criteria Decision Making. Edited by J. L. Cochrane and M. Zeleny. Columbia, South Carolina: University of South Carolina Press, pp. 16-44.

McDowell, R. and Pirie C. 1985. "Benefit-Cost Analysis of Medfly Eradication in Central America." Prepared by the Policy Analysis and Program Evaluation Staff, Budget and Accounting Division, Animal and Plant Health Inspection Service, 4 March.

Marschak, J. and Radner, R. 1972. Economic Theory of Teams. New Haven and London: Yale University Press, 1972

Savage, L. J. 1954. The Foundations of Statistics. New York: Dover Publications, Inc.

Schmitz, A. and Seckler, D. 1970. "Mechanized Agricluture and Social Welfare: The Case of the Tomato Harvester." American Journal of Agricultural Economics, 54:4, pp. 569-570.

Taylor, C. R., et. al. 1983. "Aggregate Economic Effects of Alternative Boll Weevil Management Strategies." Agricultural Economics Research, 35:2 (April), pp. 19-28.

Trueman, R. E. 1974. An Introduction to Quantitative Methods for Decision Making. New York: Holt, Rinehart and Winston, Inc.

United Nations Development Programme. 1973. "Peru Fruit Fly Eradication: Report of the Evaluation Mission." May.

Wallace, M. C., et. al. 1977. "The Mediterranean Fruit Fly and its Economic Impact on Central American Countries and Panama." Prepared for the United States Agency for International Development by the UC/AID Pest Management and Related Environmental Protection Project at the University of California, Berkeley, under Contract AID/ta-C-1380. May.

DIFFERENTIAL PERCEPTIONS BETWEEN CITIZENS & SCIENTISTS REGARDING PESTICIDE USE

Glenn R. Hawkes and Martha C. Stiles
Department of Applied Behavioral Sciences
University of California, Davis

INTRODUCTION

In 1980 an event, heightened by its occurrence in California, which produces a quarter of the food grown in the United States, became an incident of international interest. The Mediterranean fruit fly, a dangerous agricultural pest, infested the densely-populated Santa Clara Valley. The State of California responded by spraying pesticides from aircraft over residential neighborhoods, resulting in a successful, but lengthy, multi-million dollar eradication program. Medfly's estimated damage exceeded millions of dollars to the agricultural industry and demanded use of pesticides, a technology perennially under public and scientific scrutiny. Aerial spraying in urban areas increased the likelihood that sensitive non-target organisms, namely the Valley's human inhabitants would be exposed to the chemicals. This provided us with the opportunity to assess perception of risk at the time residents were undergoing aerial spraying. Since that time we have seized the opportunity to obtain citizens' views of eradication programs for other voracious pest, using a variety of pesticides and application methods. This paper discusses results of the study examining attitudes of the general public and an "elite" sample of technical experts and decision-makers. The major questions addressed herein are: Do the public, technicians, decision-makers and scientists fundamentally disagree on the issue of risks, specifically those attributable to pesticides? Is there agreement among the elite respondents? Where do the major disparities and congruities of attitudes lie?

Since the 1980 appearance of the Medfly, California citizens and administrators have evidenced an ebb and flow of discourse on the urban use of agricultural chemicals. Threats of infestations from the Mexican, Caribbean and Oriental fruit flies, the gypsy moth and the Japanese beetle have resulted in aerial and/or ground application of pesticides within city neighborhoods throughout California. Coat-tailing the decisions to spray were immediate and often impassioned contraventions by community groups. In past situations, overt conflict within communities has had extreme effects on how crisis events were managed, how quickly and quietly technological developments were adopted, and how widely older technologies, like pesticides, were accepted. Indeed, when confronted

NATO ASI Series, Vol. G11
Pest Control: Operations and Systems Analysis
in Fruit Fly Management. Edited by M. Mangel et al.

with involuntary and unknown risks, the public, their elected leaders and watchdog public interest groups have few options but to respond to the threat as they see fit. A common component of the fear response is calling into question the credibility, integrity and knowledge of those in charge of public health and safety. A response by some officials to public outcry is that the people are "ignorant of the facts" and respond inappropriately to minimal-risk situations. "Public perception has come to mean in the technical literature, public **misperception**." (Sharlin & Allen, 1985). However, risk decisions have and will continue to have a major place in shaping society's future. The decisions will be shaped largely by public perceptions of risk. Since the potential for social conflict pervades hazard management and risk evaluation processes, it is important to isolate patterns of beliefs and biases typifying those embroiled in the controversy. The magnitude and frequency of hazards from old and new technologies have increased public awareness of risk potential. Therefore, information on how and why people perceive risk has become a paramount concern of decision-making entities.

Even though research examining expert and citizen perspectives on technological risks is rare, the existing data show divergent views and judgmental biases on the part of scientists, government officials and the general public. (Slovic, et al 1979). Other studies have established correlations among various risk factors (i.e. magnitude of catastrophe, levels of uncertainty, voluntariness of exposure) and acceptability. Some have identified socio-demographic characteristics of groups who tend to be more risk averse than others. This study will describe perceptions of risk and risk acceptance of six groups involved in pest eradication programs using chemical pesticides. Patterns of beliefs distinguishing high risk and low risk groups will be identified.

In his pioneering work on social benefit and technological risk, Starr (1969) suggested that acceptable levels of risk varied with the degree of voluntariness of the event. As voluntariness or control decreases, acceptance decreases. However, Green (1978) reported correlations between high levels of perceived safety with increased involuntariness of participation in an activity. These polar conclusions may indicate that attitudes and behaviors toward risk are event-specific. That is, risk research should be done within the context of each activity.

A study by Hawkes, et al, (1984 a,b) shed light on how citizens undergoing aerial spraying of pesticides perceive the uncertainty of risk. It also described the relationship between perceived risk and reported precautionary behaviors. They found that women were more risk averse than men, and younger participants were more concerned about health and environmental hazards of spraying than the older respondents. This trend was demonstrated also among those having college degrees versus the less formally educated. In addition, level of confidence in hazard management experts, media exposure, and political orientation directly influence degree of perceived risk and risk acceptability. Perceived benefit directly influenced how risky respondents considered a situation. Slovic, et al (1979) concluded that perceived risk is influenced by imaginability and memorability of the hazard. Fischhoff, et al (1978) found that experts are just as prone to overconfidence, as are lay people. If hazard management experts must look beyond available data and depend on their own determinations and estimations of risk, data show they consistently underestimate the probability of hazard. Cited examples include the Three Mile Island incident, the Teton Dam failure, the Hyatt Regency collapse in Kansas City, Union Carbide in Bhopal, and, of course, chemicals such as DDT, EDB, and DBCP. Neihaus and Swaton (1981) reported findings that attitudinal variations are primarily due to diverse beliefs about the technologies, and to a lesser degree due to basic value differences among groups.

Several questions remain unanswered. Are there belief differences among the numerous agents (i.e. experts vs. citizens) interacting to oppose or support a potentially risky and controversial event? If so, how do those differences influence interactive behaviors?

"One would like to believe that, as evidence accumulates and the public and the experts come to share a common experience, their perceptions would converge toward one 'appropriate' view. Unfortunately, this is not likely to be so. A great deal of research indicates that once informed, people's beliefs change very slowly, and are extraordinarily persistent in the face of contrary evidence" (Nisbett and Ross, 1980). Previously-established systems of beliefs or impressions funnel current and future information into tried and tested molds. Veridicality and trustworthiness of subsequent input is increased if it complements initial beliefs. Otherwise the new evidence is suspect of being false or unreliable (Slovic, 1981).

This study examines the alignment and/or incongruity of attitudes between citizens, scientists and other special groups toward one technology, pesticides. Citizens from cities involved in California's urban pest eradication programs using aerial and ground application of pesticides were surveyed. The elite groups were involved in pesticide application, regulation, research, and information dissemination. Many were directly involved in eradication programs.

THE STUDY

Two independent samples were drawn: one representing the general public and the other representing elite groups. To sample the **public**, names of subjects were systematically drawn from four California phone directories for Stockton, Milpitas, Orangevale, and the Concord/Clayton area. Each of these communities had been involved in a state pest eradication program. Stockton and Milpitas were aerially treated for Medfly. Orangevale had Japanese beetle; and Concord/Clayton experienced gypsy moth.

The **elite** sample consisted of individuals involved in: (1) policy making or implementation of pesticide regulations; (2) research and scientific advisement on agricultural issues including pesticides and (3) technical aspects of chemical use such as application. Most elite specialists were located in counties recently threatened by pest infestations. Five sub-groups were used for this analysis (1) elected leaders (city, county, state); (2) government agency administrators (county, state and federal); (3) applicators, chemical representatives and growers; (4) selected university scientists and specialists (entomologists, toxicologists, and other agriculturalists); and (5) public interest groups. This sample was systematically drawn from directories of university, state, county, and city officials; membership rosters from the Council of California Growers, Western Agricultural Chemical Association, California Agricultural Production Consultants Association.

The public was contacted by phone first to solicit participation. A **questionnaire** was mailed to all participants (public and elite) with return envelopes. Items in the questionnaires addressed risk, acceptability, costs, benefits, and trust in hazard management entities. Four-point Likert-like responses were included for each item. A portion of the questionnaire was also devoted to demographic characteristics. Measures were constructed by scaling items and summing scores across each individual variable. Level of significance and frequency of responses are reported.

Degree of perceived **risk** of pesticides was measured by several items describing various settings and modes of application. Respondents were asked to evaluate: level of perceived harm or safety attributable to ground and aerial spraying of pesticides, risk to home users, field workers in cities and on farms, for agricultural use and neighborhood pest infestations.

Acceptability of the risk addressed the public's willingness to endure the risk of the chemicals. The expert questionnaire was worded to measure degree of risk experts **expect** citizens to accept. Extent of acceptance was measured by expressed necessity or desirability to continue extensive use of the chemicals, preference for more benign pest control methods, and direct report of how much hazard the public should tolerate.

Benefit and Cost included perceived qualitative consequences of using pesticides (i.e. economic betterment, more jobs, better food, health dangers, pollution, etc.)

Trust or confidence in decision-making entities was reported for local, state and federal governments, industry, media, and special interest groups. Trustworthiness reflected reliability and accuracy of information relayed to the public to assist in protection or to communicate level of hazard.

The Samples

The **(citizen) public sample** was 54% male. The average median age across the four sites was 46 years, and the range was 17-88 years old. Respondents reported being well-educated with a median level of 14 years. Eighteen percent of this group described themselves as politically liberal, 56% moderate and 26% conservative.

An overwhelming majority of the **"elite"** sample was male (83%), Average age of elites matched the public at 46 years; however, they reported having more education (17-year average) than residents. Moderates also dominated the elite sample (55%), and almost one-third (31%) were conservatives. Liberals were, again, in the minority with 14% characterizing themselves with this political bent.

To examine similarities and differences in attitudes and beliefs about pesticides the five variables described above were utilized: degree of perceived risk, risk acceptability, perceived costs and benefits of a potentially risky event, and level of trust in hazard managers.

RESULTS

Using one-way analysis of variance mean responses of the five elite categories were compared with those of the general public. Differences were significant between elite groups and citizens on perceptions of risk, benefits and costs, permissible levels of risk exposure, and confidence in risk managers. Differences **among** the five elite categories are also significantly noteworthy. It appears that groups more closely aligned with the regulation and/or protection of agribusiness or frequent usage of agricultural chemicals and those removed from the industry respond in extremely diverse fashions. Results are presented first describing statistical variations between attitudes of the general public (citizens) and each of the five elite groups. Second, all six groups (five elite and the public) are compared to show direction and degree of response to individual items.

Risk and Acceptability

All respondents (public and elite) were asked to describe to what extent they agreed or disagreed with five variables measuring degree of perceived risk and five on risk acceptance. Acceptability measured how much pesticide risk the public will tolerate and how much elite groups expect them to tolerate under various conditions. Scores were summed across each set of items for risk and acceptability, and one-way analysis of variance was employed in comparing means for each group. A significance level of .01 was selected. The first phase of analysis was a comparison of risk perception by the elite sample and the public. All elite respondents, but elected leaders, differed significantly from the public in level of perceived pesticide hazard ($p<.01$). In order to determine direction of differences or similarities among all six groups elite + public), the Tukey B post hoc comparison was used at the .05 confidence level. This also yielded a scale of who perceived the greatest to the least risk.

Applicators reported the lowest level of risk perception, and public interest groups (PIRGs) the highest, of all the groups including the public. PIRGs also expressed significantly greater risk aversion ($p<.05$) than other groups tending to perceive higher hazard levels (i.e. the public and elected leaders). Applicators, those who consult, mix, and spray, contended there is next to no risk to using the substances. This response was so extremely confident in pesticide **safety** that other low risk groups (government and university) closely aligned with agricultural industry differed substantially ($p<.05$). Therefore, on the continuum applicators fell on the extreme side of minimal risk. Government and university groups landed somewhere on the low side of the scale. Public interest groups were at the extreme maximum risk pole, while elected leaders and the public fell on the high to moderately-high side of perceived risk.

Since past research has shown that perceived hazard influences how willing respondents are to endure risks, we expected an inverse relationship between expressed risk and acceptability. This relationship is graphically shown in the matrix (Figure No. 1).

Examination of individual unscaled risk and acceptability items revealed where major variations occurred. For example, respondents were asked to evaluate the risk of using pesticides at home and on farms. (Figure No. 2) In assessing the risk or safety of the two user groups, a majority of citizens (59%) believed home users were very safe in applying pesticides, yet thought agricultural applicators were extremely **unsafe** (61%). Elected leaders (60%) viewed home use as safer than agricultural use (54%). Fifty-eight percent (58%) of PIRG elites perceived a

RELATIONSHIP BETWEEN RISK AND ACCEPTABILITY

	HIGH RISK	LOW RISK
LOW ACCEPTANCE	Public Interest Groups Elected Officials The General Public	
HIGH ACCEPTANCE		University Scientists Government Administrators Applicators

FIGURE 1

substantial **danger** to home applicators and slightly more hazard (64%) to farm applicators. Most of the scientists (89%) and government administrators (81%) said household consumers were free from harm. Government (90%) and university groups (86%) also felt agricultural workers were safe, while applicators were 97% certain that neither home users nor field workers are at risk from using pesticides.

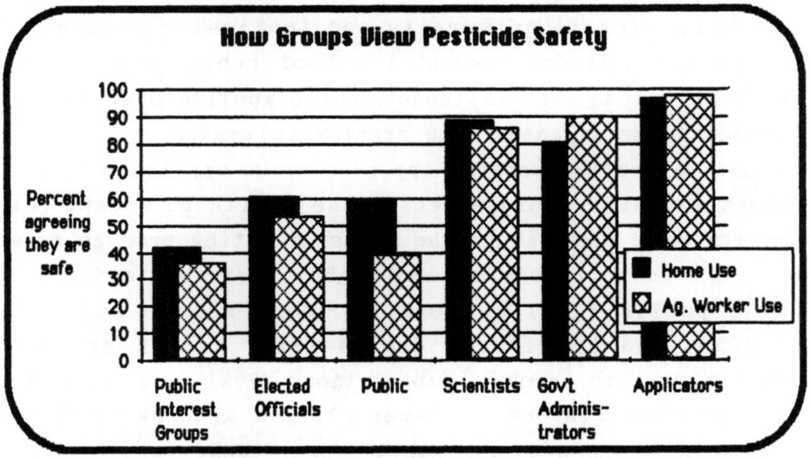

Figure 2

Type and setting of pesticide application influences risk perception. Aerial more than ground application of pesticides even on farms heightened a sense of danger for respondents. The public sample was more skeptical about aerial agricultural application than lower risk percep-tion groups, i.e. scientists, applicators and government administrators. Barely half of the citizens and elected leaders and fewer public interest representatives were confident in the safety of aerial application on farms. Eighty-two percent or more of government administrators and university specialists were convinced agricultural use of aircraft or ground rigs is not dangerous. Applicators viewed spraying farmland from either aircraft or ground rigs as 100% safe. Surprisingly, two-thirds (66%) or more of university, government, and applicator groups perceived **little or no danger** in pesticide application by aircraft in **urban areas.** The high-risk participants, however, were extremely concerned about aerial application in cities. Seventy percent (70%+) or more of the public, elected leaders, and public interest groups said it was a highly risky activity.

Acceptability was a measure of levels of pesticide risks citizens were tolerant of and levels of tolerance elites expected of citizens. Significant ($p<.01$) variations in degrees of acceptable risk between **all** elite groups and the general public were found using the one-way analysis of variance. As shown in the previous risk-acceptability matrix (Figure No. 2), applicators were characterized by high acceptance and public interest representatives by a comparably low degree of acceptability. Government and university specialists were notably more tolerant than elected officials and the general public. All high risk elite groups differed significantly ($p<.05$) from one another, while government responses were similar to university and applicator groups'. Elected leaders' and public interest groups' level of public risk acceptability was substan--tially less than citizens', while applicators, university and government specialists expected much greater tolerance.

Individual variables describe types of risks applicators, scientists, and government administrators expected the public to accept (Figure No. 3). Respondents were asked about the risks of eating produce treated with pesticides. This use of pesticides was highly desirable and acceptable to the consuming public (77%) as long as there are no bugs. However, government (93%) and applicator groups (98%) expect greater public willingness. University experts' expectations (79%) were the most closely aligned with citizens' responses. Fewer elected leaders (49%) and public interest groups (27%) accepted the public risk. Most felt citizens should not have to eat the sprayed food. Even though eating the produce was tolerable, actually spraying the crops situated near residences reduced

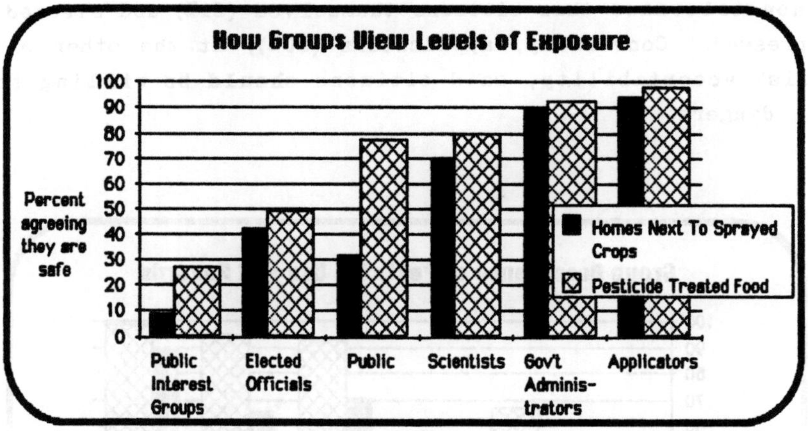

Figure 3

public acceptance dramatically. About two-thirds (66%) of the public sample expressed fear from spraying the nearby cropland. Ninety-one percent (91%) of the PIRGs were highly concerned about citizens safety under those conditions. On the other hand, 90%+ of the applicators and government administrators expressed no concern over the public living adjacent to sprayed farmland.

While many elite respondents were confident in pesticide safety in various settings, many also opted for using more benign methods such as biological controls before resorting to pesticides. Over 80% of the public and elected officials and 100% of the PIRGs did not want the chemicals used as a rule of thumb, but as a last resort. Many scientists (64%), government administrators (48%) and applicators (43%) also considered biologicals as a favorable option. In the absence of such biological methods a majority of all categories, except public interest groups, viewed extended and continued use of agricultural chemicals as necessary and acceptable (Figure No. 4). Not surprisingly, applicators (99%), government (95%) and scientific groups (93%) had the highest acceptance rate on this item. Fewer elected leaders (67%), citizens (57%), and public interest representatives (39%) believed pesticides should be so imperative to agriculture's future.

All groups were asked generally and directly if the public should be willing to accept the hazards inherent in pesticide use. In their risk-conservative vein, public interest groups (80%) maintained that the citizens have **no obligation** to accept pesticide dangers. This is much greater **non-tolerance** than citizens themselves (61%) and elected leaders (58%) expressed. Conversely, applicators (90%), at the other extreme of public risk acceptability, said citizens **should be** willing to accept potential dangers.

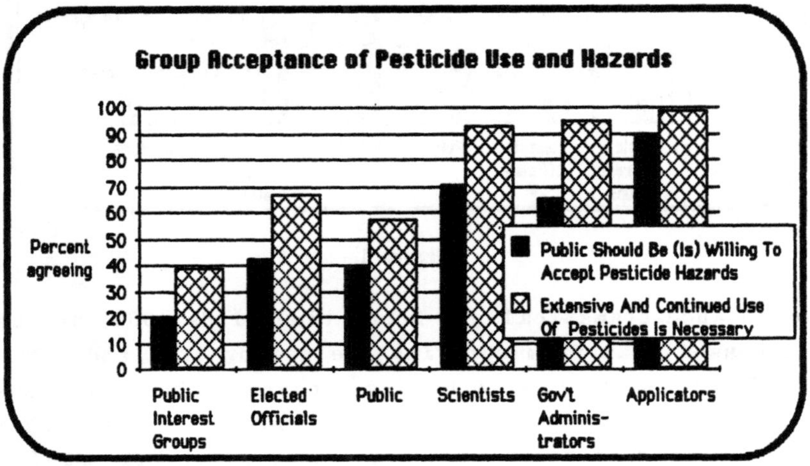

Figure 4

Costs and Benefits

Perceived qualitative consequences (gains or losses) of an activity such as, pesticide use are known to influence level of risk and acceptability. Elite groups and citizens were presented cost and benefit lists and asked to assess the probability that each cost and benefit would result from using the chemicals (Figure No. 5). As expected, the response patterns were very similar to perceived risk. The low-cost, high-benefit cohort consisted of applicators, government administrators, and scientists. Public interest groups, elected leaders, and the general public were high-cost, low-benefit respondents. One-way analysis of variance yielded significant ($p<.01$) variations between citizens' high degree of perceived **COST** of pesticides and the low level perceived by the low-risk groupings. No substantial differences occurred between the citizens and elected officials and PIRG representatives. Degree of **BENEFIT** derived from using agricultural chemicals was similarly distributed across the five elites and the general public with one exception. Like the applicators, government administrators, and university scientists, elected leaders were notably ($p<.01$) more optimistic about the positive consequences of pesticides than were the public.

% AGREEING
HIERARCHY OF COSTS & BENEFITS

Costs	Appli-cators	Govt Admin.	Univ. Scient.	PIRGs	Elect Lead.	THE PUBLIC
Air & Water Pollution	12.0	16.7	29.6	84.0	64.9	72
Endanger Lives	2.4	9.5	25.9	76.9	59.3	62
Create Large Gov't Bureaucracies	32.5	21.4	19.2	30.0	30.3	58
Increase Dependence on Small Groups of Experts	39.0	30.0	38.4	60.0	43.4	47
Deplete Natural Resources	3.6	4.8	3.7	66.7	26.8	37
Benefits						
Higher Quality Food	100.0	97.6	96.3	45.5	70.7	63
Provide More Jobs	91.5	80.5	70.6	33.3	57.3	58
Increase Calif. Tax Base	85.0	62.5	61.5	40.0	60.0	37
Increase Econ. Well-Being of All*	97.5	96.8	96.3	74.0	87.7	53

*For elites, frequencies are combined responses to three items addressing increased well-being of consumers, agribusiness, and chemical companies.

FIGURE 5

Air and water pollution was considered the greatest cost by citizens, PIRGs, and elected officials. Applicators, government agencies, and scientists placed pollution low on the cost scale. Over sixty percent (60%+) of the general public, PIRGs, and elected leaders viewed loss of life as highly likely. They were considerably less concerned about more dependency on specialized groups. Endangerment to human lives was perceived as a less likely cost by government, university and applicators groups than was pollution or the cost of increasing dependency on specialists. Less than 5% of applicators, government, and scientists saw depletion of natural resources as notable cost of using agricultural chemicals. Groups believing there is little risk to pesticide use and few qualitative costs also reported greater benefits. The government, university and applicator specialists ranked increased quality of produce as the greatest benefit of using pesticides. Citizens, too, felt better food was the most important gain. Betterment of the economic well-being of groups (i.e. consumers, chemical companies, and agribusiness), was also rated at the top of the benefit list by all expert groups. PIRGs and elected officials were less likely to recognize the benefits to consumers from the use of the chemicals. By a slight margin, university scientists, PIRGs, and elected officials felt chemical companies were the primary beneficiaries. Government administrators viewed chemical companies and the public as benefiting equally. Provision of jobs was ranked highly by (92%) applicators, (81%) government administrators, and (71%) university scientists. Fewer members of the general public (58%), elected officials (57%) and PIRGs (33%) were confident that the chemicals reduce unemployment.

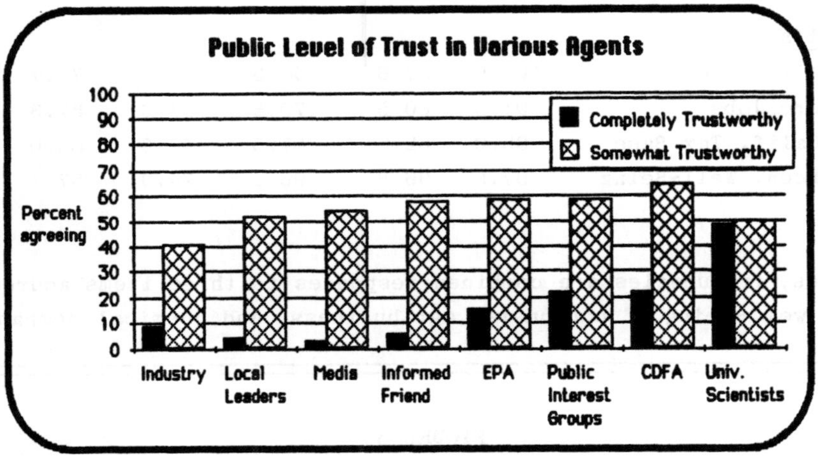

Figure 6

Who Trusts Whom

How did perceptions of risks, benefits, costs and level of acceptability influence confidence in agents dealing with aspects of hazard management? Citizens and elite categories assessed how trustworthy or reliable certain groups are during a crisis event particularly where public protection is of paramount interest (Figure No. 6). University scientists were trusted most completely by the general public. State agencies (like California Dept. of Food and Agriculture) and public interest groups were rated equally dependable, second to university experts.

LEVEL OF TRUST IN RISK MANAGEMENT AGENCIES

HIGH RISK GROUPS		Industry	PIRGS	Fed. Gov. (EPA)	State Gov. (CDFA)	Comm. Leaders	Univ. Scient.	Media
Public %	complete trust	9	22	15	22	4	48	3
	some trust	41	59	59	65	52	50	54
Elected Leaders	complete trust	8.5	13.6	13.8	28.8	1.7	44.1	0.0
	some trust	50.8	62.7	72.4	61.0	64.4	50.8	44.8
PIRGs	complete trust	0.0	23.1	0.0	0.0	0.0	15.1	0.0
	some trust	23.1	69.2	76.9	38.5	81.8	84.6	53.8
LOW RISK GROUPS		**Industry**	**PIRGS**	**Fed. Gov. (EPA)**	**State Gov. (CDFA)**	**Comm. Leaders**	**Univ. Scient.**	**Media**
Appli-cators	complete trust	32.1	0.0	7.1	34.5	0.0	32.1	0.0
	some trust	61.9	14.3	79.8	64.3	26.2	61.9	9.5
Govt. Admin.	complete trust	16.3	0.0	16.3	53.5	2.3	46.3	0.0
	some trust	79.1	45.2	76.7	44.2	39.5	53.7	16.3
Univ. Scien.	complete trust	3.4	0.0	10.3	31.0	0.0	72.4	0.0
	some trust	62.1	27.6	75.9	62.1	27.6	27.6	24.1

FIGURE 7

Almost all elite categories rated themselves as the agent they would trust most during hazard management situations. Elected officials followed their constituency's lead with having the greatest confidence in university scientists, yet placed themselves as one of the least reliable hazard management groups. Even though public interest groups, too, placed highest trust in themselves, they were very economical with those they did trust. PIRGs reported having total confidence in only themselves and university experts. State organizations (i.e. CDFA) received the highest marks from applicators, and listed themselves equally with university scientists.

It is worth mentioning that none of the lower risk perception categories (applicators, government administrators and university scientists) considered public interest representatives as **completely** trustworthy. In addition, of these three categories, **state** institutions reported the **highest** level of confidence in PIRGs. This is ironic in light of the historical adversary nature of the relationship between these two groups. The **media** (TV, newspapers, radio) unanimously received the lowest score, as no one in the elite sample indicated "complete faith" in this group either. The general public trusted only about 3% of media messages.

SUMMARY AND DISCUSSION

As is apparent from this data review, respondents (citizens and decision-makers alike) entertain ideas about risk that congeal into firm and consistent belief systems. Two groups emerge with divergent beliefs about the riskiness of pesticides. One group consists of the more optimistic respondents who see little risk, want others to be less skeptical, have great confidence that the benefits outweigh the costs, and have faith that most hazard management agencies are doing a reasonable job. The three elite groups conforming to these beliefs are the applicators, government officials, university scientists. It is not a coincidence, too, that these participants are closely aligned with, if not directly involved in, agriculture, pesticide use, regulation or policy-making. The other cohort consists of the public, elected leaders, and public interest groups. These are the skeptics who believe there is an inherent level of risk from pesticides and these risks are often beyond acceptable levels. Being more cautious about the risk of the chemicals, these respondents also view the potential costs as somewhat catastrophic to humans and the environment, and the benefits as slightly inequitable. Not surprisingly, these respondents are not involved directly with the agricultural industry and have lived in areas where state pest eradication programs were implemented. Even though the two categories, consisting of three sub-samples each, emerged intergroup attitudinal differences are evident. Applicators consistently responded in an extremely favorable tone to all items on pesticide safety, risk, acceptability, costs and benefits. This level of favorability exceeded others' who tended to play a protagonistic role in pesticide-agricultural issues. Many of the single responses indicated that the applicators were so confident that a carte-blanche approach to use prevailed.

On the other hand, the public interest groups may be described as the antagonists in the pesticide and agriculture debate. This group consistently differed from the others reporting a similarly cautious attitude toward pesticide risks. Many responses by the PIRGs demonstrated an highly-protective attitude, and the carte-blanche approach appeared to be directed toward reducing the use of pesticides or instituting stricter regulations. The elected officials, too, were cautious regarding the potential for hazard, more so than were citizens. Yet they were also the most closely aligned to the general public in their responses. This study confirms the diversity of belief systems influencing attitudes and behaviors of six groups involved in an activity with many known and unknown hazards and proven benefits. It is noteworthy that the respondents who are less concerned about risks are those associated with pesticides and agriculture. Some previous data has shown that education may effect caution or concern. However, all groups reported high levels of formal schooling, 14 or more years. It is more likely that Mazur (1979) was on the right track in his hypothesis that previous training, education, and orientation tends to bias or otherwise reinforce attitudes and related behaviors.

During stressful events within communities, factions tend to align with those closest to their own ideology. It should also be recognized from this study, that the high- and low-risk groups consist of those most often at loggerheads during a controversial event. It is also demonstrated here that not all groups aligned on one side of an issue or another agree or disagree with the same intensity. The general public described herein, tends mostly toward the middle-of-the road, yet still on the risk-averse side. Where do the various agents described in this study agree? They agree that there are known benefits, yet they do not agree on which are the most likely to occur. Most agree that there is a certain level of risk, but they do not agree on the intensity or probability of the risk. And a majority agree that safer methods should be explored. For those interested in examining controversy, these data present an obvious pattern which is much more pervasive in social conflict events than previously known. The scenario will not change much from case to case: those in control, that is, experts, decision makers and technicians tend to view pest eradication and general use of pesticides as safe, necessary and beneficial to everyone. They enter a community with the best of intentions and explain to citizens all is well, that the chemicals have been used for years with no known ill effects. The citizens, those not directly participating in decisions, the special interest groups, and community leaders have a considerable degree of skepticism. They cannot deny the known benefits of using a pesticide on their gardens or in their homes. But they can certainly question the use of aircraft in their city neighborhoods, or introducing a chemical not extensively used. This skepticism increases as daily news coverage of "**SAFE**" technologies such as preservatives, pesticides, automobiles, medicine, and so on, being banned or restricted. Along with this coverage are debates among the scientists and other experts about the safety of the substances, testing procedures, etc. thus perpetuating the cycle of skepticism or resignation.

The belief and knowledge gaps between citizens and hazard management agencies are not insignificant. Government and private industry often face the problems of angry citizen groups who believe they are acting in the best interest of their community. Government and industry then ask what they can do to allay fear or what does the public want. Citizens participating in the study offered a wide range of suggestions. Some recognize the need for accepting a level of risk, but they also observe the hazards in continually doing so. These are some of their ideas:

> "More money should be granted to universities, etc. for research into the side-effects of the chemicals used--dioxon, etc., before they are put on the commercial market. Obviously, we cannot control pests without some use of pesticides; however, there is a need for more thorough research to establish safety before pest infestation occurs so that a crisis situation doesn't force the use of a pesticide that could later be found harmful." (Milpitan)

> "An ever-increasing population makes the need for more sources of food and power extremely urgent. The use of man-made chemical compounds is probably the most efficient way to meet these needs, but only if they can be made safe, not only to use, but also to insure that they leave no poisonous residue to contaminate the earth, the sea and the atmosphere." (Stocktonite)

Others were more adamant about "over-reaction" by public agencies.

> "Twenty-nine (Japanese) beetles caught and a megabuck eradication program. Much like using an atomic bomb in a street fight. Dioxon, TCE, DDT....Let's not expand the list of expensive mistakes!!!" (Orangevalean)

Many expressed the need for more information on all aspects of eradication, before events reach crisis proportions.

"It is my opinion that the public as a whole is not fully educated on the matter of pesticides. Only in a crisis, such as the Medfly, is the public informed." (Stocktonite)

"I feel more action should be taken to educate the general population on the real hazards of pesticides and other chemical substances the public comes in contact with." (Milpitan)

"I'm sick of discrepancies of facts among the professionals. What constitutes an infestation? 31 dead Japanese beetles?" (Orangevalean)

At the practical level, respondents proposed programs to allow for increased exposure to the desired information.

"There should be a regular program on T.V. or in newspapers in a language we can understand. This could be information on chemicals, pests, the latest experimental programs, etc.; we could then be able to put pressure on our public officials before crises develop." (Stocktonite)

"Distribute and display at shopping centers samples of pests along with information on who to contact (like a county agricultural agent). There should be more displays of pests in addition to the one in the county offices." (Milpitan)

"I think stores that sell produce should display signs stating that the produce has been sprayed with pesticides and to thoroughly wash the product before consumption. This would increase public awareness." (Concordian)

Comments were to an extent positive, with citizens offering suggestions on how best to inform them. They are willing to hear and learn more about the problems threatening their food supply. Even though many times government is perceived as unresponsive, the public has confidence that matters are not beyond redemption, and are consistent in their desire to be active participants in decision-making.

Knowing that beliefs differ, and recognizing that all perceptions are valid may not, in itself, assist in reducing the heat of an event. However, (1) being cognizant of how one group differs from another, (2) being receptive to suggestions for improving the system within which we all must operate, and (3) acknowledging the degree of overconfidence to which all mortals are prone, will certainly help achieve a less tempestuous present and a much safer future.

REFERENCES

Fischhoff, B., Slovic, P., Lichtenstein, S. (1978). Fault Trees: Sensitivity of Estimated Failure Probabilities to Problem Representation. Journal of Experimental: Human Perception and Performance 4, p. 342-355.

Green, C. H. (1981). Risk: Attitudes and beliefs. In D.V. Canter (Ed.), Behavior in Fires. Chichester: Wiley.

Hawkes, G.R., Pilisuk, M., Stiles, M.C., Acredolo, C. (1984a). The Medfly Crisis: Citizens' Response to Eradication Risks. California Agriculture, 38 (1 and 2).

Hawkes, G. R., Pilisuk, M., Stiles, M.C. and Acredolo, C. (1984b). Assessing Risk: A Public Analysis of the Medfly Eradication Program. Public Opinion Quarterly, 48, p. 443-451

Mazur, A., (1973). Disputes Between Experts. Minerva IX, 2.

Neihaus, F., and Swaton, E. (1981). Acceptability and Acceptance of Risk by the Public. International Atomic Energy Agency Presentation, January 1981.

Nisbett, R. and Ross, L. (1980). Human Inference: Strategies and Shortcomings of Social Judgment. Englewood Cliffs, N.J.: Prentice-Hall.

Sharlin, H. I. and Allen, D. (1985) EDB: A Case Study in the Communication of Health Risk. Unpublished report submitted to the Environmental Protection Agency.

Slovic, P., Fischhoff, B., Lichtenstein, S. (1979). Rating the Risk Environment, 21(3), p. 14-20, 36-39.

Slovic, P., Fischhoff, B., Lichtenstein, S. (1981). Perceived Risk: Psychological Factors and Social Implications. Proceedings in Rural Sociology. London A376, p. 17-34.

Starr, C. (1969). Social Benefits Versus Technological Risk. Science, 165, p. 1232-1238.

*Copies of the questionnaire may be obtained by contacting the authors.

FEASIBILITY ASSESSMENT OF PEST MANAGEMENT

G.A. Norton,
Silwood Centre for Pest Management,
Imperial College,
Silwood Park, Ascot, Berks SL5 7PY, U.K.

INTRODUCTION

The concern of this paper is with attempts to improve pest management. Such improvements can be sought by developing and implementing new technology - such as resistant varieties, controlled droplet application, or sterile insect techniques, - and/or by changing management - through the adoption of regional control strategies, designing integrated pest management, or by modifying insect trap layout. In the public sector, these attempts to improve pest management are made on the grounds that proposed changes will -

(i) result in more efficient pest management,
(ii) reduce environmental/health risks,
(iii) increase the sustainability of crop production, for instance, by delaying pesticide resistance, and/or
(iv) allow an improved response to outbreaks.

In recognising that intentions to improve pest management are frequently not realised, it is argued here that a different approach, involving more feasibility assessment, is required. The paper consists of 4 parts:

1. A brief description of a programme to control leaf cutting ants in Trinidad by means of aerial baiting. This illustrates the general problem of implementing changes in pest management.
2. An outline of a research procedure appropriate for practical pest management.
3. Discussion of case studies to illustrate two important concepts involved in this procedure - decision models and development pathways.
4. Some general conclusions.

LEAF CUTTING ANTS IN TRINIDAD

Over ten years ago a scheme was devised for controlling leaf cutting ants in Trinidad, a pest which attacks citrus, cocoa, and other

NATO ASI Series, Vol. G11
Pest Control: Operations and Systems Analysis
in Fruit Fly Management. Edited by M. Mangel et al.

important crops on the island (Lewis and Phillips, 1973; Lewis, 1973; Lewis and Norton, 1973). The technique developed involves a locally made bait - of citrus pulp and aldrin, applied by aircraft. It was tested on an island off Trinidad and found to be more effective than current practices, where workers had to search to find the nests and then treat each nest individually. Indeed, aerial baiting has the advantage that small nests, which are unlikely to be detected by workers, would also be killed out with aerial baiting, reducing next year's problem. Thus, aerial baiting was found to be technically feasible.

On examining the cost of aerial baiting relative to the conventional method of control, it was found that aerial baiting was cheaper. In other words, aerial baiting would be more cost effective. Although some redistributional effects were likely to result from full-scale adoption of aerial baiting - for instance, manual searching for nests would no longer be needed - there would also be a considerable saving in foreign exchange since the citrus bait would be locally made. Thus, on balance, aerial baiting appeared to be economically desirable.

Examination of the environmental and safety aspects of aerial baiting indicated that less active ingredient will be applied to the environment and that operator hazard will be no worse than previously, and probably far less hazardous. Thus, aerial baiting appeared environmentally acceptable, at least compared with existing practices. Since aircraft were available for aerial baiting at a time when they were not needed for spraying sugar cane in Trinidad, this technique also seemed practically feasible.

Despite all these advantages, aerial baiting of leaf cutting ants was not implemented in Trinidad: there seem to be two main reasons. First, as the Trinidad oil industry increased, the relative importance of crops like citrus and cocoa declined. Therefore, the urgency for controlling leaf cutting ants was reduced. The second reason appears to be the budgetary cost and political risk associated with implementing aerial baiting. Thus, what in fact did happen was the government produced the citrus pulp bait and then gave it to the farmers, courtesy of the Department of Agriculture. In political terms this appeared to be far more attractive, although probably much less effective, than aerial baiting.

The lesson I personally learned from this experience was the importance of determining what is likely to be feasible in political terms as well as in all the other economic, environmental, and practical aspects. This message is expressed in a diagrammatic form in Fig. 1.

In general, when attempting to develop and implement improvements in pest management, what is needed is a strategic research approach that first identifies the target improvement and then works towards meeting it. Industry, especially Japanese industry, is very good at this; identifying what the customer needs and working towards providing it. In this case, the marketing and commercial sectors give direction to scientific input.

By contrast, public sector research often appears to be carried out in relative isolation, both in a discipline sense and in an institutional sense - with research often separated from extension activities. The result is that natural scientists tend to predominate

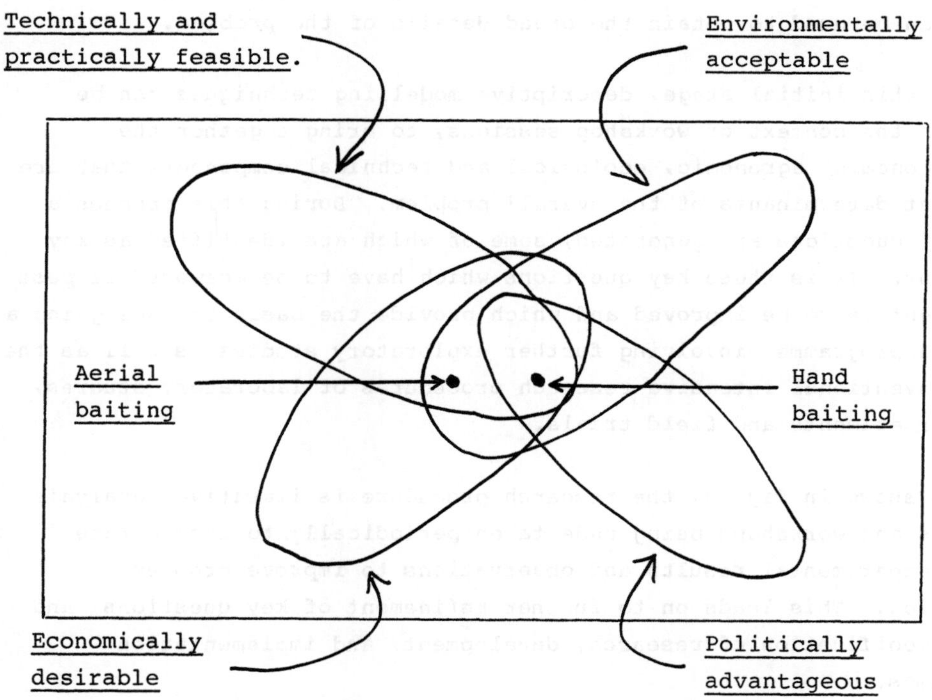

Fig. 1. The components of feasibility assessment

in public sector programmes and developments tend to be technologically driven rather than being designed to best meet a need.

A STRATEGIC RESEARCH PROCEDURE

One means of tackling this problem is to develop research procedures for pest management that are more explicit and comprehensive (Norton, 1982). It is ironic that the need for this has been recognized for developing country problems but is much less appreciated, or at least implemented, in developed countries.

The suggested procedure is outlined in Fig. 2. As with certain other research procedures (Byerlee et al., 1980; Collinson, 1984), the first task in this procedure is to define the problem in terms of its major components and relationships. In doing this, exploratory investigations will be necessary to answer such key questions as who the target farmers are to be, what the distribution of the pest is, what the major information gaps are and where they occur, and in which direction agricultural development is headed. To this end, interview surveys with farmers and the use of technology forecasting techniques (e.g. Smith, 1985) can be used to obtain the broad details of the problem.

At this initial stage, descriptive modelling techniques can be used, in the context of workshop sessions, to bring together the socio-economic, agronomic, ecological and technical components that are important determinants of the overall problem. During this process a range of questions are generated, some of which are identified as key questions. It is these key questions which have to be answered if pest management is to be improved and which provide the basis for designing a research programme, involving further exploratory studies as well as the more conventional intensive research procedures of laboratory studies, field experiments and field trials.

As shown in Fig. 2, the research procedure is iterative, analysis sessions and workshops being undertaken periodically to incorporate recent experimental results and observations to improve problem definition. This leads on to further refinement of key questions, and to the modification of research, development, and implementation programmes.

Clearly, a variety of systems analysis techniques can be of value throughout this research procedure. In workshops, we have found

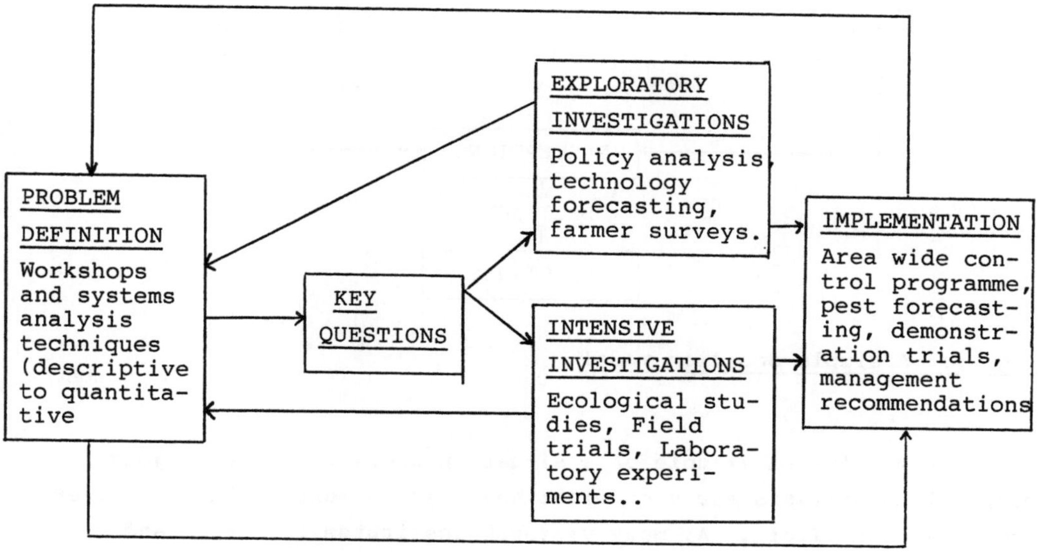

Fig. 2. A Strategic Research Procedure

descriptive analysis techniques, using time profiles and interaction matrices, for example, to be extremely useful. Simulation and other modelling techniques can also complement the research effort in helping to identify key questions. These and other modelling techniques can also be used to design recommendations or to implement pest control programmes (Conway, 1984).

In this paper, the focus is on descriptive analysis techniques and their value in providing on-going feasibility assessment. Two key concepts are important for this-decision models and development pathways.

DECISION MODELS

The decision making process can be viewed in terms of the model outlined in Fig. 3. Decision makers attempt to choose the option they think will best meet their objectives in the face of their perceived problem (Norton & Mumford, 1983). In pest control, the decisions that farmers make, or which public agencies make, can be viewed in the same way in as much as the three variables shown in Fig. 3 will determine their choice. If scientists are to produce an alternative option in pest control, the extent to which it will be implemented will depend on the decision makers' assessment of its performance relative to other options. Thus, at an early stage a feasibility assessment of new pest control options needs to be undertaken.

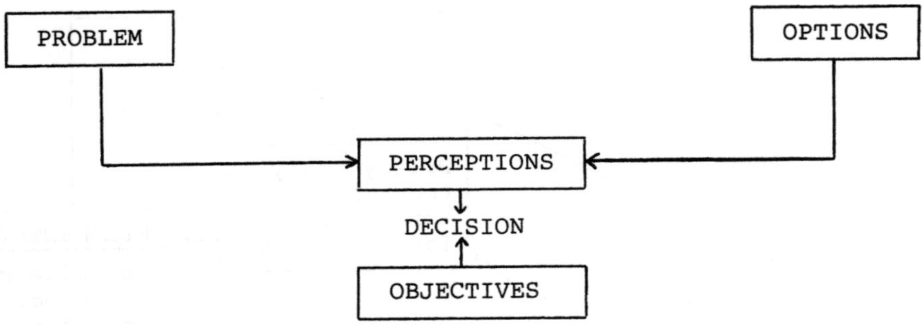

Fig. 3. A simple decision model

Let me give an example of a situation where I believe a good feasibility analysis early on might have led to more efficient research and extension effort. At many research institutes in Europe and in North America a good deal of effort has been devoted to the development of supervised insect control (or scouting) in apples. After fifteen years, there is still little evidence that supervised control has been adopted to any degree by apple growers.

A key question that should have been asked at an early stage is - What factors will affect whether growers switch from current calendar spraying to supervised control of insect pests? Put another way - What constraints do we need to break or overcome in order for supervised control to be adopted?

In a recent analysis, Fenemore & Norton (1985) found there to be four major constraints -

1. Calendar spraying of mildew fungicides makes calendar spraying of insecticide much cheaper.
2. Economic threshold decision rules are difficult to design and employ in English apple orchards.
3. The high cost and/or lack of trained labour for scouting may be prohibitive, and
4. The potential risks of lower quality fruit associated with supervised control may be unacceptable.

A second example concerns public decision making in pest control and the choice that faces the Malaysian Department of Agriculture in responding to outbreaks of brown planthopper that periodically occur on rice. My colleague, K.L. Heong, together with Department of Agriculture staff, has attempted to explicitly set out the range of options and the objectives the Department wishes to meet (Table 1.). In setting out the

Table 1. Possible Policy options and objectives for public pest management. The initial problem is to assess how each option is likely to meet each objective.

	OBJECTIVES						
OPTIONS	Improve farmers' ability to control pest	Improve farmers' incentives	Strengthen political support	Keep Dept's cost low	Ensure prevention of pest loss	Ensure prevention of pest spread	Reduce future pest outbreaks
SHORT TERM 1. Warn and advise farmers							
2. Advise and provide credit							
3. Advise and subsidise pesticides							
4. Advise, subsidise and supervise spraying							
5. Moss treat and charge farmers							
6. Moss treat at Dept's cost							
MEDIUM TERM 7. Intensive pest surveillance							
8. Implement area wide cultural/ biological control							
9. Training courses for farmers							

problem in this way, this descriptive decision model provides a focus for discussion, helping to identify what is likely to be politically feasible, what further information is required, and what constraints need to be overcome if a particular alternative is to be adopted.

DEVELOPMENT PATHWAYS

The idea here is that in attempting to assess the feasibility of proposed changes in pest management it is important to know where agricultural and other developments are headed, and how this will affect pest problems and the attributes of various control strategies.

The process of agricultural development can be viewed as crossing a landscape whose topography is determined by a variety of factors, particularly agricultural policy, economic forces, and technological capability. The way in which development proceeds can affect pest management in the following ways -

(a) It can affect the structure of the industry, such as the proportion of part-time farmers and the agricultural wage rate, both of which have implications for control strategies, such as labour intensive supervised control.

(b) Agricultural developments, in influencing the cropping practices adopted, can affect pest status by making the cropping system more favourable or susceptible to pest attack, thus increasing potential crop loss.

(c) It produces developments in crop protection technology, and

(d) Influences the way in institutional development proceeds.

Three examples are now given to illustrate some of the concepts involved and the importance of appreciating how specific pest problems have developed.

Developments in Sugar Beet in the U.K.

The adoption of precision drilling and monogerm seed in the U.K., and throughout Europe, has allowed sugar beet growers to drill their crop to a stand, the driving force being to reduce the labour involved in thinning. However, since damage at the seedling stage cannot be compensated for by adjacent plants when drilled to a stand, this practice increases the potential for damage caused by seedling pests. One consequence of adopting these technologies therefore is that growers have found it necessary to adopt soil applied insecticide practices to protect the more vulnerable seedling stage.

This particular example illustrates how development can be "locked in" to a particular pathway by the adoption of particular technologies. In this case, the main "locking-in" factors are economies of scale and complementarity of different technologies. Other "locking-in" factors include policy precedents and institutional development associated with operating a particular technology.

The Development of Cotton Pest Problems in the Sudan Gezira

This example illustrates how understanding the way in which pest problems have developed can give important clues to how they might best be resolved.

At present, whitefly is the major pest problem in cotton in the Gezira. While it is important to investigate such direct solutions as the application of granular insecticides, this particular solution is both costly and potentially dangerous to operators. By understanding how the whitefly problem has arisen, and in appreciating the wider aspects of cotton management in the Gezira, alternative means of improving the situation might be derived.

An analysis of this problem (Griffiths, 1984) identified the major variables that have shaped developments in the Sudan Gezira over the past 60 years. In particular, important changes have occured at an institutional level including changes in objectives and in cost sharing arrangements; also changes have occured in cropping practices (particularly in varieties, crops grown, and rotations), and in pesticide use.

The analysis revealed a number of key questions concerning the origins of the whitefly problem and possible solutions. For instance, changes in the official crop rotation in the 1960's resulted in an increased area of groundnuts and a reduction in fodder crops. This affected the whitefly problem in at least two ways. First, since groundnuts provide an alternative host for Heliothis, attacks by this pest on cotton increased, leading to increased spraying and the emergence of whitefly as an induced pest. The second effect arose from the reduction in fodder crops. Since tenants still maintained livestock, they irrigated cotton after harvest, as well as fallow land, to provide feed but at the same time making the whitefly problem worse. By taking this more comprehensive view, more sustainable long term solutions to the problem might be revealed.

Technology Forecasting for Pesticide Application in the U.K.

The previous examples have shown how an ex-post study of developments can reveal important features of pest development and indicate the type of "knock-on" consequences that can result. This example illustrates how a more positive approach can be taken to future developments, in attempting to forecast what is likely to happen in the next five years or so. Smith (1985) sent a postal questionnaire to over 200 experts in pesticide application machinery both in government and the commercial sector. The aim was to identify those developments in the U.K. likely to occur by the year 1990 on which the experts were largely agreed and those on which they differed. An indication of the type of statement the respondants were asked to agree or disagree with is given in Table 2. In obtaining some idea of the rate of development of particular technologies one can then consider how they will promote or constrain other pest management options, such as alternatives to pesticide, or alternative management strategies.

Table 2. Statements concerning future developments of application technology with which experts were asked to agree or disagree (Smith, 1985).

Statement	% Agree	% Disagree	% No Reply
By 1990, the number of sprayers on farms allowing farmers to control droplet size accurately will have increased by at least three times.	86	13	1
By 1990, with increased use on farms of spinning disc and electrostatic sprayers, the acreage sprayed by hydraulic pressure sprayers will decline by at least 50%.	24	74	2

CONCLUSION

It has been argued that a feasibility analysis of potential changes in pest management should be undertaken at an early stage in research and development programmes and should be re-assessed at major points in project development. To implement this approach, attention needs to be paid to two major issues -

1. The need for a more explicit and comprehensive research procedure where descriptive models in particular are used to identify key questions.
2. To find ways in which development can be prevented from becoming unnecessarily "locked in". Some means must be found of maintaining flexibility in physical system development as well as in the development of scientific and extension institutions.

REFERENCES

Byerlee, D., Collinson, M., Perrin, R., Winkelmann, D. and Biggs, S. (1980). Planning Technologies Appropriate to Farmers - Concepts and Procedures. Centro Internacionale de Mejoramiento de Maiz y Trigo, Mexico. 71pp.

Collinson, M. (1984). Diagnosing the problem of small farmer needs. In Proceedings of the Fourth Agricultural Sector Symposium, World Bank, New York. pp. 124-146.

Conway, G.R. (Ed.) (1984). Pest and Pathogen Control: Strategic, Tactical and Policy Models. John Wiley and Sons Ltd., New York. 488pp.

Fenemore, P.G. and Norton, G.A. (1985). Problems of implementing improvements in pest control: a case study of apples in the U.K. Crop Protection 4, 51-70.

Griffiths, W.T. (1984). A Review of the Development of Cotton Pest Problems in the Sudan Gezira. M.Sc. thesis, University of London.

Lewis, T. (1973). Aerial baiting to control leaf-cutting ants (Formicidae, Attini) in Trinidad. II. Field application, nest mortality and the effect on other animals. Bulletin of Entomological Research 63, 275-287.

Lewis, T. and Norton, G.A. (1973). Aerial baiting to control leaf-cutting ants (Formicidae, Attini) in Trinidad. III. Economic Implications. Bulletin of Entomological Research 63: 289-303.

Lewis, T. and Phillips, F.T. (1973). Aerial baiting to control leaf-cutting ants (Formicidae:Attini) in Trinidad. I. The bait, its production and the effects of weathering on attractants and persistence of toxicants. Bulletin of Entomological Research 63: 263-273.

Norton, G.A. (1982). A decision analysis approach to integrated pest control. Crop Protection 1, 147-164.

Norton, G.A. and Mumford, J.D. (1983). Decision making in pest control. In Advances in Applied Biology, volume 8. Academic Press, New York. pp. 87-119.

Smith, A.K. (1985). Forecasting future developments in application technology. Crop Protection 4: 121-129.

III. BIOLOGY

EVALUATION OF COLOR AND FOOD-ODOR TRAPPING METHODS IN THE OLIVE FRUIT FLY

A. P. Economopoulos
"Demokritos" National Research Center
15310 Aghia Paraskevi Attiki, Greece

Summary

Traps based on visual (color) and olfactory (food) attraction have been tested for monitoring or control purposes. The correlation between trap catches and population size or fruit infestation has been studied in certain cases. Lure combination often resulted in catches substantially higher than those of individual lures. Control applications have shown that even powerful, combined-lure traps must be applied at high densities for substantial infestation prevention. In years of very high population densities, even large numbers of such traps may not be adequate for effective control.

Further experimentation is needed to study the correlation between trap catches and actual population density at different seasons and weather conditions. Further work is also needed on the development of powerful traps which could keep infestation low at low cost. The effect of such traps on non-target insects should be explored carefully.

Introduction

Trap development received much attention in the olive fruit fly, Dacus oleae (Gmelin) (Diptera: Tephritidae), in the recent years. Trapping techniques for both monitoring and control purposes have been studied extensively. In both cases, the use of insecticides could be reduced either because of better timing of sprays or because some or all sprays could be replaced by effective trapping systems. Thus, traps based on color visual lures, food olfactory lures, pheromones and combined lures have been developed. The combination of lures having different effectiveness at different climatic conditions, or affecting differently the two sexes or individuals at different stage of development or physiological stage, was sought particularly in cases where traps aimed at control applications.

NATO ASI Series, Vol. G11
Pest Control: Operations and Systems Analysis
in Fruit Fly Management. Edited by M. Mangel et al.

Also, considerable work was done on trap lifespan extension, which is very important in control applications.

Color Traps

The discovery that certain colors attract certain insect species strongly, led to the use of the most powerful ones in trapping devices. Prokopy et al. (1975) found that the yellow color, in particular the one that reflects highest amount of light between 520-580 nm and very little below 520 nm, it attracts more than any other color the olive fly. Although yellow traps were found considerably less powerful than olfactory ones (Economopoulos 1979, Delrio et al. 1983), they could be useful since they usually last longer. This is especially true when, instead of sticky material, a long lasting insecticide is applied on the trap surface, preventing thus insects from covering up the trap surface. When used at high densities, yellow traps have been reported to result in considerable infestation reduction (Economopoulos 1979), but beneficial insect destruction also (Neuenschwander 1982). The high number of traps required per tree, especially in cases of large or dense trees and/or high population densities (Economopoulos 1979), increases the cost and makes the method no practical.

Yellow traps have been tested for forecasting purposes and infestation levels were predicted on captured females (Ballatori et al. 1980). Correlation between olive fly catches on yellow color sticky panels and population size has been attempted by using a system of six yellow traps per middle size olive tree (Delrio et al. 1983). The six yellow traps had been found to trap most of the flies present in the tree (Economopoulos unpublished). The yellow trap caught fewer flies than those present in the tree throughout the year. The difference was more pronounced in fall - beginning of winter. In the same work, the range of attraction of yellow trap was found not to extend beyond the canopy of the trap tree.

Food-odor Traps

The McPhail odor trap (McPhail 1937) has been for a long period practically the sole trap used in practice for monitoring purposes. Water solutions of ammonium salts or proteinaceous substances have been used. The above trap is of short duration, no more than 4-5 days under summer warm-dry weather, breaks easily, and it is rather difficult to handle. Protein hydrolyzate solutions (e.g. Buminal$^{(R)}$) have been usually found much more powerful than ammonium salt solutions. Prokopy and Economopoulos (1975) discovered that about 2/3 of flies attracted do not enter the trap to drown in the lure solution. They land on its exterior surface flying away thereafter. This led to the use of sticky material on the trap surface so that all arriving flies are caught. Nevertheless, use of sticky material makes trap handling even more difficult and unpractical.

The number of insects trapped in the McPhail trap has been correlated to population size and infestation level either empirically or through a mathematical formula based on trap catches and temperature (Kapatos and Fletcher 1983). The empirical approach is usually applied by spraying campaign authorities in large scale bait sprays. Correlation of McPhail as well as pheromone trap catches to population size has been also attempted by using the six yellow trap system mentioned before. Thus, the McPhail trap baited with Buminal water solution caught similar numbers of flies to the six yellow trap system in spring and summer and much smaller in fall-beginning of winter. In the same work the phermomone trap was found to attract many more males than those present in the tree in spring and fall, while in summer it attracted few to zero flies. The distance of attraction was estimated to extend at least up to 80 m in the phermomone trap and 20-40 m in the McPhail protein trap (Delrio et al. 1983).

Combined-lure Traps

In several instances it has been shown that lure combination increases total trap catches considerably (Economopoulos in press). For example, when the yellow trap is combined with an ammonium acetate dispenser it catches about three times more olive flies than the yellow trap alone (Economopoulos and Stavropoulou-Delivoria 1984). Thus, for increased trapping rates, devices combining two or more different lures have been developed. This was attempted especially in cases of traps developed for mass-trapping and infestation prevention. When a trap combining a yellow color sticky panel with an ammonium acetate dispenser was used at one trap per tree density, it gave substantial infestation protection in years of low population density (Economopoulos et al. in press). In high population years the above method gave unsatisfactory results. Delrio (in press) combined in the same trap yellow color sticky panel, ammonium salt, Buminal protein and phermomone. Again, when one to two such traps were applied per olive tree, substantial infestation reduction was obtained only in years of low population density. In both experiments the olive groves used had considerable degree of isolation from other olive groves. The combination of phermomone with yellow sticky panels (one per nine olive trees) gave rather poor control results (Broumas et al. 1983).

References

Ballatori, E., Cl. Pucci and C. Ricci. 1980. Relation entre l'infestation des olives et les captures d'adultes de Dacus oleae (Gmel.) par pieges chromotropiques. Redia LXIII:417-439.

Broumas, Th., P. Katsoyannos, C. Yamvrias, C. Liaropoulos, G. Haniotakis and F. Strong. 1983. Control of the olive fruit fly in a pest management trial in olive culture. Proc. Symp. on "Fruit Flies of Economic Importance" (Athens, Nov. 1982), A.A. Balkema/Rotterdam:584-592.

Delrio, G. Biotechnical methods for olive pest control. Proc. Meeting EEC, FAO and IOBC on "Integrated Control in Oleiculture" (Pisa, April 1984):in press.

Delrio, G., A.P. Economopoulos, P.V. Economopoulos, G.E. Haniotakis and R. Prota (alphabetically). 1983. Comparative study on food, sex and visual attractants for the olive fruit fly. Proc. Symp. on "Fruit Flies of Economic Importance" (Athens, Nov. 1982), A.A. Balkema/Rotterdam:465-472.

Economopoulos, A.P. 1979. Attraction of Dacus oleae (Gmelin) (Diptera, Tephritidae) to odor and color traps. Z. ang. Ent. 88:90-97.

Economopoulos, A.P. 1979. Application of color traps for Dacus oleae control: olive groves with different degree of isolation, tree-size and canopy density. Proc. Symp. IOBC/WPRS on "Integrated Control in Agriculture and Forestry" (Vienna, Oct. 1979):552-559.

Economopoulos, A.P. Use of traps based on color and/or shape. In, "Fruit Flies: Biology, Natural Enemies, Control" Eds. A. Robinson, G. Hooper, Elsevier Science Publishers: in press.

Economopoulos, A.P. and A. Stavropoulou-Delivoria. 1984. Yellow sticky rectangle with ammonium acetate slow-release dispenser: an efficient long-lasting trap for Dacus oleae. Entomologia Hellenica 2:17-23.

Economopoulos, A.P., A. Raptis, A. Stavropoulou-Delivoria and A. Papadopoulos. 1986. Control of Dacus oleae by yellow sticky traps combined with ammonium acetate slow-release dispensers. Ent. exp. appl. 40: in press.

Kapatos, E. and B.S. Fletcher. 1983. Seasonal changes in the efficiency of McPhail traps and a model for estimating olive fly densities from trap catches using temperature data. Ent. exp. appl. 33:20-26.

McPhail, M. 1937. Relation of time of day, temperature and evaporation to attractiveness of fermenting sugar solution to Mexican fruit fly. J. Econ. Entomol. 30:793-799.

Neuenschwander, P. 1982. Beneficial insects caught by yellow traps used in mass-trapping of the olive fly, Dacus oleae. Ent. exp. appl. 32:286-296.

Prokopy, R.J. and A.P. Economopoulos. 1975. Attraction of laboratory-cultured and wild Dacus oleae flies to sticky-coated McPhail traps of different colors and odors. Environ. Entomol. 4:187-192.

Prokopy, R.J., A.P. Economopoulos and M.W. McFadden. 1975. Attraction of wild and lab-cultured Dacus oleae flies to small rectangles of different hues, shades and tints. Ent. exp. appl. 18:141-152.

SOME ISSUES IN FRUIT FLY ECOLOGY

Brian S. Fletcher,
CSIRO, Division of Entomology,
55 Hastings Road, Warrawee, NSW, 2074, Australia.

SUMMARY

This paper outlines some of the important topics in fruit fly ecology related to basic ecological issues, ecological issues of both academic and practical interest and ecological contributions to pest management programmes. Topics covered include mechanisms of speciation, life history strategies, foraging behaviour, population regulation, bioclimatic potential, interspecific competition, host range and the importance of ecological data in the optimization of pest management procedures.

INTRODUCTION

For convenience, issues relating to fruit fly ecology can be grouped into three main categories, 1) basic ecological issues; 2) ecological issues of both academic and practical interest; 3) ecological issues concerning pest management procedures.

In this paper I will discuss some of the topics which fall into each of these three categories but the main emphasis will be on those related to the planning and implementation of management programmes aimed at the suppression or eradication of fruit fly populations, as these are the ones of most interest in the current context.

BASIC ECOLOGICAL ISSUES

There are a number of basic ecological issues related to studies on fruit fly ecology which are of considerable academic interest to biologists but in most cases are not considered of immediate concern by individuals or agencies responsible for the implementation of quarantine procedures or management programmes against pest species. On the

NATO ASI Series, Vol. G11
Pest Control: Operations and Systems Analysis
in Fruit Fly Management. Edited by M. Mangel et al.

)ther hand, many of the issues are of general ecological interest and not confined only to researchers on fruit flies. Some good examples are:-

1. Studies on mechanisms of speciation and host race formation in Rhagoletis spp. (Bush, 1969, 1975) and the contribution these findings have made towards the understanding of speciation mechanisms and the debate regarding the possibility of sympatric speciation.
2. The investigation of life history strategies of different tephritids and the contribution of those findings to the current theories about the ways in which life history traits are related to survival and reproduction in different habitats (Bateman, 1979; Zwolfer, 1983; Fletcher, 1986).
3. Studies on foraging behaviour in Rhagoletis and other fruit flies in relation to food, mates and larval resources and the contribution of these data to foraging theory (Prokopy and Roitberg, 1984).
4. Detailed quantification of the population dynamics of some species and the examination of which factors regulate fruit fly populations in time and space (e.g. Bateman 1968; Kapatos and Fletcher, 1986; Fitt, 1986), including the roles of inter- and intra-specific competition and the role of other density-dependent mechanisms in population regulation, subjects which have been of concern to ecologists for many years (Andrewartha and Birch, 1984).

Although these issues are not of particular importance to people involved primarily in the planning and implementation of regulatory or control procedures against fruit flies, this does not mean that basic studies have not made significant contributions to fruit fly management procedures. There are a number of good examples of how basic research findings have been of direct benefit by providing better means of monitoring or controlling fruit fly populations, or may lead to the development of better control procedures in the future.

For example, studies on possible host races in Rhagoletis cerasi (Boller et al. 1976) revealed that there are two geographical races of the cherry fly in Europe with a high degree of undirectional sexual incompatibility between them. A situation which it may be possible to exploit in control programmes by using the incompatible insect technique, which has already been used with some success against mosquitoes (Davidson, 1974).

Studies on the role of colour and shapes as foraging stimulii in a number of Rhagoletis spp. and Dacus spp. (Prokopy, 1977) have facilitated the development of more efficient traps for monitoring and even suppressing populations of some pest species (Prokopy, 1975; Remund and Boller, 1983; Economopoulos, 1977).

Also, as I will discuss in more detail later, ecological data on the phenology, dispersal, host relationships and population dynamics of a number of pest species have been used to good effect in optimizing pest management procedures against them.

ECOLOGICAL ISSUES OF BOTH ACADEMIC AND PRACTICAL INTEREST

Among the most important issues which fall into this category are those concerning the reasons for the current distribution and host preferences of the major pest species of fruit fly and the extent to which we might predict these will change in the future.

Historically, the four major genera of fruit infesting tephritids, i.e. Anastrepha, Ceratitis, Dacus and Rhagoletis were restricted to specific and largely separate regions of the world. Anastrepha in the neotropical regions of South and Central America and the Carribean, Rhagoletis in the holoartic regions of America, Europe and Asia with a few species extending into the mountain regions of Central and South America, Ceratitis in Africa and the Mediterranean basin, and Dacus mainly in the tropical and subtropical regions of Africa, Asia, Australia and the Pacific Islands.

During the past century a number of species have been able to colonize new areas as a result of accidental introductions, e.g. C. capitata which is now established in Hawaii, South and Central America and Western Australia (Back and Pemberton, 1918; Hendrichs et al., 1983; Fisher, 1981); Dacus dorsalis and D. cucurbitae in the Ogasawara Islands of Japan and Hawaii (Koyama, 1980; Back and Pemberton, 1917); Dacus tryoni in New Caledonia and the Society Islands (Bateman et al., 1973); Anastrepha suspensa in Florida (Swanson and Baranowski, 1972) and Rhagoletis cerasi in Crete (Neuenschwander et al., 1983).

As the amount of international travel and trade increases, it seems inevitable that without adequate safeguards the number and frequency of introductions will increase. The ability to be able to predict in advance which regions of the world are likely to be subject to invasion and colonization by specific fruit fly species is clearly of some significance in this context and there are a number of ecological issues which have a bearing on this.

Bioclimatic potential

Many species of fruit fly seem capable of colonizing geographic areas where they do not currently occur. Information regarding temperature thresholds, development rates and response to other climatic factors (e.g. rainfall) can be used to predict the areas that particular pest species might have the potential to colonize if suitable host fruits are available.

This approach was used in the early 1950s (Messenger and Flitters, 1954; Messenger, 1960), to determine which parts of mainland USA were potentially at risk from *D. dorsalis* after it became established in Hawaii during World War II. Meats (1981) carried out a similar type of study on the bioclimatic potential of *Dacus tryoni* in Australia which enabled him to predict which areas are most and least at risk.

Another factor relevant to the probability of establishment of fruit flies in new areas is the biological characteristics that make a species a good colonizer. Bateman (1979) concluded that the highly mobile species (e.g. *D. dorsalis* and *D. tryoni*) may be relatively poor colonizers because their dispersive habits would lead to a rapid dilution of the founder individuals with the result that they and/or their offspring would have great difficulty finding mates. In the case of less mobile species like *C. capitata* and *Rhagoletis* spp. the more concentrated aggregation of adults during the critical period following introduction of a founder colony into a new area would enhance the probability of individuals finding mates.

This difference in colonizing ability may help explain why, by and large, *Dacus* spp. have only successfully colonized islands, whereas *C. capitata* has colonized not only islands but also large parts of continental South and Central America and Australia.

Interspecific competition

A contentious issue in fruit fly ecology concerns the importance of interspecific competition, or other biological interactions, in limiting the establishment of certain species in areas which seem bioclimatically suitable.

Thus, although medfly has colonized large parts of South and Central America, despite possible interactions with Anastrepha sp. (Pavan and Souza, 1979), it has never become established in India despite early records of introductions (Kapoor, 1970), or in South East Asia and the South Pacific. Furthermore, although it became established in both Eastern and Western Australia before the turn of the century it was only able to maintain itself in Western Australia.

Possibly there are good historical reasons related to trading routes and travel patterns to account for this. However, there is also the possibility that medfly has been effectively excluded from many areas due to the presence of indigenous Dacus spp. This hypothesis has some support from two directions. Firstly medfly, which early this century was the major fruit fly pest in the Sydney region, disappeared at the same time as Dacus tryoni became the dominant fruit pest in that region (Andrewartha and Birch, 1984). In Western Australia on the other hand, where there are no native Dacus sp. that infest the commercial fruit which are the major hosts of medfly, it has persisted and expanded its range. Unfortunately, no studies on the reasons for the change in status of medfly in the Sydney region were carried out at the time, so it is not possible to say exactly what was the cause of its disappearance. Secondly, the less dramatic but significant decrease in status of medfly as a fruit pest that is reported to have occurred in the lowlands of Hawaii following the introduction of D. dorsalis (Kaiser et al., 1974), may indicate that the polyphagous Dacus sp. have a depressant effect on medfly populations, either because of competitive interactions on or within fruit, or possibly as a result of other biological interactions, as suggested by Andrewartha and Birch (1984).

Even more contentious than the possibility of C. capitata - Dacus interactions are those involving the polyphagous Dacus spp. Could D. dorsalis for example colonize an area where D. tryoni was already established, or vice versa? If so, would one always exclude the other, or could they co-exist by dividing up the resources in some way? At the moment we have no information about this and currently there seems little chance of carrying out any controlled experiments to examine the situation.

Interestingly, however, the recent discovery of D. latifrons in Hawaii (Vargas and Nishida, 1985), indicates that when a suitable 'niche' is available, a species which appears to have a limited biopotential in terms of host range and rate of increase compared with D. dorsalis and D. cucurbitae (the two Dacus spp. that were already present) can, nevertheless, become established.

Host range

A further issue related to the distribution and pest status of fruit flies concerns the range of new hosts a species will infest. The highly polyphagous species like D. dorsalis, D. tryoni and C. capitata seem to be able to switch quite rapidly to new hosts and genetic selection if it occurs at all appears to be minimal. In monophagous and stenophagous (= hosts limited to a few closely related plants of the same family) species host-range expansion is much slower and when it occurs is normally restricted to plant species closely related to current hosts. The expansion in host range of Rhagoletis pomonella in North America from its native host hawthorne onto introduced species of apple, plum, sour cherry, apricot, pear and rose hips is well documented (Prokopy and Berlocher, 1980). Bush and Diehl (1983) have suggested that the adoption of permanent new hosts arises through changes in host recognition and larval survival genes and may eventually lead to speciation.

Sometimes the host range of a fruit fly 'species' varies in different parts of its geographic range. This, depending on the circumstances, many indicate differences in phenotypic response due to differences in host availability, some genetic selection for different hosts, the existence of different host races or even the existence of two or more members of a species complex that are difficult to distinguish morphologically.

A good example of the last situation occurred in Australia in the mid-seventies when some flies with the taxonomic characteristics of D. dorsalis were caught in methyl eugenol baited traps in Northern Australia. It was only after ecological studies indicated that they were not behaving like D. dorsalis, and instead of being polyphagous were breeding in a single native host, that detailed taxonomic studies were carried out and the fact that they belonged to a previously undescribed species of the dorsalis complex, subsequently named D. opiliae, confirmed (Drew and Hardy, 1981). If its ecological characteristics had been more similar to D. dorsalis, the fact that it was a separate species would probably have gone unrecognised for a much longer period.

ECOLOGICAL ISSUES RELATED TO THE PEST MANAGEMENT OF FRUIT FLIES

Perhaps the central issue in this context is the extent to which ecological studies can benefit fruit fly control and eradication programmes. Clearly ecological data about the distribution, host range, phenology, population levels and dispersive behaviour of a pest fruit fly are useful in planning control or eradication strategies against it. There are, however, more specific ways in which ecological studies can improve or optimize pest management practices whether it be for suppression or eradication of an established population. Here I will discuss three examples.

The first concerns the sterile insect technique. The other two illustrate ways in which ecological studies have or can assist in the optimization of management programmes, using as examples two studies I have been directly involved in. Both examples relate to the olive fly, D. oleae. One concerns the development of a pest management system for D. oleae in Corfu, designed to minimise the use of insecticides, based on information about the insect's ecology and economic thresholds. The other illustrates how once comprehensive data on the population dynamics of a species have been collected, they can be used to develop simulation models. These enable us to examine not only the population dynamics of the species under different conditions but also the likely effects of various pest management procedures.

Sterile insect technique

Ito and Koyama (1982) have discussed the benefits of ecological studies to their successful S.I.T. campaign against D. curcurbitae in Okinawa. These included estimates of the density and survival rates of the target populations, the sexual competitiveness of sterile mass reared insects in the field, and the ability to realistically simulate the field situation in order to determine minimum release rates to achieve eradication. Cirio (1974) and Hendrichs et al. (1983) have also discussed the role of ecological studies in relation to S.I.T. programmes against C. capitata. The benefits to be gained by incorporating ecological studies into such programmes appears unquestionable.

Development of a pest-management system for D. oleae in Corfu

Almost all the 3.5 million olive trees, which are the sole host of D.

oleae on Corfu, belong to the single variety 'lianolia'. They produce large numbers of small fruits used almost exclusively for oil production. The fruit start to form in mid-June and when free from olive fly infestation may remain on the trees until the following May or early June. Harvesting starts in mid-November and the fruit are allowed to fall naturally onto nets beneath the trees, from which they are collected regularly. The harvesting period thus extends from November until the end of fruit fall in early June.

Long term ecological studies on D. oleae in Corfu were started in 1973 as part of a joint Greek Government/FAO project and this provided the quantitative data on survival and development rates of the immature stages, survival, maturation rates and reproductive potential of adults, the efficiency of the McPhail traps, and the crop losses due to infestation, that were used to develop the pest management system (Kapatos and Fletcher, 1983 a,b,c; 1986).

The phenological studies indicated that on Corfu, D. oleae has four main generations per year (Kapatos and Fletcher, 1984). Adults, the only stage to survive from the previous season, pass the fruitless early summer period in a reproductive diapause. They mature and oviposit the eggs of the first generation in the new fruits in July and early August. During the period from September until February-March there are two partly overlapping generations with peak numbers of immature stages in the fruit in late September/mid-October and November respectively. Most of the population overwinter as pupae in the soil and these give rise to adults which emerge in February-March. They mature and oviposit in the fruit still remaining on the trees in April-May. This gives rise to the fourth generation of adults which emerge in late May and June and pass the summer in an immature state until the new fruits of the following season become ripe enough to be infested.

It became clear from these data that effective suppression of the adult population in July just prior to the start of the new infestation should drastically reduce infestation levels right up to the start of harvesting, after which infestation ceases to be an economic problem. In areas where suppression of adults in July was not effective, subsequent infestations could be controlled either by aerial bait sprays or cover sprays applied from the ground.

To ensure the success of this system while restricting insecticide application to a minimum, two different sets of criteria had to be used to determine when to spray, both of which relied heavily on ecological data.

The first requirement was to develop an effective method of timing the application of the initial aerial bait spray in July so that it coincided with the period when females were maturing their ovaries and thus highly responsive to the protein hydrolysate/insecticide bait, but had not yet started to oviposit in the fruit (Kapatos and Fletcher, 1982). The second was to develop meaningful economic thresholds so that the decision on whether to apply treatments later in the season could be made. These thresholds needed to be expressed both as % infestation of fruit (for cover sprays which kill the immature stages in the fruit) and females trapped per week (for aerial bait sprays which are only effective against adult populations).

The model used to forecast the time of application of the initial aerial bait spray in July is shown in Fig. 1. It is based on the relationship between fruit size and ovarian maturation rates. Economic thresholds for cover sprays and aerial bait sprays (based on 1980 data for control costs per tree and olive oil revenues) are shown in Tables 1 and 2. The way these parameters were derived has been described in detail in early papers (Kapatos and Fletcher, 1983 a,b).

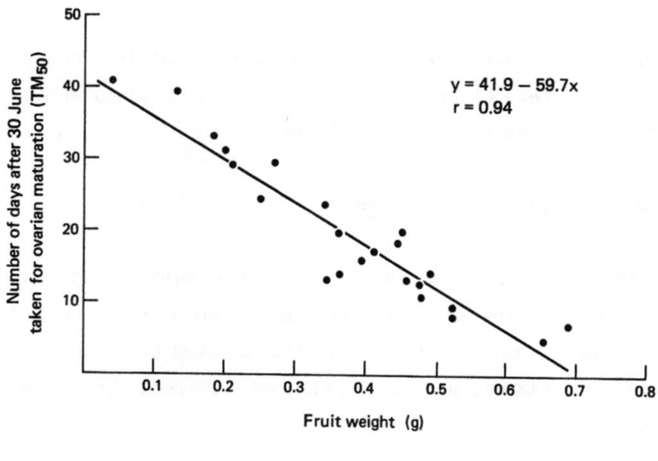

Figure 1. The correlation between the number of days after the 30th June taken for ovarian maturation (TM_{50}) and fruit weight on the 30th June (from Kapatos and Fletcher, 1983a).

Table 1. Calculated economic injury levels for D. oleae infestation, expressed both as number of infested fruits per tree and percentage infestation, in relation to the cost of cover sprays for late July-August, September and October, the periods when these control measures would be normally carried out. The calculations refer to trees bearing an average of 30,000 olive fruits each (from Kapatos and Fletcher, 1983b).

Late July-August		September		October	
No. of fruits infested	% infestation	No. of fruits infested	% infestation	No. of fruits infested	% infestation
2277	7.59	1848	6.16	3094	10.31

Table 2. Calculated economic injury levels for air bait sprays, expressed as weekly number of females caught per McPhail trap, in relation to the cost of the spray and the proportion of trees with fruits in the area, for September and October (from Kapatos and Fletcher, 1983b).

Proportion of trees with fruits (A)	Number of females per trap/week	
	September	October
0.25	16	6
0.50	8	3
0.75	5	2
1.00	4	1

Implemented properly, this pest management system for olive fly can not only reduce the overall costs of control but also ensure that the use of insecticides is kept to a minimum.

Simulation of management procedures

Another use of ecological data is in the development of tactical simulation models which can be used to investigate the population dynamics of target species and examine the effects of various management procedures, e.g. spray applications and release of sterile males, on their population dynamics.

In the case of D. oleae we were able to use the ecological data collected in Corfu to develop a biologically realistic simulation model which has enabled us not only to examine various population processes but also investigate the effects of various control strategies.

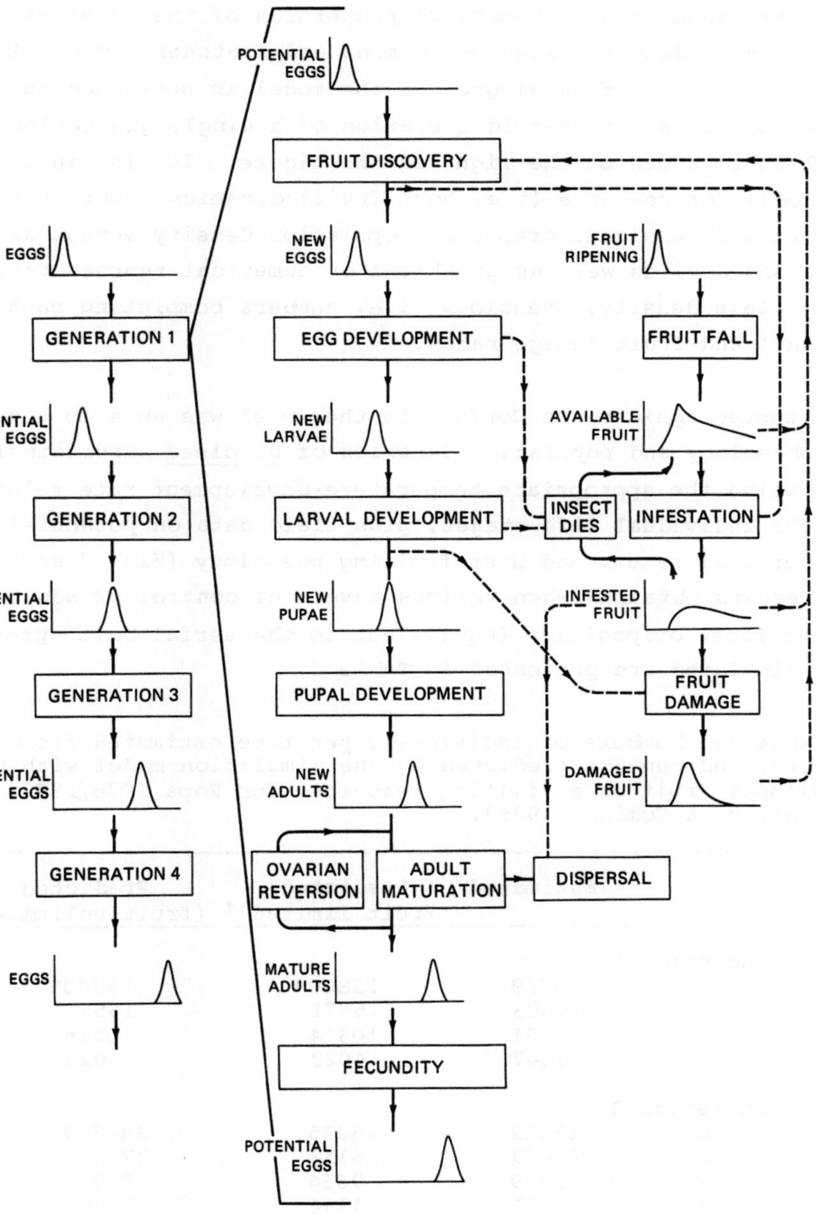

Figure 2. Flow diagram showing the components of the model which relate the population fecundity distribution in one generation to the population fecundity distribution in the next (the fruit availability sub-model can be omitted if desired). Solid lines indicate population flows; dashed lines controlling effects; rectangles indicate submodels, and graphs state variables.

The development and mathematical properties of the model have been described in detail elsewhere (Comins and Fletcher, 1986, Fletcher and Comins, 1985). A flow diagram of the model is presented in Fig. 2. As shown, it is a four-fold iteration of a single generation sub-model which is expanded on the right of the figure. It also includes a sub-model for resource (i.e. host fruit) dynamics. Output is in the form of 3-dimensional graphs of population density versus development class and day, as well as graphical or numerical representation of stage-class density, phenology (i.e. numbers completing each stage per day) and fruit damage rates.

When tested against the Corfu data the model was able to simulate both the phenology and population dynamics of D. oleae surprisingly accurately using the appropriate temperature-development rate relationships for the individual life stages, plus field data on potential fecundity, survival rates, and host fruiting phenology (Fig. 3 and Table 3). The results obtained when various levels of control of adults prior to the first oviposition (equivalent to the aerial bait spray in July) were simulated are presented in Table 4.

Table 3. Numbers of individuals per tree estimated from field data, and numbers predicted by the simulation model with and without fruit as a limiting resource, for Ropa 1976/1977, (from Fletcher & Comins, 1985).

	Estimated	Predicted (fruit limited)†	Predicted (fruit unlimited)
Generation 1			
E	15769	15800*	15800*
L	14083	15571	15571
P	9351	10314	10314
A	3647	4023	4023
Generation 2			
E	25422	28035	288839
L	23879	26353	271220
P	8709	9654	99354
A	4531	4994	51664
Generation 3			
E	3518	4200	2458933
L	3430	4125	2308938
P	3157	3794	2128804
A	650	794	447048

† Fruit numbers initially set at 55000.
* Number of eggs input at the start of the simulation
E = eggs; L = larvae; P = pupae; A = adults.

These data indicate that over 90% of females need to be killed before there is any lasting suppression of the population. In fact, at lower levels of suppression, by the fourth generation population levels have returned to or even exceeded those that occur in the absence of control, due to the initially slower rate at which the host fruits are utilized. This does not mean, however, that some economic advantage would not be gained by lower levels of suppression because there would still be some reduction in fruit loss due to infestation prior to the start of harvesting in November.

This model can also be used to investigate other pest management procedures, including more sophisticated spray regimes and the sterile insect technique (Comins and Fletcher, 1986).

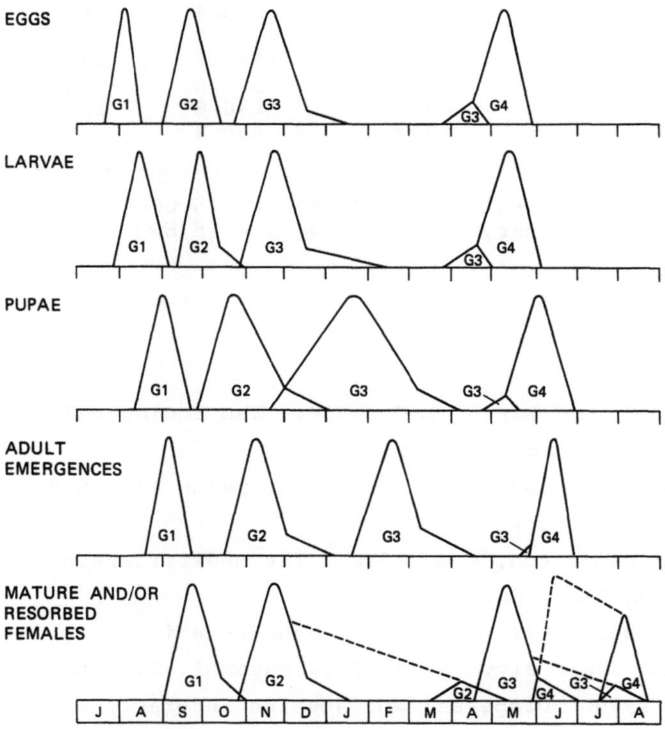

Figure 3. The phenology of D. oleae predicted by the computer simulation model using temperature data for the 1976/77 period in Corfu. Dotted lines indicate the periods when the females have resorbed ovaries.

Table 4. Numbers of D. oleae completing each stage for the four generations per year after different levels of suppression of adults prior to the first oviposition period in July.

		Level of Suppression			
Generation	Stage	Normal	90%	99%	99.9%
	E	14512.2	1462.3	137.9	14.2
1	L	9613.0	968.6	91.4	9.4
	P	3749.1	377.6	35.6	3.7
	A	2924.3	294.7	27.8	2.9
	E	23039.0	5586.8	582.6	60.7
2	L	8439.8	2046.6	213.4	22.2
	P	4388.7	1064.2	111.0	11.6
	A	877.7	212.8	22.2	2.3
	E	3673.7	3185.8	413.9	44.4
3	L	3387.1	2937.2	381.6	40.9
	P	711.3	616.8	80.1	8.6
	A	355.6	308.4	40.1	4.3
	E	1231.6	1489.7	225.7	25.0
4	L	1077.1	1302.8	197.4	21.9
	P	1055.1	1196.8	194.0	21.5
	A	527.6	598.4	97.1	10.8

+ Initially 55,000 fruit with additional 15,000 fruit becoming available in February. Initial eggs = 15,800

REFERENCES

Andrewartha, H.G. and Birch, L.C. 1984. The Ecological Web. The University of Chicago Press, 506p.

Back, E.A. and Pemberton, C.E. 1917. The melon fly in Hawaii, USDA Bull. No. 491, 64p.

Back, E.A. and Pemberton, C.E. 1918. The Mediterranean fruit fly in Hawaii USDA Bull. No. 536, 118p.

Bateman, M.A. 1968. Determinants of abundance in a population of the Queensland fruit fly. Symp. Roy. Entomol. Soc. Lond. 4: 119-131.

Bateman, M.A. 1979. Dispersal and species interaction as factors in the establishment and success of tropical fruit flies in new areas. Proc. Ecol. Soc. Aust. 10: 106-112.

Bateman, M.A., Insunza, V. and Arretz, P. 1973. The eradication of the Queensland fruit fly from Easter Island. F.A.O. Plant Protection Bull. 21:114-115.

Boller, E.F., Russ, K., Vallo, V. and Bush, G.L. 1976. Incompatible races of European cherry fly, Rhagoletis cerasi (Diptera: Tephritidae), their origin and potential use in biological control. Ent. Exp. & Appl. 20: 237-247.

Bush, G.L. 1969. Sympatric host race formation and speciation in frugivorous flies of the genus Rhagoletis (Diptera, Tephritidae) Evolution, 23: 237-251.

Bush, G.L. 1975. Modes of animal speciation. Ann. Rev. Ecol. Syst. 6: 339-364.

Cirio, U. 1974. Basic ecologiche per un programme di lotta control la ceratitis capitata Wied. nell'isola di Procida. Redia, LV: 159-201.

Comins, H.N. and Fletcher, B.S. 1986. Simulation of fruit fly population dynamics with particular reference to the olive fruit fly, Dacus oleae. Ecol. Modelling, In press.

Davidson, G. 1974. Genetic Control of Insect Pests. Academic Press, New York.

Drew, R.A.I. and Hardy, D.E. 1981. Dacus (Bactrocera) opiliae, a new sibling species of the dorsalis complex of fruit flies from northern Australia (Diptera: Tephritidae). J. Aust. Ent. Soc. 20: 131-137.

Economopoulos, A.P. 1977. Controlling Dacus oleae by fluorescent yellow traps. Ent. Exp. & appl. 22: 183-190.

Fisher, K. 1981. Fruit fly under attack from the sterile insect technique. J. Agric. West. Aust. 2: 51-52.

Fitt, G.P. 1983. Factors limiting the host range of tephritid fruit flies, with particular emphasis on the influence of Dacus tryoni on the distribution and abundance of Dacus jarvisi. Ph.D. thesis. Univ. of Sydney, Australia.

Fitt, G.P. 1986. The importance of interspecific interactions among tephritids. In Robinson, A. and Hooper, G.H.S. (eds.). Fruit Flies: Their Biology, Natural Enemies and Control. Elsevier Press, Amsterdam. In press.

Fletcher, B.S. 1986. Life history strategies of tephritid fruit flies. In Robinson, A. and Hooper, G.H.S. (eds.). Fruit Flies, Their Biology, Natural Enemies and Control. Elsevier Press, Amsterdam. In press.

Fletcher, B.S. and Comins, H. 1985. The development and use of a computer simulation model to study the population dynamics of Dacus oleae and other fruit flies. Atti XIV Cong. Naz. Ital. Ent. Palermo, 561-575.

Hendrichs, J., Ortiz, G., Liedo, P. and Schwarz, A. 1983. Six years of successful medfly program in Mexico and Guatemala. In Cavalloro, R. (ed.). Fruit Flies of Economic Importance. A.A. Balkema, Rotterdam. 353-365.

Ito, Y. and Koyama, J. 1982. Eradication of the melon fly: role of population ecology in the successful implementation of the sterile insect release method. Protection Ecol. 4: 1-28.

Kapatos, E.T. and Fletcher, B.S. 1982. Development of a method for determining the time of air treatments against adults of D. oleae (Gmel.) (Diptera: Tephritidae). (In Greek with English Summary). Agric. Res. (Athens), 6: 403-416.

Kapatos, E.T. and Fletcher, B.S. 1983a. Development of a pest management system for Dacus oleae in Corfu by utilizing ecological criteria. In Cavallaro, R. (ed.). Fruit Flies of Economic Importance. A.A. Balkema, Rotterdam. 593-602.

Kapatos, E.T. and Fletcher, B.S. 1983b. Establishment of economic injury levels and intervention thresholds for Dacus oleae infestations in Corfu. Ent. Hellenica, 1: 37-45.

Kapatos, E.T. and Fletcher, B.S. 1983c. Seasonal changes in the efficiency of McPhail traps and a model for estimating olive fly densities from trap catches using temperature data. Ent. Exp. & Appl. 33: 20-26.

Kapatos, E.T. and Fletcher, B.S. 1984. The phenology of the olive fly, Dacus oleae (Gmel. (Diptera: Tephritidae), in Corfu. Zeitschrift. Angew. Ent. 97: 360-370.

Kapatos, E.T. and Fletcher, B.S. 1986. Mortality factors and life budgets for the immature stage of the olive fly, Dacus oleae (Gmel.) (Diptera, Tephritidae) in Corfu. Zeitschrift. Angew. Ent. In press.

Kapoor, V.C. 1970. Indian tephritidae with their recorded hosts. Oriental Insects, 4: 207-251.

Keiser, I., Kobayashi, R.M., Miyashita, D.H., Harris, E.J., Schneider, E.L. and Chambers, D.L. 1974. Suppression of Mediterranean fruit flies by oriental fruit flies in mixed infestations in guava. J. Econ. Entomol. 67: 355-360.

Koyama, J. 1980. The Okinawa project of eradicating fruit flies. In Proceedings of Symposium on Fruit Fly Problems. Nat. Inst. Agric. Sci. Japan. 99-106.

Meats, A. 1981. The bioclimatic potential of the Queensland fruit fly, Dacus tryoni, in Australia. Proc. Ecol. Soc. Aust. 11: 151-163.

Messenger, P.S. 1960. California climatic factors and the oriental fruit fly. Calif. Dept. Agr. Bull. 49: 235-241.

Messenger, P.S. and Flitters, N.E. 1954. Bioclimatic studies of three species of fruit flies in Hawaii. J. Econ. Entomol. 47: 756-65.

Neuenschwander, P., Michelakis, S., Russ, K. and Hobans, E. 1983. Ecological studies on Rhagoletis cerasi L. in Crete for the use of the incompatible insect technique. In Cavallaro, R. (ed.). Fruit Flies of Economic Importance. A.A. Balkema, Rotterdam, 41-51.

Pavan, O.H.O. and Souza, H.M.L. 1979. Competition between Ceratitis capitata and Anastrepha fraterculus. In Kommendahl (ed.). Proc. Symp. IX. International Congress of Plant Protection, Washington, D.C. 158-161.

Remund, U. and Boller, E.F. 1983. Pieges visuels pour lutte biotechnique et prevision de lu mouche de la cerise, Rhagoletis cerasi L. In Cavalloro, R. (ed.). Fruit Flies of Economic Importance. A.A. Balkema, Rotterdam. 490-491.

Prokopy, R.J. 1975. Apple maggot control by sticky red spheres. J. Econ. Entomol. 68: 197-198.

Prokopy, R.J. 1977. Stimuli influencing trophic relations in Tephritidae. In Intl. Symposium on Trophic Relations in Insects. Colloques Internationaux du C.H.R.S. 265: 305-334.

Prokopy, R.J. and Roitberg, B.D. 1984. Foraging behaviour of true fruit flies. Amer. Scientist, 72, 41-49.

Swanson, R. and Baranowski, R.M. 1972. Host range and infestation by the Caribbean fruit fly, Anastrepha suspensa (Diptera: Tephritidae), in South Florida. Proc. Fla. St. Hort. Soc. 85: 271-274.

Vargas, R.I. and Nishida, T. 1985. Survey for Dacus latifrons (Diptera, Tephritidae), J. Econ. Entomol. In press.

Zwolfer, H., 1983. Life systems and strategies of resource exploitation in tephritids. In R. Cavalloro, (ed.). Fruit Flies of Economic Importance. A.A. Balkem, Rotterdam. 16-30.

MEDITERRANEAN FRUIT FLY ASSOCIATED BACTERIA: TRANSMISSION AND LARVAL SURVIVAL

Vincenzo Girolami

Istituto Entomologia Agraria dell' Universita' di Padova
Via Gradenigo n.6 35100 Padova

Summary

The Med fly can conserve larval micro-organisms and seems to hereditarily transmit them to its offspring. Furthermore, micro-organisms may be acquired from the environment, adult food and fruit.

The presence of bacteria is essential to larval growth and perhaps it is also important for adult survival.

The chemicals involved are still unknown but are not linked to the presence of unspecific mixtures of vitamins or hydrolysed proteins.

Micro-organisms which cause the fruit to rapidly rot, may kill larvae.

Therefore the survival of the Med fly seems to depend on the presence of a balanced micro-flora.

The yield of mass rearing can be influenced by micro-organisms.

Inherited bacterial species which are useful for survival, may be considered true symbionts.

Symbiotic association was first observed among _Tephritidae_ , in the olive fly (_Dacus oleae_ Gmelin) at the beginning of the century (Petri 1909).

Although symbionts have since been found in all true fruit flies (_Dacinae_ and _Trypetinae_), in some respects the olive fly remains a unique case due to the presence of specific bacteria inside the "larval blind sacs", the shape of the adult pharyngeal bulb and other aspects that will be treated later.

The adult pharyngeal bulb.

The pharyngeal bulb, first found in the olive fly (Petri 1909), is an organ common to all fruit flies, (Girolami 1973) consisting of a stomodeal diverticulum placed at the end of the cibarial pump in front of the Cerebrum.

In all true fruit flies, with the exception of the olive fly and some african _Dacus_ the pharyngeal bulb's most common shape is round with a short stalk connecting it to the pharynx (Fig. 1). The cells at its apex are elongated. This kind of pharyngeal bulb, was described as the " _C. capitata_ type" (Girolami 1973).

The elongated cells, whose submicroscopic structure was first shown in _Rhagoletis pomonella_ are called columnar epithelial cells (Ratner et Stoffolano 1984).

The symbiotic bacteria in adult fruit flies multiply inside the pharyngeal bulb, and are released in compact masses (bacterial masses) and cross the foregut entering the midgut where they can be seen inside the peritrophic membrane. These masses dissolve at the beginning of the hindgut where single bacteria only can be found.

NATO ASI Series, Vol. G11
Pest Control: Operations and Systems Analysis
in Fruit Fly Management. Edited by M. Mangel et al.

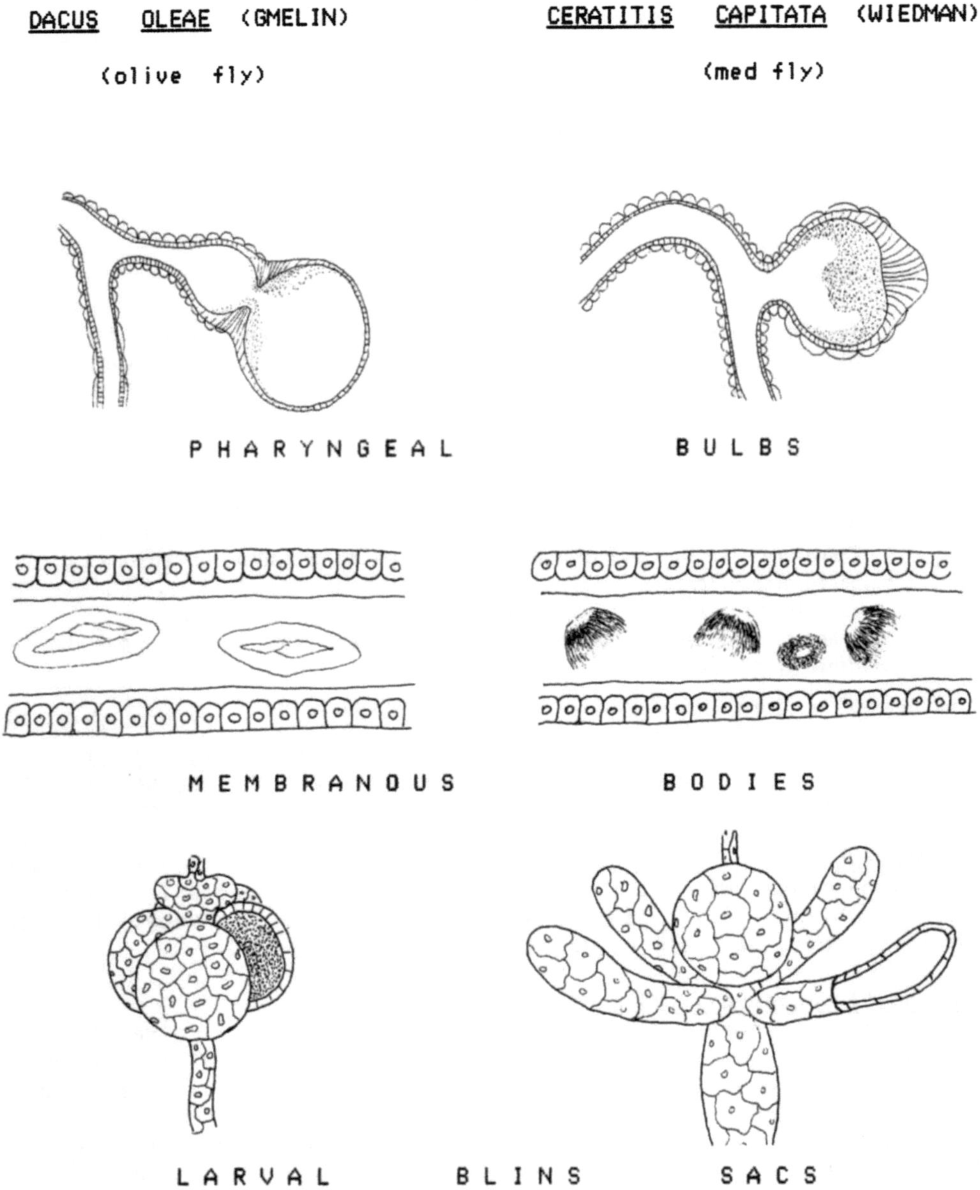

Fig. 1 - Morphological aspects of the symbiosis of the medfly and the better known olive fly.

The pharyngeal bulb is smaller in the Medfly (and contains few bacteria). The elongated "wall cells" are in the apex of the bulb in the medfly (and other fruit flies) but at the base of the bulb in the olive fly.

The "membranous bodies" that leave the bulb (and form the core of the "bacterial masses"), enter the Mesenteron and conserve the internal shape of the bulb inside the perytrophic membrane of the bacteria free Med flies.

The intestinal blind sacs of larvae are elongated and empty in the medfly, but roundish and full of specific bacteria in the olive fly.

Pharyngeal bulb secretion.

The core of the bacterial masses is produced by the fruit flies: adults free of bacteria and reared in sterile conditions, show in place of the bacterial masses, numerous transparent bodies (membranous bodies or agglomerates) (Fig.1) inside the peritrophic membrane (Girolami 1973). In Ceratitis capitata the bodies conserve the shape of the thickened intima of the pharyngeal bulb and are similar to a medusa. Columnar Epithelyal cells seem responsible of the secretion of membranous bodies. The bulb of Ceratitis capitata can produce numerous membranous bodies per day even in old unfed adults.

Transmission apparatus and larval bacteria.

In the olive fly finger-like invaginations, full of bacteria, are observed in the ovipositor (Petri 1909). During oviposition bacteria are released from the invaginations to the egg micropilar area. Symbionts penetrate the larvae before hatching.

Other fruit flies are lacking in these special structures for bacteria transmission (Stammer 1929).

The blind sacs of olive fly larvae are roundish in shape and full of bacteria whereas those of other fruit flies including Ceratitis capitata (Fig 1) are more elongated and empty; bacteria are free inside the intestinal lumen.

Symbiont identity.

Pseudomonas savastanoi has been considered the symbiotic bacteria of the olive fly, having been isolated from the blind sacs of larvae. Today there is no concrete reason to consider Pseudomonas savastanoi the symbiotic bacteria but the olive fly possesses a specific unidentified bacterial species difficult to multiply in vitro (Girolami at cavalloro 1972).

Other fruit flies seem to possess less specific bacteria as can be observed under microscope. It also seems that the bacteria are easy to cultivate. The most investigated insect-associated bacteria are those of Rhagoletis pomonella where the most common species is shown to be Klebsiella oxytoca (Howard et al. 1985) and not Pseudomonas melophthora as previously believed (Allen et al. 1934). However, it appears that symbiosis can be linked to more than one bacterial species in different specimens or populations of the same fruit fly.

The possibility that different bacteria can occupy symbiosomes (or special organs) of the same species renders the very concept of symbiosis questionable. The existence of symbiosis in Dacus tryoni Froggatt has been denied (Drew et al. 1983) and all bacteria were considered to come from food, especially from leaf surface bacteria.

The ease with which the medfly can be reared on artificial media, permits experimental demonstration of the transmission possibility of a specific bacteria. The importance of symbionts on growth and survival in the laboratory can also be investigated; we can therefore speculate on the importance of micro-organisms in particular on the successful colonisation of new host fruits in nature and on the yield of production in mass rearing.

MATERIALS AND METHODS

Rearing conditions.

Adults were kept in cubic net cages (10 cm), according to the techniques developed for the laboratory rearing of D. oleae (Girolami 1979).

An old laboratory strain "Ispra" (Cavalloro and Girolami 1969) was commonly used; insects collected from oranges (Naples, December 1984) were also investigated.

In order to lower bacterial presence, the whole cage with flies can be immerged in 1% formalin for some minutes, and is then placed in a sterile laminar flow cabinet to dry. Sterilised food is placed in droplets on the cage and water is added with 0.5% formalin.

Egg collection and larvae sterilisation.

Freely laid eggs were left to fall through the bottom of the net cages into a water and 0.5% formalin solution. The cages were kept in a laminar flow cabinet. The eggs were collected by filtering the water through a sterile stainless steel net (0.11 mm mesh), and placed in a stainless steel net dish, (0.15 mm mesh) inside a sterile Petri dish containing a small quantity of 0.5% formalin. When the larvae hatched they passed through the mesh and separated from the dead or empty eggs.

The newly hatched larvae were collected with a sterile "pipette Pasteur" and divided into stainless steel net dishes (0.11 mm mesh). The larve could be rinsed with sterile water inside the dishes and placed in Petri dishes with water.

After at least one hour of rinsing in water, some drops of broth containing bacterial colonies (or broth only for controls) can be added to the dishes.

After a further hour, the larvae are transferred to the larval growth media inside the net dishes. These are placed upside down on the larval medium inside the Petri dishes (5 cm diam.). The whole operation can be carried out in sterile conditions after autoclaving the media inside the Petri dishes.

Larval rearing.

Traditional larval medium (Cavalloro e Girolami 1969) was routinely employed. For reseach purposes an autoclavable medium was employed : Water: 100; cellulose powder: 60; glucose: 25; brewers yeas: 15; HCl 1N: 4 ml (final pH 4.0 ca).

Benzoic acid: 0.5; brewers yeast "autolisate" (Fratelli Piccioni Brescia Italy) or soy hydrolyzate (N. B. C., Cleveland, Ohio): 3; were added or not in different experimental media.

Filtered sterile vitamin solution (suspension) (vitamin diet fortification mixture (N.B.C.): 10; tween 80: 3; water: 100) was added after media sterilisation, in the ratio of 1 part of solution every 100 of medium.

Pupae handling.

The Petri dishes containing media and larvae are put inside larger, sterile Petri dishes, which in their turn are placed inside a third dish. This nest of dishes is useful to avoid contamination and dehydration. Care must be taken to avoid contamination of the external surface of the smaller dish when inserting the larvae. It is not essential that the outer dish is kept sterile.

The mature larvae are able to raise the lid of the smaller (5 cm diam.) Petri dish full of larval medium, and pupate on the sterile bottom of the second dish.

In order to observe bacteria transmission from pupae to adults, the pupae can be collected and placed in short sterile test tubes and covered with sterile sand. The test tubes are stoppered by a stainless steel mesh and vertically placed in a tall Petri dish.

When adults emerge, they reach the surface leaving pupae exuviae in the bottom. The layer of sterile sand thus minimises contamination from bacteria present on the exterior of the pupae after emergence. (In nature adults have no further contact with the exuviae.)

Emerged adults can be fed by putting some drops of sterile water and food (usually 10% sterile glucose solution) on the screen.

Field collected pupae can be put in 5% formalin stirred solution for many hours, and then placed in test tubes as above.

The pupae can also be placed on agar media in order to observe the growth of bacteria colonies. However, the pupae, which become covered by a watery film, usually die.

RESULTS AND DISCUSSION

Larvae Sterilisation.

Newly-hatched larvae collected in non-sterile water were placed in different concentrations of formalin for different periods and were then placed in Petri dishes containing agar culture medium in order to observe the development of bacterial colonies. The larvae must remain in 0.1% formalin for 24 hours or 0.5% formalin for 12 hours in order to prevent bacteria growth.

If the eggs are left to fall directly into a 0.5% formalin solution and are placed in the same solution after collection, the newly emerged larvae are free of micro-organisms. Nearly all the eggs hatch, and therefore this method was generally adopted.

Bacterial innoculation of larvae.

If colonies are put on freshly-rinsed larvae, micro-organisms can be killed by residual formalin inside the intestine; one hour's immersion in renewed water may be sufficient to prevent this.

Some hours are necessary for the micro-organisms to occupy the larval intestine. Therefore, for some hours contact of the innoculated larva with medium must be avoided. The low pH and preservatives of larval medium can kill bacteria outside larvae.

SPEED OF LARVAL GROWTH (IN DAYS)

	LARVAl MEDIUM				LARVAL TREATMENT		
	plane	+ preserv.	+ vitamins	+ protein hydrolys.	non disinfected (tap water)	disinfected with formalin: rinsed only	disinfected with formalin: innoculated with bacteria or yests
NOT STERILISED	+	+			6 - 8	6 - 8	6 - 8
	+	+	+	+	6 - 8	6 - 8	6 - 8
	+		+	+	?	6-8 (?)	6-8 (?)
	+				?	6-8 (?)	6-8 (?)
STERILISED	+	+			8 - 10	> 15	8 - 10
	+	+	+	+	8 - 10	> 15	8 - 10
	+		+	+	?	> 15	8 - 10
	+				?	> 15	8 - 10

TAB. 1 - Average length of larval growth on different artificial media of the Medfly larvae with or without different micro-organisms (T > 25 C).

When non-autoclaved media are used, the presence of micro-organisms in the larvae does not affect larval growth which is always rapid. Bacteria may be assumed from the media.

When autoclaved media are used, the presence of micro-organisms in the larvae improves larval growth. From initial experimentation, it seems that similar results can be obtained when micro-organisms derive from females (or the environment) or when pure colonies are added.

Vitamin mixtures or protein hydrolysate in larval growth media do not influence growth even when micro-organisms are absent.

When non autoclaved preservative free media are used the multiplication of bacteria and fungi can kill larvae. If sterilized larvae are used, the slow multiplication of micro-organisms of the media, does not normally kill larvae. The presence of some micro-organisms is essential for larval growth even if it is not clear which chemical is involved.

Larval growth and the presence of bacteria.

Symbiont free larvae grow very slowly on sterile media (more than 20 days at 25 C), and transform into small pupae . If micro-organisms are innoculated as above, larval growth is improved; pupation occurs with only a few days delay in comparison with traditional rearing methods where formalin is not used and the larval growth medium is not sterilised (Tab. 1).

Similar results were obtained by using colonies of lactobacillus acidophilus or yeasts. One yeast (not yet identified) was isolated from the intestine of an adult fly and probably derived from ingested food.

Larval growth is normal, both when sterile media are employed with non-sterilized larvae and when sterilized larvae are put onto non-sterilized media.

On microscopic examination of larval intestine content, few bacteria can be found even if preservatives are not employed. Nevertheless colonies are obtained around the pupae placed on agar culture media. Of course, colonies are not obtained when pupae deriving from bacteria free larvae are used.

Non-sterilized larvae cannot be put on preservative free growth media because this leads to the multiplication of bacteria and fungi that can kill larvae.

This may also occur with fruit. When larvae are transferred from entirely rotten fruit, the new fruit also rots very quickly. This may depend on the amount of micro-organisms or on the presence of some specific micro-organism responsible for the rapid deterioration of the fruit.

Micro-organism free larvae were not able to survive on surface sterilized pieces of fruit pulp in Petri dishes.

Symbiont transmission from larvae to adults.

As mentioned above, pupae can preserve larval bacteria even if in small quantities. It is not known where they survive.

The pharyngeal bulbs of emerging adults may contain larval bacteria or yeasts even if they are kept in sterile conditions and fed sterile food.

In fact adults that derive from larvae grown on non-autoclaved media or fruit, contain micro-organisms in most insects observed. Adults that derive from larvae grown on autoclaved media and innoculated with micro-organisms, do not seem to contain symbionts but further experimentation is needed since this could be true for some bacterial species and not others. Lactobacillus acidophilus improved larval growth but was later absent in adults even if preservative free larval growth media was used.

Symbiont transmission from adults to eggs.

When females are offered artificial agar fruits covered with parafilm (Boller 1968), colonies may be observed in the space between the egg and the parafilm along the closed channel of penetration made by the ovopositor. It is almost certain that the micro-organisms derive from the ovopositor since the agar balls, containing culture medium, were sterilized and the surface parafilm was immersed in different disinfectant solutions. Nevertheless micro-organisms are not always found; perhaps this is because some symbionts cannot multiply, due to the anaerobic conditions inside the agar medium.

Eggs, collected in sterile water under cages placed in a laminar flow cabinet, always contain bacteria. Contamination can happen in water and therefore it is not certain that all newly-laid eggs contain bacteria.

Adult bacteria.

The micro-organisms found inside the pharyngeal bulb and the intestine can derive from larval micro-flora. In fact, if larvae are reared on non sterile media or on fruit, one day after the adults emerge in sterile test tubes (and are fed sterile food and water), numerous micro-organisms can be observed. However, they are not visible if 0.5% formalin is added to the water.

If food containing a specific bacteria or yeast is offered to adults, the same species of micro-organism is found inside the bulb and the intestine. Micro-organisms are conserved by adults even if the food is later changed or sterilized.

Pupae emerging from medfly larvae reared on fruits which are all contaminated by micro-organisms preserve many bacterial species after drastic surface disinfection. Numerous colonies can be obtained by rinsing the pupae after disinfection and placing them on agar culture medium.

CONCLUSIONS

Symbiont transmission.

From experimental data it emerges that some (but not all) bacterial species or yeasts may be transmitted from eggs to larvae, pupae and finally to adults.

Adults are able to conserve larval bacteria or to take them from food and to store them.

Micro-organisms may be transmitted from females to eggs and larvae but it is not yet clear if they are (as in the olive fly) always transmitted.

Symbiont free larvae can take micro-organisms from the exterior environment as shown by their intake of larval medium bacteria. It is obvious therefore that they can also use surface micro-organisms on fruit or wind-born spores that reach the pulp of the fruit through the ovipositional hole (in the olive fly fruit bacteria cannot substitute symbionts since symbiont free larvae die).

Flies, reared for many generations on artificial media, can lay viable eggs inside fruit and the larvae grow normally. It is not yet known if the bacteria are inherited or found in the fruit.

Larval survival.

a) Artificial media. Micro-organisms are necessary for larval survival even with nutrient rich growth medium. It is obvious therefore that they are fundamental for larval development inside fruit.

It is not clear what nutrients the bacteria supplies to the larvae. From initial experimental data it emerges that they do not supply common aminoacids or vitamins as supposed for the olive fly (Hagen 1966).

A small quantity of micro-organisms seems sufficient to improve larval growth, in fact the micro-organisms multiply at a very slow rate in the intestine even if preservatives are not present.

This matter merits further basic investigation above all to improve the yield of laboratory rearing and stabilize mass production.

b) Fruit. The fruit pulp around the larvae is rotten but larvae prefer to lacerate new unrotten pulp and therefore ingest fresh fruit. The micro-organisms that invade the pulp are probably not essential for larval growth since they are abundant in the intestinal micro-flora. Micro-organism free larvae do not seem able to survive in fruit.

Micro-organisms which cause the fruit to rapidly rot, may kill larvae that need fresh pulp to eat.

Therefore larval growth is linked to the presence of a balanced micro-flora and some micro-organisms may be dangerous to larvae which need the presence of certain micro-organisms.

Both adults which are collected in the field and which are reared in the laboratory without disinfectants, lay eggs on fruit offered and produce larvae with a similar survival rate (on apples, pears or oranges).

Adult symbiosis.

Inside pharyngeal bulbs of the *C. capitata* , there is a smaller quantity of bacteria than inside the bulb of the olive fly. The bacteria can be observed in thin layers between the membranes of the "bodies" and not in compact masses as in the olive fly. Nevertheless they are always present in adults collected in nature or in the laboratory strains reared without disinfectants.

The presence of complex and specific organs, and continuous secretions in adult fruit flies, strongly suggests that bacteria, either inherited or acquired from food or environment, are important for adult survival or fecundity (Girolami 1983). The presence of the above organs in males suggests that they are not only linked to transmission and larval survival.

Inoculating larvae with *lactobacillus acidophilus* it is possible to obtain large pupae and fertile females which are bacteria free.

These adults allow us to test the influence of different species of micro-organisms on fecundity and survival. Future research will be undertaken in this direction.

Possibility of bacteria intake from environment (only med flies)	Hereditary Transmission	Possibile loss of micro-organisms
	ADULT	
From food ——> From environment ——>	↓	
From fruit surface ——>	EGG	<—— Lack of Transmission ?
	↓	
	LARVA	
From rotten pulp ——>	↓	<—— Competition with other micro-organisms ? <—— Disinfectants in food (e.g. phenols) ?
	PUPA	
	↓	<—— Selection of suitable microbial species ?
	ADULT	

TAB. 2 - General view of possibilities of micro-organisms intake in the different stages of development of the medfly (on the left). Bacteria can be inherited or assumed from the environment and food. The highly specialised symbionts of the olive fly can only be hereditarily transmitted.

The hypothesis on bacteria loss (on the right) are not yet confirmed in nature but are suggested by laboratory observations.

Symbiont identity and survival.

Unlike the olive fly, bacterial colonies are always obtained from dissected pharyngeal bulbs of the Med fly which show the presence of bacteria under the microscope. Therefore the bacteria of the med fly are less specialized than those of the olive fly.

Micro-organisms may be inherited from larve and transmitted from the mother to the larvae, even if the specialized transmission devices of the olive fly are absent.

Med fly micro-organisms may derive from the environment, from fruit or from adult food and (someone) may be transmitted under laboratory conditions irrespectively of their origin (Tab.2).

If the same thing occurs in nature, different bacterial species may be inherited and transmitted to offspring when larval growth is succesful. On the other hand the success of larval survival may depend on the inherited micro-organisms.

This means that if larval mico-oganisms are not suitable for larval growth, they are negatively selected.

Since the med fly is poliphagous, the successful colonization of a different species of fruit can depend on the micro-organisms derived from the mother and therefore on the species of fruit used by the previous generation.

Caution is needed when generalising but it may be that in *Ceratitis capitata* some bacterial species are preserved and transmitted. In this case larval growth on a particular fruit could be influenced by the maternal bacteria.

Inherited and transmitted micro-organisms which are useful for survival, may be considered true symbionts even if they exsist independently of insect presence and even if different species are found in different populations or individuals.

Since micro-organisms result fundemental for larval growth on artificial media it is reasonable that they play an important role in the yield of mass rearing. It is well known that the number of flies produced can change drastically even if all the climatic conditions are controlled and the larval medium is identical. Both micro-organisms of eggs and newly hatched larvae derived from adult rearing and micro-organisms of media can be fundamental.

It could be important to revise micro-biological aspects of all mass rearing stages.

ACKNOWLEDGEMENTS

I thank Dr. F. Rensi and Dr.Maria Fattori for their collaboration in research and Dr.B. Biolati for his advice and supplying bacterial strains. I also thank Dr. L. Furlan for insect field collections.

This work was partially supported by the Italian Research National Council (IPRA).

REFERENCES

Allen T.C., Pinckard J.A., Riker A.J., 1934. Frequent association of Phytomonas melophthora With various stages in the life cycle of the apple maggot, Rhagoletis pomonella . Phytopathol., 22: 228 - 238.

Boller E.,1968. An artificial oviposition device for the European cherry fruit fly, Rhagoletis cerasi . J. Econ. Entomol. 61: 850-852.

Cavalloro R., Girolami V., 1969. Miglioramento nell'allevamento in massa di Ceratitis capitata Wied. (Diptera : Trypetidae). Redia, LI: 315-327.

Drew R.A.I., Courtice A.C., Teakle D.S., 1983. Bacteria as a natural source of food for adult fruit flies (Diptera : Tephritidae). Oecologia, 60: 279-284.

Girolami V., Cavalloro R., 1972. Aspetti della simbiosi batterica di Dacus oleae (Gmel.) in natura e negli allevamenti di laboratorio. Ann. Soc. Ent. Fr. (N.S.), 8 (3): 561-571.

Girolami V., 1973. Reperti morfo-istologici sulla batteriosimbiosi del Dacus oleae (Gmel.) e di altri Ditteri Tripetidi, in natura e negli allevamenti su substrati artificiali. Redia, LIV: 269-294.

GirolamiV., 1979. Studi biologici e demoecologici sul Dacus oleae (Gmelin). I. Influenza dei fattori ambientali abiotici nell' adulto e negli stadi preimmaginali. Redia, 62: 147-191.

Girolami V.,1983. Fruit fly symbiosis and adult survival: general aspects. Proc. Int. Symp. CEC-IOBC, Athens, Nov.1982, pp: 74-76. In: Cavalloro R.. Fruit flies of economic importance. A.A. Balkema/ Rotterdam, pp: 1-642

Hagen K.S. 1966. Dependence of the olive fruit fly, Dacus oleae , larvae on the symbiosis with Pseudomonas savastanoi for the utilisation of the olive. Nature, 204: 423-424.

Howard D.J., Bush G.L., Breznak J.A., 1985. The evolutionary significance of bacteria associated with Rhagoletis . Evolution, 39 (2): 405-417.

Petri L., (1909). Ricerche sopra i batteri intestinali della mosca olearia. Mem. Staz. Pat. Veg., Roma, pp: 1-129.

Ratner S.S., Stoffolano J.G, (1984). Ultrastructural changas of the esophagel bulb of the adult female apple maggot, Rhagoletis pomonella (Walsh) (Diptera : Tephritidae). Int. J. Insect Morphol. & Embryol., 13: 191-208.

Stammer H.J. (1929). Die Bakteriensymbiose der Trypetiden (Diptera). Z.Morphol. Oekol. Tiere, 15: 481-523.

THE ROLE OF SEX PHEROMONES IN THE CONTROL OF THE OLIVE FRUIT FLY, DACUS OLEAE: PRESENT STATUS - PROSPECTS

George E. Haniotakis
Biology Dept, N.R.C. "Demokritos"
P.O. Box 60228
GR-153 10 Aghia Paraskevi, Greece

ABSTRACT

Since 1981, when synthetic sex pheromones of Dacus oleae became available, much effort has been made toward their use in the control of this pest. A brief review of background information pertaining to this subject is presented here. Attempts at practical use of D. oleae pheromones include: monitoring, mass trapping and mating disruption. Although definite correlation of pheromone trap catches and olive fruit infestation has been reported, pheromone traps are not used yet for monitoring wild populations. Mass trapping, after its 5-year developmen4 process, today shows good possibilities of success. Although mass trapping in its present form cannot replace current control practices, it is expected that improvement of trap efficiency and establishment of optimal trap densities will make this goal possible in the near future. Mating disruption is still at the preliminary stage, though some encouraging results have been reported.

INTRODUCTION

The olive fruit fly, Dacus oleae Gmelin, is a major pest of olives in the Mediterranean region and Near East, where olive growing is one of the most important components of agriculture and where 98% of table olives and olive oil is produced. The damage is caused by larvae which bore and feed inside the olive fruit. The losses, both quantitative and qualitative, are due to the consumption of part of the fruit as well as to fruit drop, and can be as high as 80% or more of production, if no control measures are taken. Present control measures include:

a) Preventive method: This consists of bait sprays applied from the ground or the air. Baits consist of protein hydrolysates and

NATO ASI Series, Vol. G11
Pest Control: Operations and Systems Analysis in Fruit Fly Management. Edited by M. Mangel et al.

organophosphorus insecticides. Sprays are effective against the adult stage only. Timing is therefore of great importance for elimination of the maximum number of females prior to oviposition. Timing is based today on catches of McPhail traps baited with ammonium salt solutions. The economic threshold level for bait sprays is one fly per trap per day except for the first application in the early summer when other parameters are taken into consideration, including sex ratio, female fecundity and fruit development. Bait sprays are not effective at the orchard level. Area-wide applications are required which are usually undertaken by governmental or cooperative organizations.

b) Therapeutic method: This consists of cover sprays from the ground with organophosphorus insecticides. It is effective against all insect stages with a residual activity of 10-15 days depending on the insecticide used. Economic threshold level for cover sprays is considered a 5% fruit infestation for Greece and 10-15% for Italy. Accurate fruit sampling is important in this case. This method is effective at the orchard level and is usually used by individual olive growers for their own protection.

The ecological and toxicological consequences of prolonged use of wide-spectrum insecticides over large areas, as is the case with air bait sprays and cover sprays as well as the scarcity and high cost of farm labor, in the case of ground bait and cover sprays, have encouraged scientists in all interested countries to seek alternative control methods. The discovery of a powerful sex attractant for male olive fruit flies which became readily available after its identification and synthesis, has stimulated research aiming at its use for monitoring and control of this pest.

The work which has been done on this subject up to the present will be reviewed here, as well as its present practical status and future prospects.

BACKGROUND INFORMATION

The utilization of pheromones in the mating process of D. oleae was suspected by Schultz and Boush (1971) and Economopoulos et al. (1971). The presence of a sex attractant pheromone in this species was demonstrated in laboratory (Haniotakis, 1974) and field (Haniotakis, 1977) experiments. It is released by female flies and acts as a potent male attractant. It was found to consist of a mixture of four components (Mazomenos and Haniotakis, 1981, 1985), three of which (a-pinene, n-nonanal and ethyl dodecanoate) were known substances and were identified rather quickly, while the fourth substance (1,7-dioxaspyro [5.5] undecane) was new and its identification, based on the work of Francke and Reith (1979) and Francke et al.(1979a, 1979b), was delayed. Two research groups working independently identified and synthesized almost simultaneously this last compound (Baker, 1980; Mazomenos et al., 1981)

Although this latter compound is very active by itself and is considered the major component of the blend, the activity of the full blend is higher than that of the major compound alone (Mazomenos and Haniotakis, 1981, 1985). The identification of the major component of the female pheromone blend from wild male rectal glands (Mazomenos and Pomonis, 1983) raised doubts as to its exact function. Recent work (Haniotakis et al., 1985), however, demonstrated that males respond only to the R-(-)- enantiomer of the major pheromone, or olean, as it is referred to in that work, while the S-(+)- enantiomer sems to act as an arrestant or aphrodisiac on female D. oleae flies. In the same work it was shown that females produce and release olean in the racemic form. Two additional substances of female origin (E-6-nonon-1-ol and p-cymene) have been reported to act as female attractants and aphrodisiacs (Gariboldi et al., 1982). De Marzo et al. (1978) reported that D. oleae male flies strongly attract females by a sex pheromone at the period of the day which coincides with the sexual activity of this species but no attempt was made to identify the substance(s) involved.

Studies on other aspects of the sexual behavior of D. oleae have also been made in an attempt to understand the mechanisms involved and their mode of operation. These are as follows: Biological rhythms associated with mating and factors affecting them (Loher and Zervas, 1979); rhythmicity in sex pheromone production, release and response to it; factors affecting pheromone production, release and response to it (Haniotakis, 1979b, Mazomenos, 1984); fine structure of antennal sensilla (Hallberg et al., 1984), electrophysiology of pheromone perception (van der Pers et al., 1984); and biosynthesis of major pheromone (Mazomenos, 1983). Despite this considerable body of information pertaining to sex pheromones, the complete picture of the sexual behavior of this species is far from clear. Additional work is required to tie all information together, reveal the mechanisms involved and their sequence, and explain the exact function of each of the substances involved.

USE OF PHEROMONES

Considerable effort has been devoted toward the development of technologies for the use of sex pheromones in the control of D. oleae. Work was thus done on the following: finding suitable pheromone dispensing systems and establishing optimal pheromone concentration ranges (Mazomenos et al., 1983; Jones et al., 1983); developing pheromone traps for monitoring and/or control (Haniotakis, 1979a; Haniotakis et al., 1982; Jones et al., 1983); relative attraction of pheromone, feeding and color(attractants and their combinations (Haniotakis and Skyrianos, 1981; Haniotakis et al., 1982; Delrio et al., 1983; Jones et al., 1983; Fiori et al., 1983; Quaglia et al., 1983; Delrio, 1984; Haniotakis, 1985); field evaluation of pheromone traps (Haniotakis, 1981; Delrio et al., 1983; Ramos et al., 1983; Crovetti et al., 1982); effect of trap positioning on insect catches (Haniotakis et al., 1982; Jones et al., 1983; Quaglia et al., 1983); effective range of pheromone traps (Delrio et al., 1983); pheromone trap efficiency and factors by which efficiency is affected, including trap density, insect population density, orchard size, tree size and density, wind velocity and temperature (Haniotakis and co-workers, unpublished data).

On the basis of all this information, several schemes for practical use of D. oleae sex pheromones were designed and tested from 1980 till today. These involve the use of pheromones alone or in combination with other attractants on various types of traps under different trap densities and under different orchard conditions to cover specific situations of the olive culture in this area. Each new design was based on experience obtained from previous tests, as well as on new technological developments, for example, new types of traps or dispensers of the various attractants. The results of these tests will be reviewed below. Based on these results, future developments or need for further work will be discussed.

Monitoring

Accurate monitoring of wild D. oleae populations is not easy. Various traps have been tested as monitoring devices with the following results:

McPhail trap catches are strongly affected by biotic factors both of pest and host origin and non-biotic factors (wind velocity, temperature, relative humidity), so that correlation between trap catches and population density or fruit infestation levels is difficult (Neuenschwander and Michelakis, 1979; Kapatos and Fletcher, 1982). Definite correlation was found between level of fruit infestation and yellow sticky trap catches (Ballatori et al., 1980; Ricci et al., 1979). In similar two-year studies with five different traps (Crovetti et al., 1983), good correlations were found between fruit infestation and the following: a) males or females captured in delta yellow traps raited with 100 mg of the major pheromone; b) males or females or males and females caught in delta yellow traps baited with 500 mg of microencapsulated ammonium acetate; and c) males and females caught in a bottle trap baited with an ammonium acetate/water solution (500 mg/50 cc). No definite conclusions could be drawn, however, due to high variability of the results. Delrio et al. (1978) were able to establish a satisfactory economic threshold level of yellow sticky trap captures and cover sprays. For bait sprays, however, yellow sticky traps are not reliable (Delrio, 1982a).

Results of two-year studies on the relative captures of sticky pheromone and McPhail traps and fruit infestation levels in different locations throughout Greece (20 traps of each type per location) are awaiting analysis. It is expected that working economic threshold levels for bait sprays will be established (Haniotakis, unpublished data).

Sticky pheromone traps of a neutral wood color are very useful in accurately timing the first bait sprays in early summer. These traps clearly show the reproductive activation of D. oleae, following a reproductively inactive period of about 1.5 months starting in mid-May (Haniotakis et al., 1982).

Mass trapping

In the efforts towards practical use of pheromones in D. oleae, emphasis was given to mass trapping as having high likelihood of success. Since ground bait sprays (a form of mass trapping in which a small olive branch every other tree is sprayed with about 300 g of bait, with a residual activity of no more than a week) give satisfactory crop protection, a permanent insect attracting and killing station, such as a trap, should have at least equal possibilities of success. The efficacy of mass trapping was tested at orchard level in isolated and non-isolated orchards and at area-wide level, with different trap densities and different trap types, alone or integrated with bait sprays as follows:

1) In 1980 and 1981 a mass trapping trial was conducted in an isolated 400-tree orchard near Athens, Greece, in which insecticides had never been used before. Yellow sticky traps at a density of 1 trap/2 trees (yellow is a mild attractant for both sexes of D. oleae), baited with pheromone alone, were used (Haniotakis et al., 1983). In 1980, a year with low pest population density, crop

protection was considered by the owner of the orchard as satisfactory, since early (summer-early Fall) fruit infestation was below 2%, and late (late Fall) infestation remained below 10% as compared to 60% in the control orchard. The distinction between early and late fruit infestation is important because early infestation usually results in complete loss of the infested fruit by dropping on the ground, while fruit infested later on may not drop, meaning that the loss amounts only to the portion of fruit actually consumed by the insect. In 1981, a year with high pest population density, crop protection was not satisfactory. Early fruit infestation was around 20% in the test orchard as compared to about 50% in the control and late infestation about 30% as compared to about 60% in the control. It was then decided to repeat the test in 1982, a year with intermediate pest population density, with a trap density of 1.5 traps/2 trees and the addition of an ammonia (an intermediate attractant for both sexes of the olive fly) releasing dispenser on the trap (Haniotakis, 1984). Early fruit infestation was below 2% in the test orchard vs about 45% in the control and late infestation about 10% vs about 70% in the control. Crop protection was satisfactory.

2. In 1981 a mass trapping trial was conducted in a 10,000 tree section at the edge of a large orchard at Ksironomi Boeotias, Greece (Broumas et al., 1983). Yellow sticky traps baited with one pheromone and one ammonia-releasing dispenser were used at a density of 1 trap/9 (nine) trees. The orchard outside the test area was protected by three air bait sprays. In the test area an aerial bait spray was also applied on October 4th. Early fruit infestation in the test area reached the 2.8% level compared to 0.8% in the sprayed orchard and late infestation 8.1% compared to 0.3% respectively. Despite the differences in fruit infestation levels, harvest data showed no income differences.

3. In 1981 a mass trapping trial was conducted in an orchard of 500 young trees east of Heraklion, Crete, Greece. The orchard was isolated from all sides except the south, where the nearest orchard (non-protected) was at a distance of about 500 m. Yellow sticky traps baited with one pheromone and one ammonia-releasing dispenser were used at a density of 1 trap/2 trees. One ground bait spray was applied on August 22-24 (Haniotakis et al., 1983). Early fruit infestation in the test area reached the 50% level compared to 90% for the control and late infestation 30% and near 100%, respectively. Crop protection was considered unsatisfactory.

4. In 1981 a mass trapping trial was conducted in a 600-tree section of a large olive orchard west of Heraklion, Crete, Greece.

Yellow sticky traps with pheromone and ammonia-releasing dispensers were used at a density of 1 trap/2 trees following an aerial bait spray to reduce high early summer pest population. An adjacent large olive orchard was not protected during 1981 due to low fruit load (Haniotakis et al., 1983). Early fruit infestation in the test section of the orchard was around 60% compared to about 40% in the control and late infestation about 50% compared to 40%, respectively. Increased pest population density was observed inside the test area compared to the outside orchard, due probably to the effect of the attractants used. This higher population density resulted in higher fruit infestation levels. Crop protection was unsatisfactory.

5. In 1982 two mass trapping trials were conducted in Tanagra Boeotias as follows: The main olive orchard of the Tanagra plain extends inside a small valley formed by the foothills of a mountain toward the south. Two small orchards, one of 700 trees and another of 150 trees, about 300 m apart, are separated from the main orchard by an open field approx. 500 m wide. In the 700-tree orchard, yellow sticky traps baited with pheromone and ammonia-releasing dispensers were installed at a density of 1 trap/2 trees. In the 150-tree orchard, yellow traps with the same attractants, sprayed with a 3% deltamethrine (Decis, Hoechst, Greece) and 3% honey water solution, were installed (Broumas et al., 1984; Haniotakis, 1984). The main orchard was protected by aerial bait sprays except for a strip of about 1000 trees near the experimental orchards which was sprayed from the ground to avoid insecticide drift. Maximum early fruit infestation was as follows: in the orchards with sticky traps 0.2%, insecticide traps 0.2%, ground sprayed 0.6%, and air sprayed 0.6%. Late infestation was 2.7%, 0.8%, 1.2% and 2.6%, respectively. Crop protection was satisfactory.

By this time it had become obvious that sticky traps were not practical for mass trapping of D. oleae The efficiency of these traps falls significantly with time and pest population density increase. Trap placement is therefore required several times during the olive growing season (about 6 months) which makes the method non-economic. For this reason, the insecticide-treated traps were introduced here. These traps, however, lose their effectiveness in killing landing insects when it rains, which may occur during the olive season, requiring replacement after each rainfall. Ammonia-releasing dispensers containing ammonium salts replaced the ones containing aqueous ammonia solutions, which were used in earlier tests, as more practical.

6. In 1983 a mass trapping trial was conducted in two olive

orchards of 400 trees each in Tanagra Boeotias. These two orchards were among the few in the area carrying a full crop, with fruit load in the general area being very low and for which no protective measures were taken. No isolation existed between test orchards and the surrounding orchards. Plywood rectangles 15x20x0.4 cm of natural wood color were used as traps at a density of 1 trap/2 trees. Traps were immersed for 15 min in 0.05% (a.i.) solution of deltamethrine (Decis). These traps have a long residual activity and do not require replacement throughout the olive growing period (Haniotakis et al., 1985). Pheromone and ammonia-releasing dispensers were used as attractants (Broumas et al., 1984). The use of yellow traps was discontinued at this point because it was established that such traps are very destructive to the beneficial fauna of the olive ecosystem (Neuenschwander, 1982). The combination of sex attractant pheromone with ammonia was shown to result in a decrease of male captures compared to pheromone alone but an increase on female captures compared to ammonia alone, an outcome highly desirable for mass trapping D. oleae (Haniotakis, 1985). Maximum early fruit infestation in the two test orchards was 2.3% and 0.8% vs 3.1% and 3.0% in the controls, respectively. Late infestation was 16.9% and 18.8% vs 47.6% and 55.8% in the controls. Crop protection was not satisfactory.

7. In 1983 a mass trapping trial was conducted in about 800 trees located on a peninsula of the island of Ithaki, Greece (Haniotakis, 1984). Traps and trap density were as in the case described above. Nearby orchards of the island were sprayed by air three times during the olive season of 1983. No sprays were applied in the test orchard. Early fruit infestation in the test area was below 2% and slightly less in the nearby sprayed orchards. Late infestation was around 10% and 9%, respectively. Crop protection was satisfactory.

8. From 1980 to 1983 mass trapping trials were conducted in 100-tree orchards in Sorso, Sardinia, Italy (Delrio, 1982b; 1984). Sticky Rebell traps (Swiss Federal Research Station, Wadenswil, Switzerland) baited with one pheromone dispenser, one cotton wick impregnated with protein hydrolysate and with ammonium carbonate salt spread on the sticky substance, were used at densities of 1 or 2 traps/tree. The results are summarized below (from Delrio, 1984):

Year	Olive Yield kg/tree	Grove 1 Unprotected	Grove 2 1 trap/tree		Grove 3 2 traps/tree	
		% wormy olives	No. of	% wormy	No. of	% wormy

			flies caught	olives	flies caught	olives
1980	40	36.4	6,135	9.1	20,402	4.8
1981	15	64.5	84,714	62.5	163,725	73.5
1982	34	27.5	52,344	10.2	99,022	3.5
1983	44	15.8	16,339	4.8	44,586	3.7

An evaluation of the results up to this point led to the conclusion that adequate olive protection cannot be achieved by mass trapping in situations where insect immigration into the protected area is not excluded, or in cases of high pest population densities. The problem of pest immigration into the protected area could possibly be solved by increasing the size of this area, and the problem of high pest population density by the installation of traps before populations reach high densities. The development of a long-lasting trap (active life of a full year) appeared to give new prospects to this method since continuous operation of the traps in the orchards could gradually decrease insect population density, theoretically to the eradication point. It was thus decided to apply the method the following year on an area-wide scale, as described below.

9. In 1984 a mass trapping trial was conducted in a 30,000 tree section at the edge of an extended olive orchard at Mochos, Heraklion, Crete, Greece (Haniotakis et al., 1985). Traps were of the plywood type described above and attractants were also the same as above. Trap density was approx. 1 trap/3 trees. Four ground bait sprays were applied to the nearby orchards during 1984. Early fruit infestation in the test orchard was below 2%, while infestation at harvest time was 2.4%, compared to 0.7% of the sprayed areas. Infestation of 2.4% resulted in an estimated income loss of about US$0.04/tree. Crop protection was considered satisfactory. In this trial an edge effect was observed, while the effect of immigrating insects in the entire test orchard was impossible to know. It was thus decided that the following year the method would be tested in an area large enough so that any possible influence from outside orchards would be excluded. This test is described below.

10. During 1985 the mass trapping method against D. oleae is being tested in three olive growing areas on the island of Crete and one area on the island of Euboea, Greece. The experimental areas were selected on the basis of their degree of isolation. All four areas were more or less separated from the remaining olive growing areas of

the two islands by mountain ranges or forested areas for a considerable distance. The sizes of the test areas were, in thousands of trees, 800, 650, 300 (Crete) and 350 (Euboea).

Traps and attractants were as in the previous case. Pheromone and ammonium bicarbonate salt formulations were made on an industrial scale and their attractiveness seems slightly lower compared to those prepared in the laboratory. Traps were dipped in a 0.075% (a.i.) solution of deltamethrin (as opposed to 0.05% used earlier) for 30 min (as opposed to 15 min used before). This stronger insecticide concentration on the trap appears to have a slight repellent effect on D. oleae.

Although the tests are still in progress and the final conclusions cannot be drawn before the end of November, 1985, a few remarks will be presented here based on the information available up to this point:

The size of the mass trapping area did not change the efficiency of the method as established in previous tests. Pest population density and rate of reproduction remain the most important factors which regulate efficiency. Additional factors were identified which affect efficiency of the method, as follows: Tree density inside the orchard, more specifically, long distances between trees reduce trap efficiency by restricting inter-tree insect movement. Insect movement is also restricted by high temperatures and high wind velocity. Although trap installation took place during a period of low adult pest population density (early Summer), sudden adult emergence in high numbers from pupae in the ground from the previous (Spring) generation resulted, in certain areas, in an increase of adult population which required spraying. Installation of traps prior to the spring generation would have been more appropriate. In other areas and during periods of high rates of reproduction, slow but steady increases of pest population occurred, which therefore required spraying when presently accepted economic threshold levels were reached. There were areas, however, which did not require any sprays. It is obvious that new economic threshold levels for intervention must be established for mass trapping areas. Implementation of such large-scale mass trapping tests is of vital importance, as this alone can determine the success or failure of the method.

In conclusion, mass trappping in D. oleae is effective under certain conditions. In its present form it cannot replace bait sprays in all cases. More work is required to improve trap efficiency and to determine optimal trap density. Since a similar mass trapping method was successful in another Dacus species (D. dorsalis) (Koyama et al., 1984), it is probable that it can be made effective for D. oleae as well.

Mating disruption

The major pheromone component was used in a one-hectare experimental lot in Spain in 1981 and 1982 to test the possibility of controlling D. oleae by mating disruption. One or 2 dispensers per tree were loaded with 55 mg of the pheromone (Jones et al., 1983). Although there were highly significant differences between trap catches in treated and untreated plots, significant differences in fruit infestation levels were not established. The experiments were repeated in 1983 (Montiel and Simon, 1984). In the latter case, significant differences in fruit infestation levels between treatment and control were found. More specifically, fruit infestation in the last week of October was 35.20% in the test orchards and 66.09% in the control.

Experiments in Greece showed significant reduction of pheromone trap catches at densities of 1 trap/2 trees (Haniotakis et al., 1982). Two preliminary experiments, however, on mating disruption in experimental plots of sizes equivalent to those utilized in Spain, surrounded by orchards protected either by mass trapping or by bait sprays, with 5 dispensers per tree loaded with 50 mg of the complete pheromone blend or 25 mg of the major pheromone component, resulted in no reduction of pest population density and fruit infestation. On the contrary, both population density and fruit infestation in test orchards was higher than that in controls (Haniotakis and coworkers, unpublished data). Further experiments in Greece have indicated that for mating disruption of D. oleae with sex pheromones to the level of adequate crop protection, a large number of release sites per tree with release rates yet to be determined will be required. With up to 12 dispensers per tree loaded with 50 mg of the pheromone blend or 16 dispensers per tree loaded with 20 mg, complete disorientation was not achieved in small experimental plots. To establish efficacy of this method for D. oleae control, much work is still needed.

REFERENCES

Ballatori, E., Pucci, C., and Ricci, C. 1980. Relation entre l'infestation des olives et les captures d'adultes de Dacus oleae (Gmel.) par pieges chromotropiques. Redia. 63:417-39.

Baker, R., Herbert, R., Howse, P., Jones, O., Francke, W., and Reith, W. 1980. Identification and synthesis of the major sex pheromone of the olive fly (Dacus oleae). J.C.S. Chem. Comm. 1:52-53.

Broumas, T., Haniotakis, G., Liaropoulos, C., and Yamvrias, C. 1984.

Experiments on the control of the olive fruit fly by mass trapping. Proc. Internat. Joint Meeting, CEC/FAO/IOBC on the Integrated Control in Olive Groves, Pisa, Italy, April 3-6, 1984. In press.

Broumas, T., Liaropoulos, C., Katsoyannos, P., Yamvrias, C., Haniotakis, G., and Strong, F. 1983. Control of the olive fruit fly in a pest management trial in olive culture. Proc. Internat. Symp. of CEC/IOBC on Fruit Flies of Economic Importance, Athens, Greece, Nov. 18-19, 1982. pp. 584-92.

Crovetti, A., Quaglia, F., and Fornasazi, L. 1982. Prove di laboratorio e pieno campo per valutare l'attivita biologica di sostanze feromoniche del Dacus oleae Gmelin. Frustula Entomologica, N.S. 4:357-64.

Crovetti, A., Quaglia, F., Malefatti, P., and Conti, B. 1983. Research on different methods for the evaluation of the level of olive fruit fly infestation - Results obtained in the experiments carried out in Tuscany in the biennium 1980-1981. Proc. Internat. Symp. of CEC/IOBC on Fruit Flies of Economic Importance, Athens, Greece, Nov. 16-19, 1982. pp. 330-36.

Delrio, G. 1982a. Le esche avvelenate nella lotta al Dacus oleae (Gmel.) Frustula Entomologica, N.S. 4:277-95.

------ 1982b. Esperienze di lotta integrata in olivicoltura in Sardegna. Proc. CEC Meeting on "Etat d'Avancement des Travaux et Echange d'Information sur les Problemes Poses par la Lutte Integree on Oleiculture", Antibes, France, Nov. 1981. pp. 73-85.

------ 1984. Biotechnical methods for olive pest control. Proc. Internat. Joint Meeting, CEC/FIO/IOBC on Integrated Control in Olive Groves, Pisa, Italy, April 3-6, 1984. Io press.

Delrio, G., Economopoulos, A., Haniotakis, G., and Prota, R. 1983. Comparative field studies on food, sex and visual attractants for the olive fruit fly. Proc. Internat. Symp. of CEC/IOBC on Fruit Flies of Economic Importance, Athens, Greece, Nov. 16-19, 1982. pp. 465-72.

Delrio, G., Luciano, P., Ortu, S., and Prota, R. 1978. Variazioni delle popolazioni di Dacus oleae Gmel. e programmazione della lotta nella olivicoltura sarda. Atti Giornale Fitopatologiche. 1978:269-76.

DeMarzo, L., Nuzzaci, G., and Solinas, M. 1978. Studio anatomico, istologico, ultrastrutturale e fisiologico del retto ed osservazioni etologiche in relazione alla possibile produzione di feromoni sessuali nel maschio di Dacus oleae Gmel. Entomologica. 14:203-66.

Economopoulos, A., Giannakakis, A., Tzanakakis, M.E., and Voyadjoglou,

A. 1971. Reproductive physiology of Dacus oleae: Anatomy of the adult rectum and odors emitted by adults. Ann. Entomol. Soc. Amer. 64:1112-116.

Fiori, G., Ricci, C., and Ballatori, E. 1983. Preliminary evaluation of the use of a combined "Prokoboll" sex pheromone trap for Dacus oleae (Gmel.) Proc. Internat. Symp. of CEC/IOBC on Fruit Flies of Economic Importance, Athens, Greece, Nov. 16-19, 1982. pp. 473-82.

Francke, W., Hindorf, G., and Reith, W. 1979a. Alkyl-1,6-dioxaspiro[4.5]-decanes - A new class of pheromones. Naturwissenschaften. 66:618-19.

------ 1979b. Mass-spectrometric fragmentation of Alkyl-1,6-dioxaspiro[4.5] decanes. Naturwissenschaften. 66:619-20.

Francke, W., and Reith, W. 1979. Alkyl-1,6-dioxaspiro[4.4] nonane; eine neue Klasse von Pheromonen. Liebigs Annln. Chem., pp. 1-10.

Gariboldi, P., Jommi, G., Rossi, R., and Vita, G. 1982. The chemical constitution and sex pheromone activity of volatile substances emitted by Dacus oleae. Experientia. 38:441-44.

Hallberg, E., van der Pers, J., and Haniotakis, G. 1984. Funicular sensilla of Dacus oleae - fine structural characteristics. Entomol. Hellenica. 2:41-46.

Haniotakis, G. 1974. Sexual attraction in the olive fruit fly, Dacus oleae (Gmelin). Envir. Entomol. 3:82-86.

------ 1977. Male olive fly attraction to virgin females in the field. Ann. Zool. Ecol. Anim. 9:273-76.

------ 1979a. Potential practical uses of the olive fruit fly sex attractant pheromone. Proc. Internat. Symp. of IOBC/WPRS on Integrated Control in Agriculture and Forestry, Vienna, Oct. 8-12, 1979. pp.471-76.

------ 1979b. Pheromone studies of the olive fruit fly, Dacus oleae (Gmel.). Final Report, NATO Research Grant No. 1352. 69p.

------ 1981. Field evaluation of the natural female pheromone of Dacus oleae (Gmelin). Envir. Entomol. 10:832-34.

------ 1984. Control of the olive fruit fly, Dacus oleae by mass trapping: Present status - prospects. Proc. VII Circum Mediterranean Plant Protection, Meeting of European and Mediterranean Plant Protection Organization, Canea, Greece, Sept. 24-28, 1984. In press.

------ 1985. Effect of combining food and sex attrctants on the capture of Dacus oleae flies. Entomol. Exp. & Appl. In press.

Haniotakis, G., Francke, W., Mori, K., Redlich, H., and Schurig, Y. 1985. Sex specific activity of R-(-)- and S-(+)-1,7-dioxaspiro[6.5] undecane, the major pheromone of Dacus oleae.

J. Chem. Ecol. In press.

Haniotakis, G., Kozyrakis, M., and Hardakis, I. 1983. Applications of pheromones for the control of the olive fruit fly. Proc. Internat. Conf. on Integrated Plant Protection, Budapest, Hungary, July 4-9, 1983. In press.

Haniotakis, G., Kozyrakis, E., and Bonatsos, C. 1985. Control of the olive fruit fly, Dacus oleae Gmelin by mass trapping: Pilot scale feasibility study. Zeit. Angew. Entomol. In press.

Haniotakis, G., Mazomenos, B., and Hardakis, I. 1982. Monitoring and control of the olive fruit fly with pheromone traps. C.R. Reunion du groupe d'experts sur l'etat d'avancement des travaux et echange d'informations sur le problemes poses par la lutte integree en oleiculture, Antibes, France, Nov. 4-6, 1981. pp. 46-55.

Haniotakis, G., and Skyrianos, G. 1981. Attraction of the olive fruit fly to pheromone, McPhail and color traps. J. Econ. Entomol. 74:53-60.

Jones, O., Lisk, J., and Baker, R. 1983. Mating disruption of Dacus oleae with the major component of its sex pheromone. Proc. Intern. Symp. of CEC/IOBC on Fruit Flies of Economic Importance, Athens, Greece, Nov. 16-19, 1982. pp. 500-05.

Kapatos, E., Fletcher, B. 1982. Development of a pest management system for Dacus oleae in Corfu by utilizing ecological criteria. Proc. Internat. Symp. of CEC/IOBC on Fruit Flies of Economic Importance, Athens, Greece, Nov. 16-19, 1982. pp. 593-602.

Koyama, J., Teruya, T, and Tanaka, K. 1984. Eradication of the oriental fruit fly (Diptera:Tephritidae) from the Okinawa islands by a male annihilation method. J. Econ. Entomol. 77:468-72.

Loher, W., and Zervas, G. 1979. The mating rhythm of the olive fruit fly, Dacus oleae Gmelin. Z. Ang. Entomol. 88:425-35.

Mazomenos, B. 1983. Biosynthesis of a sex pheromone of the olive fruit fly, Dacus oleae Gmel. Doctoral Thesis, Univ. of Ghent, Dept. of Agric., Belgium. 137 p.

------ 1984. Effect of age and mating on pheromone production in the female olive fruit fly, Dacus oleae (Gmel.). J. Ins. Physiol. 30:765-69.

Mazomenos, B., and Haniotakis, G. 1981. A multicomponent female sex pheromone of Dacus oleae (Gmelin). Isolation and bioassay. J. Chem. Ecol. 7:437-44.

------ 1985. Male olive fruit fly attraction to the synthetic sex pheromone components in laboratory and field tests. J. Chem. Ecol. 11:397-405.

Mazomenos, B., Haniotakis, G., Spanakis, I., Ioannou, A., and

Kozyrakis, M. 1983. Field evaluation of olive fruit fly pheromone traps with various dispensers and concentrations. Proc. Internat. Symp. of CEC/IOBC on Fruit Flies of Economic Importance, Athens, Greece, Nov. 16-19, 1982. pp. 584-92.

Mazomenos, B., Haniotakis, G., Tumlinson, J., and Ragousis, N. 1981. Isolation, identification, synthesis and bioassays of Dacus oleae pheromones. Panhellenic Congress of Agricult. Research, Kallithea, Chalkidiki, Greece, May 5-6, 1981. Abstr pp. 96-97.

Mazomenos, B., and Pomonis, G. 1983. Male olive fruit fly pheromone: Isolation, identification and lab bioassays. Proc. Internat. Symp. of CEC/IOBC on Fruit Flies of Economic Importance, Athens, Greece, Nov. 16-19, 1982. pp. 96-103.

Montiel Bueno, A., and Simon Mata, M. 1984. Li interruption de la communication sexuelle de la mouche de l'olive (Dacus oleae Gmel.) comme strategie de lutte integree dans l'olivier. Proc. Internat. Joint Meeting CEC/FAO/IOBC on Integrated Control in Olive Groves, Pisa, Italy, April 3-6, 1984. In press.

Neuenschwander, P. 1982. Beneficial insects caught by yellow traps used in mass trapping of the olive fly, Dacus oleae. Entomol. Exp. & Appl. 32:286-96.

Neuenschwander, P., and Michelakis, S. 1979. McPhail trap captures of Dacus oleae (Gmel.) (Diptera,Tephritidae) in comparison to the fly density and population composition as assessed by sondage technique in Crete, Greece. Bull. Soc. Entomol. Suisse 52:343-57.

Quaglia, F., Crovetti, A., and Rossi, E. 1983. Competitive comparison of the activity of five different traps for monitoring Dacus oleae adults carried out under field conditions in Tuscany in 1980 and 1981. Proc. Internat. Symp. of CEC/IOBC on Fruit Flies of Economic Importance, Athens, Greece, Nov. 16-19, 1982. pp. 457-64.

Ramos, P., Jones, O., and Howse, P. 1983. The present status of the olive fruit fly (Dacus oleae Gmel.) in Granada, Spain, and techniques for monitoring its populations. Proc. Internat. Symp. of CEC/IOBC on Fruit Flies of Economic Importance, Athens, Greece, Nov. 16-19, 1982. pp. 38-40.

Ricci, C., Pucci, C., Ballatori, E., and Forcina, A. 1979. Aleuni aspetti della dinamica delle popolazioni di adulti di Dacus oleae (Gmel.) e analysi della relazione tra infestazione e catture con cartelle chromotropichae. Notiz. Mal. Plante 100:261-82.

Schultz, G., and Boush, G. 1971. Suspected sex pheromone glands in three economically important species of Dacus. J. Econ. Entomol. 64:347-49.

van der Pers, J., Haniotakis, G., and King, B. 1984.

Electroantennogram responses from olfactory receptors in <u>Dacus oleae</u>. Entomol. Hellenica 2:47-53.

SELECTION FOR FAST AND SLOW MATING LINES IN THE MEDFLY AND ANALYSIS OF ELEMENTS OF COURTSHIP BEHAVIOUR

D.J. HARRIS, R.J. WOOD, S.E.R. BAILEY,
DEPARTMENT OF ZOOLOGY,
MANCHESTER UNIVERSITY,
MANCHESTER M13 9PL, ENGLAND.

ABSTRACT

The response to selection by single pair (SP) and mass (M) methods was greater in slow lines. SP females determined mating activity, but in M lines both sexes did so. M males started wing vibration at a closer distance and proceeded to flapping sooner than SP males. Behavioural indices showed SP fast and M slow males used pheromone signalling to slow females more often than did controls.

INTRODUCTION

It has been suggested that the effectiveness of the sterile insect technique (SIT) might be increased by selecting laboratory strains for "supernormal" performance before releasing them in the field (Chambers, 1975, Boller, 1977, Boller, Katsoyannos, Remund and Chambers, 1981). Recent work has focused on "mating activity" and "startle activity" (Harris, Wood and Bailey, 1983, Sharp, Boller and Chambers, 1983, Boller and Calkins, 1984, Wood and Harris, 1985).

The approach described in this paper uses three types of experiment: (1) selection programmes which attempt to produce fast (F) and slow (S) mating lines, (2) matings between F and S lines to assess the primary effects of selection on each sex and (3) analysis of courtship behaviour in these lines using video techiques.

NATO ASI Series, Vol. G11
Pest Control: Operations and Systems Analysis
in Fruit Fly Management. Edited by M. Mangel et al.

SELECTION PROGRAMMES

Harris et al. (1983) reported two attempts to change the mating activity of a mixed Costa Rica (CR)/Italy (I) wildtype stock at 26±2°C and 65±10% R.H. The original stock had been constructed from two reciprocal crosses between CR and I at generations F23 and F27 respectively followed by mass-mating for one further generation. The stock was bred on by mass culture as the control line (C) for these experiments. Two selection methods were used, single-pair (SP) and mass (M). Each was used in one major programme and in a replicate programme of shorter duration. The replicate programmes are not described here.

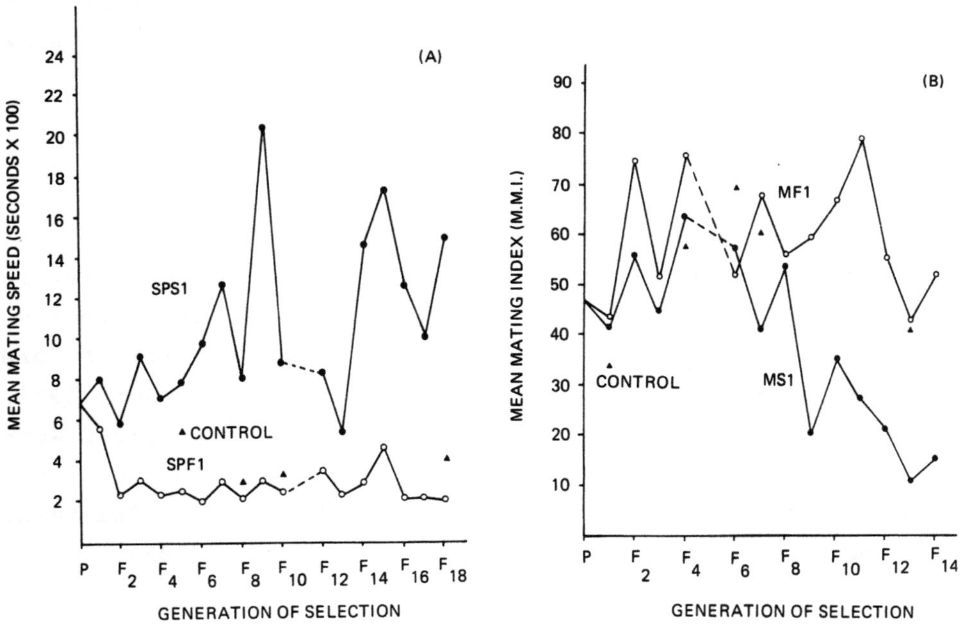

Figure 1. Programmes of (A) SP selection (mean mating speed obtained by transforming data to natural logarithms and then anti-logging) and (B) M selection (MMI after Boller et al., 1981).

Single-pair selection

In each generation 20-30 pairs per day were observed over a minimum of five days. The average selection pressure over the first ten generations was 76 $\pm$ 2% for SPF and 81 $\pm$ 3% for SPS. The mean mating speeds of F and S lines were significantly different after two generations of selection and remained so for almost all subsequent generations (Figure 1A). However, it is evident that the response to selection for slow mating was more marked than that for fast mating which remained at approximately the same level as in the control strain. There was fluctuation between generations, but in many generations there was a synchrony between the F and S lines (F1 to F5, F6 to F10 and F12 to F16 in Figure 1A) the cause of which has not been identified. Attempts to correlate mean mating speed with mean temperature and mean relative humidity proved negative.

Mass selection

The measure of mating activity in this experiment was the mean mating index (MMI) as defined by Boller et al. (1981) (Figure 1B), although there were two main differences in technique between the mating test used in this study and that used by Boller and co-workers. Harris et al. (1983) always tested one hour and twenty minutes into the photoperiod and introduced males to females under normal illumination. Boller et al. (1981) tested flies at any time during the first seven hours of the photoperiod and both sexes were first confined together for thirty minutes under red light. After this "holding period", full illumination was restored to initiate the test.

It was not until F9, after 8 generations of selection, at approximately 80% selection pressure, that F and S lines began to diverge (Figure 1B). The only clear response to selection occurred in the slow mating line, the fast mating line remaining in the range of the controls. Again, there was an apparent synchronous oscillation between lines in many generations (P to F4 and F10 onwards in Figure 1B).

MATINGS BETWEEN FAST AND SLOW LINES

Table 1

Comparison of reciprocal pairings between fast and slow lines within SP (*) and M selection programmes (after Harris *et al.*, 1983).

Mating Female x Male	SP Mean mating speed (secs)	No. of pairs (n)	M MMI (± SE)	No. of tests
SLOW x SLOW	1142 (a)	43	23.8 ± 4.2 (x)	9
SLOW x FAST	1471 (a)	77	42.5 ± 4.6 (xy)	9
FAST x SLOW	306 (b)	86	43.2 ± 6.8 (xy)	8
FAST x FAST	285 (b)	60	64.9 ± 5.5 (y)	9

(*) Data transformed to natural logarithms and the mean antilogged to give mean mating speed in seconds.
(a,b) Absence of a common letter indicates a significant difference at the 0.1% level (SP lines only).
(x,y) Absence of a common letter indicates a significant difference at the 5% level (M lines only).

Single-pair selection

Matings were carried out between SPF and SPS lines at generations F16 and F19. The data are combined in Table 1, which shows that S females mated significantly more slowly than F females irrespective of their male partner. It is concluded, therefore that the primary effect of this selection procedure was to produce females which differed in receptivity to male courtship.

Mass selection

In the M selection programme, matings were made between F and S lines in generations F13, F15, F16 and F17. The average results for 8 or 9 replicates suggest that M selection affected both sexes to a similar extent (Table 1).

COURTSHIP BEHAVIOUR

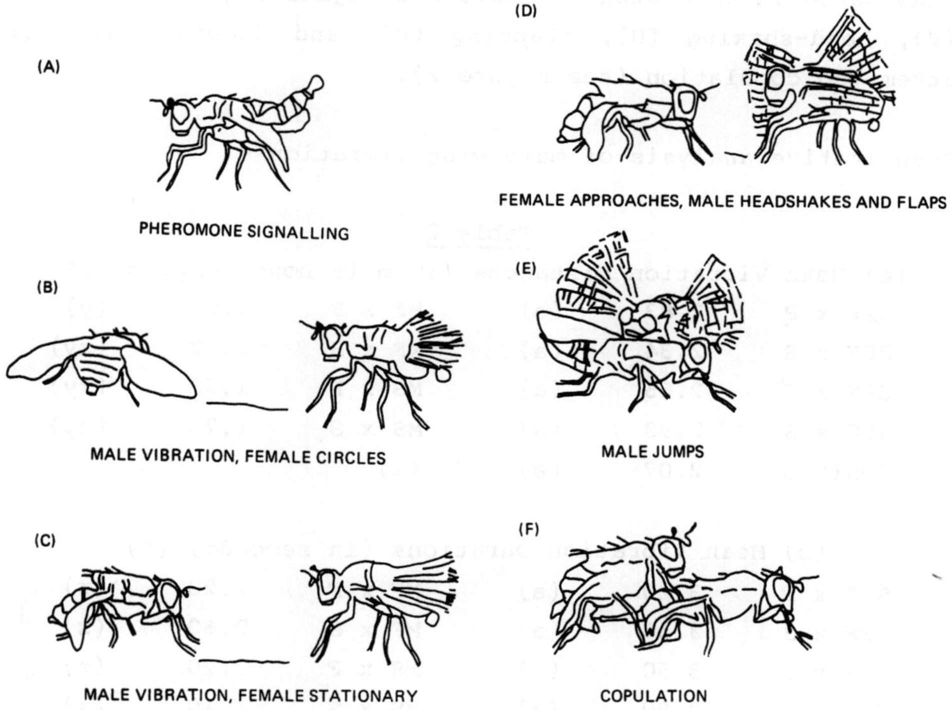

Figure 2. Medfly courtship behaviour (drawings after Feron, 1962). Stages of courtship (in sequence in two vertical columns) are labelled as referred to in the text.

To investigate the effects on courtship behaviour of the two methods of selection, single pairs were observed in a specially designed chamber controlled for temperature and humidity. Four categories of pairing (FxF, FxS ,SxF and SxS female x male) were observed in both selection experiments. These will be referred to as, for example: SPF x S (pairings of an SPF female and an SPS male) or MS x F (pairings of an MS female and an MF male). In the field, males aggregate under adjacent leaves of a host plant, one per leaf, forming a "lek" (Prokopy and Hendrichs, 1979). In this species, it is the male which attracts the female using a complex behavioural repertoire described by Feron (1962). Five elements were considered in this study - pheromone signalling (P), vibration (V), head-shaking (H), flapping (Fl) and jumping (J), i.e. attempted copulation (see Figure 2).

Quantitative analysis of male wing vibration

Table 2

(a) Mean Vibration Distances (in male body lengths) (*)

SPF x F	2.83	(a)	MF x F	1.02	(y)
SPF x S	2.34	(a)	MF x S	1.52	(xy)
SPS x F	2.65	(a)	MS x F	1.58	(xy)
SPS x S	2.93	(a)	MS x S	1.74	(xy)
CONTROL	2.07	(a) /	(x)		

(b) Mean Vibration Durations (in seconds) (*)

SPF x F	3.20	(a)	MF x F	1.20	(x)
SPF x S	3.50	(a)	MF x S	2.60	(x)
SPS x F	3.50	(a)	MS x F	3.20	(x)
SPS x S	3.80	(a)	MS x S	1.10	(x)
CONTROL	1.60	(a) /	(x)		

(*) The significance of differences between pairings was assessed using a t-test on the means.
(a,b,c) Differences concern SP and CONTROL pairings only. Absence of a common letter denotes a significant difference at the 5% level.
(x,y,z) Differences concern M and CONTROL pairings only. Absence of a common letter denotes a significant difference at the 5% level.

Several parameters were measured, but only two are mentioned here (Table 2). These are the distance of the male from the female at the start of wing vibration bouts which preceded wing flapping and the duration of vibration. For both parameters, differences between lines could be related to the method of selection. "Mean vibration distance" was measured in male body lengths (Table 2a) and "mean vibration duration" was measured in seconds (Table 2b).

Table 3

Significance of the differences between pairings using Student's t-test.

(a) Mean Vibration Distances (in male body lengths) (*)

	SPF	MS	MF
SPS	0.847 N.S.	0.059 N.S.	***
SPF		0.087 N.S.	***
MS			0.820 N.S.

(b) Mean Vibration Durations (in seconds) (*)

	SPF	MS	MF
SPS	0.495 N.S.	***	***
SPF		***	***
MS			0.823 N.S.

(*) Significance levels: *** = $p < 0.001$; N.S. = Not Significant

As Table 2 clearly shows, M selected males tended to initiate vibration at closer range and then to begin flapping earlier. In terms of vibration distance the difference between M and SP males was not always significant (Table 3a), but there was a clear separation in terms of vibration duration (Table 3b).

Behavioural indices

Flow charts of behavioural transitions were constructed (e.g. Figure 3).

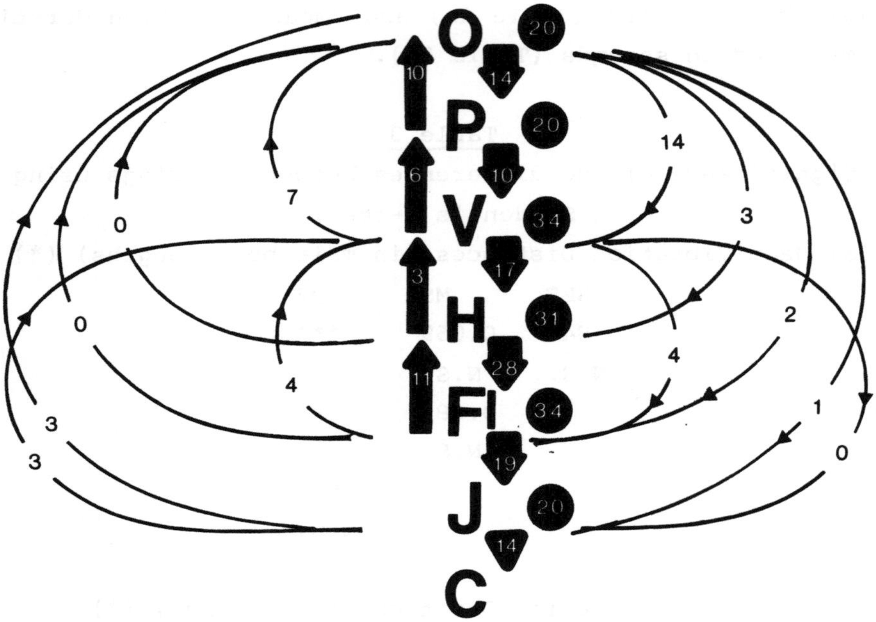

Figure 3. An example flow diagram of behavioural transitions between male courtship elements (control strain).

Key: O = no courtship; P = pheromone signalling; V = wing vibration; H = head-shaking; Fl = wing flapping; J = jumping (attempted copulation); C = successful copulation.

= number of transitions into the element

= direction of transition

= number of successful copulations

From this information, 5 indices were calculated. The first two consider the importance of pheromone signalling (P) and head-shaking (H) in proceeding to copulation. They are based on the frequency with which the next stage is attained through P or H directly, relative to all possible ways through preceding elements (terms used as in Fig.3).

$$\text{P Index} = \frac{(P \rightarrow V)}{(P \rightarrow V) + (O \rightarrow V)} \times 100$$

and

$$\text{H Index} = \frac{(H \rightarrow Fl)}{(H \rightarrow Fl) + (V \rightarrow Fl) + (P \rightarrow Fl) + (O \rightarrow Fl)} \times 100$$

Considering the P Index (Table 4a), significant differences existed in the SP programme between F and S males when they were paired to S females. When compared with the corresponding S male, an SPF male used pheromone signalling more often. In the M programme the difference between F and S males was not significant. However, when the control is considered, it is apparent that SPF and MS males both increased their use of pheromone signalling. Considering the H Index (Table 4b), the results show that within the SP lines, head-shaking tended to be less important in pairings involving SPS flies of either sex, but only in SPF x S pairings was this significantly lower than the control level. Pairings involving MF flies of either sex had a lower H Index than controls. Thus there was again a difference between selection methods.

Table 4a

P Index (*,#)

SPF x F	51.5	(a)	MF x F	45.5	(xy)
SPF x S	42.1	(a)	MF x S	37.5	(x)
SPS x F	80.0	(b)	MS x F	47.6	(xy)
SPS x S	43.9	(a)	MS x S	69.2	(y)
CONTROL	41.7	(a) /	(x)		

Table 4b

H Index (*,#)

SPF x F	96.7	(a)	MF x F	57.9	(x)
SPF x S	54.2	(b)	MF x S	46.7	(x)
SPS x F	65.0	(bc)	MS x F	42.9	(x)
SPS x S	63.8	(bc)	MS x S	69.2	(xy)
CONTROL	82.4	(ac) /	(y)		

(*) See text for explanation of how index was calculated.
(#) The significance of differences between pairings was assessed using a G-test on the raw data.
(a,b,c) Differences concern SP and CONTROL pairings only. Absence of a common letter denotes a significant difference at the 5% level.
(x,y,z) Differences concern M and CONTROL pairings only. Absence of a common letter denotes a significant difference at the 5% level.

The third index, the J Index, was calculated to assess the relative importance of male copulation attempts which were not immediately preceded by courtship. This so-called "rape strategy" was observed in field-cage tests by Prokopy and Hendrichs (1979). The index was derived as follows:

$$\text{J Index} = \frac{(O \rightarrow J)}{(O \rightarrow J) + (P \rightarrow J) + (V \rightarrow J) + (H \rightarrow J) + (Fl \rightarrow J)} \times 100$$

The strategy was found to be particularly important in MS x S pairings (Table 5).

Table 5

J Index (*,#)

SPF x F	14.8	(a)	MF x F	12.5	(xz)
SPF x S	10.5	(a)	MF x S	19.2	(xy)
SPS x F	0.0	(a)	MS x F	3.4	(z)
SPS x S	5.9	(a)	MS x S	41.7	(y)
CONTROL	5.0	(a) /	(xz)		

(*) See text for explanation of how index was calculated.
(#) The significance of differences between pairings was assessed using a G-test on the raw data.
(a,b,c) Differences concern SP and CONTROL pairings only. Absence of a common letter denotes a significant difference at the 5% level.
(x,y,z) Differences concern M and CONTROL pairings only. Absence of a common letter denotes a significant difference at the 5% level.

Two final indices were calculated to assess male success during the later stages of courtship in terms of female response to the male. The St Index measured male "holding power" in terms of female "stability" i.e. the likelihood of a jump following flapping and the R Index measured male mating success in terms of female "receptivity" i.e. the likelihood of successful copulation following a jump.

The St Index was calculated as:

$$\text{St Index} = \frac{(Fl \rightarrow J)}{(Fl \rightarrow O) + (Fl \rightarrow P) + (Fl \rightarrow V) + (Fl \rightarrow H) + (Fl \rightarrow J)} \times 100$$

The R Index was calculated as:

$$\text{R Index} = \frac{(J \rightarrow C)}{(J \rightarrow O) + (J \rightarrow P) + (J \rightarrow V) \quad (J \rightarrow Fl) + (J \rightarrow C)} \times 100$$

Table 6a

St Index (*,#)

SPF x F	70.0	(a)	MF x F	73.7	(x)
SPF x S	66.7	(ab)	MF x S	63.3	(x)
SPS x F	45.0	(b)	MS x F	61.9	(x)
SPS x S	53.4	(ab)	MS x S	53.8	(x)
CONTROL	55.9	(ab) / (x)			

Table 6b

R Index (*,#)

SPF x F	74.1	(a)	MF x F	78.1	(x)
SPF x S	84.2	(a)	MF x S	76.9	(x)
SPS x F	66.7	(a)	MS x F	62.1	(xy)
SPS x S	20.6	(b)	MS x S	41.7	(y)
CONTROL	70.0	(a) / (xy)			

(*) See text for explanation of how index was calculated.
(#) The significance of differences between pairings was assessed using a G-test on the raw data.
(a,b,c) Differences concern SP and CONTROL pairings only. Absence of a common letter denotes a significant difference at the 5% level.
(x,y,z) Differences concern M and CONTROL pairings only. Absence of a common letter denotes a significant difference at the 5% level.

S females of both programmes tended to show a lower stability (Table 6a) and a lower receptivity (Table 6b) than the corresponding F females. However in terms of the St index (Table 6a), there were no significant differences from controls. The only significant effect was that of the SPS female in decreasing the "holding power" of the SPF male. Considering female receptivity (Table 6b), the only significant difference from the control level was the reduction apparent in SPS x S courtships. This R index of 20.6 was also significantly lower than that of other SP x SP pairings. The lowering of S female receptivity was only significant within the M programme and only when compared to pairings involving a F female.

DISCUSSION

The slight response for fast mating in both programmes suggests that laboratory stocks are already near their maximum mating potential, having probably been selected unconsciously during routine laboratory rearing. In the present study, both ancestral strains had been reared more than 20 generations before the base-stock was created. The parental generation of the SP programme (Figure 1A) also exhibited an increase in mating propensity compared with CR and I strains, probably due to heterosis (Harris, Wood and Bailey, in preparation). Boller and Calkins (1984) showed that conscious selection of a newly colonised laboratory strain did not produce a higher mating propensity (as measured by its MMI) than was exhibited by the same strain exposed to unconscious selection alone. However, selection did seem to accelerate the rate at which a higher MMI was reached. It may be noted that the MMIs of both MF and MS lines (Figure 1B) fall into the same range as those reported by Boller and Calkins (1984). The synchrony of fluctuations noted in the mating propensity of different strains suggests some environmental influence is involved, but the cause is unclear. Hopefully, further investigation may reveal the source of these disturbances.

Matings carried out between F and S lines indicated that there were differences in mating behaviour between SP and M-selected flies. There was a clear female effect in the SP matings, whereas both sexes had an influence in the M matings. The selection method also affected some parameters of courtship behaviour. M-selected males tended to initiate vibration at closer range and also progressed much more quickly to flapping than did than SP-selected males. SP males differed in their pheromone signalling behaviour when paired with a (less receptive) S female. Both SPF and MS males employed pheromone signalling (P) in pairings to S females more often than control males. An SPS or MF fly of either sex reduced the occurence of head-shaking compared with the control.

These results suggest that it may be productive to investigate different selection techniques and their effect on behaviour. The high incidence of the "rape" strategy in MS x MS pairings indicates that it may be useful to select for slow mating before field release. Releasing strains exhibiting different reproductive strategies might lessen the likelihood of behavioural resistance to SIT evolving (Wood and Harris, 1985). See also Wood, this volume.

The results of F and S pairings and also the St and R indices show the importance of the female in determining successful copulation. They also point to the need for further field studies of mating activity (lek behaviour and female mate choice)(work being carried out by Kaneshiro _et al._, personal communication). It is the response of the native female which will decide the success of sterile male release.

Field tests carried out by Zapien, Hendrichs, Liedo and Cisneros (1983) showed that laboratory-reared males did poorly compared to wild males in terms of courtship and mating, when the sky was overcast and light and temperature conditions were sub-optimal. Boller and Calkins (1984) selected an Egyptian strain for laboratory mating activity at sub-optimally low temperature (19.0°C-19.8°C) and increased the MMI from 2 to 14. Another priority must be the field-testing of genetically manipulated strains in order to determine their effectiveness.

ACKNOWLEDGEMENTS

This work was carried out under MAFF licence no. PHF 250/72 issued under the Import and Export (Plant Health) and Plant Pests (Great Britain) Order 1980 and made possible by an S.E.R.C. Studentship. Thanks to Mrs. S.E. Hardman for advice on typing, to Mr. L. Lockey for taking the photographs and to Boots Company Limited and Dr. E. F. Boller for providing medfly strains.

REFERENCES

Boller, E.F. and Calkins, C.O. (1984) Measuring, monitoring and improving the quality of mass-reared Mediterranean fruit flies, Ceratitis capitata (Wied.). 3. Improvement of quality by selection. Zeit. angew. Entomol.98, 1-15.

Boller, E.F., Katsoyannos, B.I., Remund, U., Chambers, D.L. (1981) Measuring, monitoring and improving the quality of mass-reared Mediterranean fruit flies, Ceratitis capitata (Wied.). 1. The RAPID quality control system for early warning. Zeit. angew. Entomol.92, 67-83.

Boller, E.F. (1977) Sexual Activity. In: Quality Control: An Ideas Book for Fruit Fly Workers. (Eds. E.F. Boller and D.L. Chambers), IOBC/WPRS Bulletin 1977 pp. 75-76.

Chambers, D.L. (1975) Quality in mass-produced insects: definition and evaluation. In: Controlling fruit flies by the sterile-insect technique IAEA, Vienna 1975 pp. 19-32.

Feron, M. (1962) L'instinct de reproduction chez la mouche Mediterranean des fruits (Ceratitis capitata (Wiedemann)) (Diptera:Tephritidae). Comportement sexual. Comportement du pont. Rev. Path. Veg. Ent. Agric. Fr. 41, 1-129.

Harris, D.J., Wood, R.J., Bailey, S.E.R. (1983) Studies on mating in the Mediterranean fruit fly Ceratitis capitata In: Fruitflies of Economic Importance (Ed. R. Cavalloro) Balkema, Rotterdam 1983 pp. 197-202.

Kaneshiro, K., Director, Hawaiian Evolutionary Biology Program, Pacific Biomedical Research Centre, 3050 Maile Way, Gilmore 310, Honolulu, Hawaii 96822, U.S.A.

Prokopy, R.J. and Hendrichs, J. (1979) Mating behaviour of Ceratitis capitata on a field-caged host tree. Ann. Entomol. Soc. Am. 72, 642-648.

Sharp, J.L. and Boller, E.F., Chambers, D.L. (1983) Selection for flight propensity of laboratory and wild strains of Anastrepha suspensa and Ceratitis capitata (Diptera: Tephritidae). J. Econ. Entomol. 76, 302-305.

Wood, R.J. and Harris, D.J. (1985) Artificial and natural selection in the Mediterranean fruit fly Ceratitis capitata In: Fruitflies, their Biology, Natural Enemies and Control (Eds. A.E. Robinson and G.H.S. Hooper), Elsevier, Holland 1985 (in press).

Zapien, G., Hendrichs, J., Liedo, P., Cisneros, A. (1983) Comparative mating behaviour of wild and mass-reared sterile medfly Ceratitis capitata (Wied.) on a field cage host tree II. Female mate choice. In: Fruitflies of Economic Importance (Ed. R. Cavalloro) Balkema, Rotterdam 1983 pp. 397-409.

SOME OBSERVATIONS ON THE COMPETITIVENESS OF STERILIZED MEDITERRANEAN FRUIT FLY

G.H.S. Hooper
Department of Entomology,
University of Queensland,
St Lucia. Qld 4067.
Australia.

ABSTRACT

Factors which could influence the real and apparent competitiveness of sterilised Mediterranean fruit flies in suppression/eradication programs utilising the sterile insect technique are discussed.

An essential requirement of the sterile insect technique (SIT) is that the released sterilized insects effectively interact with the target population. This requires that they can disperse to find food and mating arenas, and that they have behavioural characteristics, particularly with respect to mating, that mimic those of the wild population as closely as possible. The term "competitiveness" is applied to the ability of the released males to compete with males of the target population and this can be quantified by means of competitive mating tests (Haisch 1970, Fried 1971).

There is a considerable body of knowledge on the competitiveness of the Mediterranean fruit fly (henceforth termed medfly), Ceratitis capitata (Wiedemann), irradiated with various doses of gamma radiation in the laboratory, and of factors which can influence this characteristic e.g. age of pupae when irradiated (Hooper 1971a), reduced oxygen tension in the pupae (Hooper 1971b , Ohinata et al. 1977). Competitiveness or ratio tests conducted in the laboratory can measure the outcome, in terms of egg hatch, of the interplay of a number of factors e.g. mating ability, the impact of multiple matings, sperm competitiveness. They cannot measure factors such as flight ability, courtship behaviour, pheromone production and/or release, which would be of crucial importance in the field. Furthermore, in these tests irradiated flies are compared with untreated flies of the laboratory

NATO ASI Series, Vol. G11
Pest Control: Operations and Systems Analysis
in Fruit Fly Management. Edited by M. Mangel et al.

or mass-reared strain; they do not usually involve a comparison of irradiated laboratory flies with untreated wild flies. Hence it is reasonable to assume that the competitiveness of the released flies in the field will be less than that indicated by the laboratory tests. Despite these limitations laboratory competitiveness tests have an important role in evaluating irradiation procedures.

It has been well established that while sterility is directly related to gamma dose, competitiveness is inversely related to dose (Hooper 1972). Arising from this finding Hooper and Katiyar (1971) and Hooper (1972) expressed the view that for some SIT applications, or in the initial phase of an eradication program, the use of a dose which while permitting some residual fertility in the males would enhance the competitiveness should be evaluated. Ohinata et al. (1978) also suggested this approach. Fig. 1 plots the relationship between the expected egg hatch and the ratio of released to wild (R:W) flies for various values of residual fertility (Hs) in the R males and several values for competitiveness (C). A comparison of Figs 1A and 1B indicates that C has a greater impact at low R:W ratios i.e. from 1:1 to ca 40:1, on the expected egg hatch than Hs at the levels used. (While as stated above sterility and competitiveness are inversely related, the illustrated combinations of Hs and C could be obtained in practice by irradiating pupae in air or nitrogen.)

Fig. 1C illustrates the theoretical performance of fully sterile flies with poor competitiveness (Hs = 0, C = 0.2) and flies with a low residual fertility but greater competitiveness (Hs = 2, C = 0.5). Again at low R:W ratios the flies with some residual fertility would out-perform fully sterile flies (the combinations of Hs and C used are not unreasonable). Indeed not until a R:W ratio of ca 100:1 is attained are the fully sterile flies superior. Even flies having 5% residual fertility are not markedly inferior to flies with 2% residual fertility at low R:W ratios if the competitiveness of each type of fly is comparable.

On theoretical grounds then the release of flies with a low level of fertility, but consequently having enhanced competitiveness, seems worthy of serious consideration in the initial stages of a suppression/eradication program. After reviewing the then available literature Hooper (1982) concluded that a R:W ratio of at least 100:1 is required for eradication of

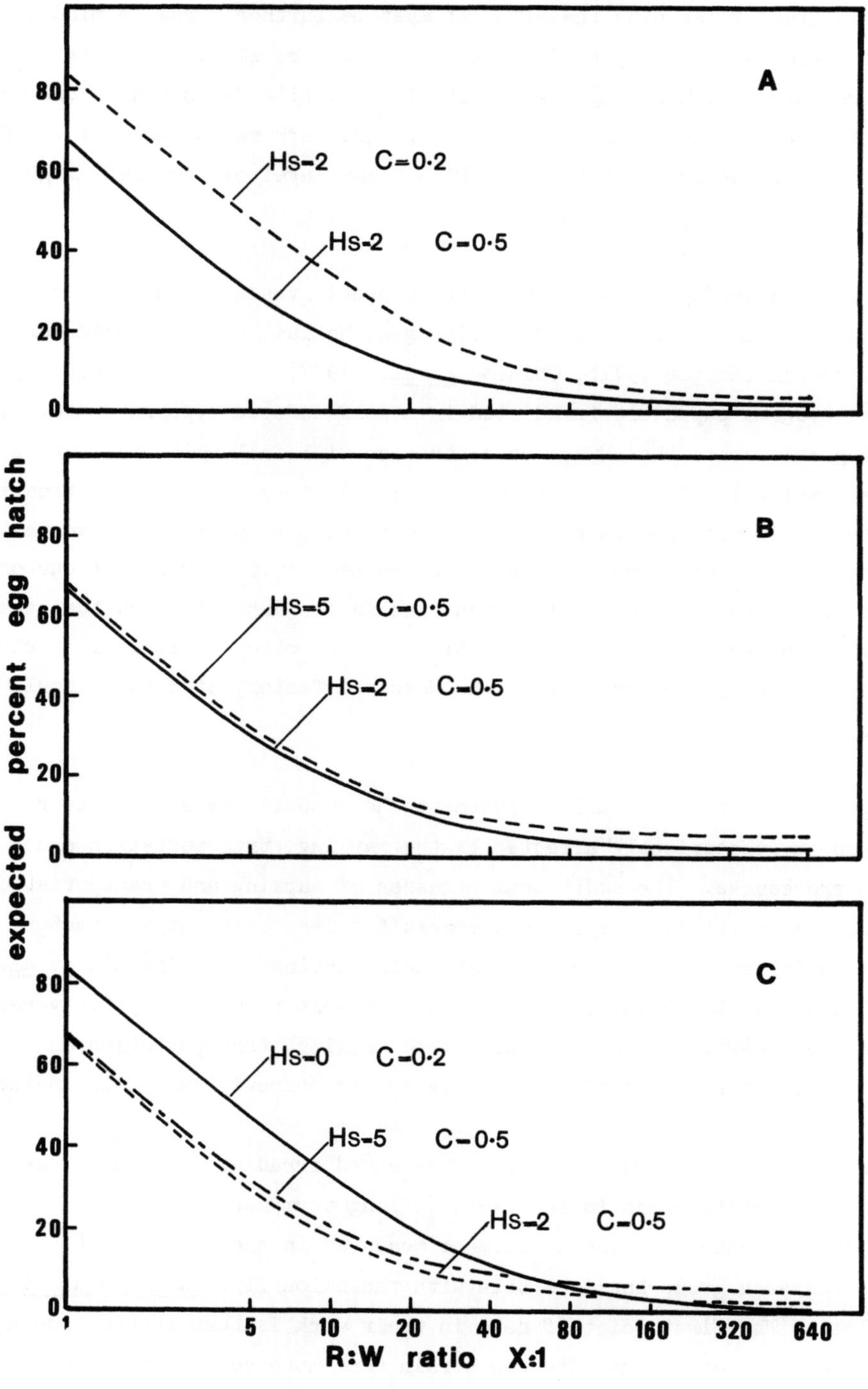

Fig. 1. Effect of level of residual fertility (Hs) and competitiveness (C) at various R:W ratios on expected egg hatch.

tephritid fruit flies with the SIT. It must be further borne in mind that the R:W ratio can normally be determined only after releases of sterile flies are made, and trap captures of sterile and wild flies can be analysed. Frequently during the initial phase of a program R:W ratios are below 100:1 and it is at these levels that flies with a low level of fertility might be expected to exert superior control.

Work with medfly has shown that the competitiveness of males can be enhanced if pupae are irradiated in nitrogen, helium, carbon dioxide or a partial vacuum (Hooper 1971b, Ohinata et al. 1977). In the Mexican Moscamed program pupae are irradiated in a self-induced hypoxic state. In achieving this state the oxygen concentration will fall while the carbon dioxide level will rise. Ohinata et al. (1977) showed that of all treatments evaluated to induce hypoxia prior to irradiation carbon dioxide gave the least enhancement of competitiveness, and Hooper (1970) found that carbon dioxide is deleterious to medfly. Hence it is the author's view that the possibility of a reduction in competitiveness and other biological attributes arising from the procedure used at the Moscamed factory at Metapa should be investigated.

Apart from the irradiation treatment the competitiveness of the released flies can be deleteriously affected by the marking, transportation and release procedures. The additional stresses of marking and transportation could be quantified by carrying out competitiveness tests at the factory after irradiation and after receipt at their destination. Ohinata et al. (1977) showed that a simulated transport treatment reduced competitiveness of flies by 16-42%. Further research into marking, transportation and release procedures to minimise deterioration in competitiveness is indicated.

Underlying the competitiveness of released irradiated flies is the effect on competitiveness in the field of long term mass-rearing. The effect of the latter can really only be measured in the field and the Okinawa research group has done this with the melon fly, Dacus cucurbitae Coquillett. The first point of note in their work is that in the laboratory irradiated males of the mass-reared strain were more competitive than untreated wild flies (Itô and Koyama 1982). This anomaly may be explained by the finding of Soemori et al. (1980) that in small cages the mass-reared males mated more than wild males, and that as cage volume per fly increased mating by mass-reared males declined while that of wild flies increased. These experiments indicate that only in large field cage experiments will a realistic indication of the competitiveness of released sterilized males be obtained.

Iwahashi et al. (1983) found that the competitiveness of released flies, based on field data, fell from ca 0.8 in the 5-10th mass-reared generation to ca 0.2 in the 18th mass-reared generation. They, and Itô and Koyama (1982) attributed this decline to the effect of continued mass-rearing. Similarly Kahinohana (1980) reported that the dispersive ability of mass-reared-flies relative to wild flies declined after 31-39 generations of mass rearing (44-52 generations from colonization). However, while the field determined competitiveness declined dramatically in 13 months so too did the proportion of wild males. While not discarding the possibility that fitness of the mass-reared flies could decrease with increased time of rearing, the alternative explanation of Meats (1983) that reduced competitiveness could apparently arise at high R:W ratios based on trap captures (i.e. when the density of wild flies is low) simply as a result of a patchy distribution of wild flies should be kept in mind.

One other possibility may warrant scrutiny. Work with Drosophila spp. has shown in laboratory experiments that the mating success of two kinds of males is dependent on the relative frequency of the two kinds, with the minority males mating more frequently than the males present at the higher frequency (Ehrman 1966). One wonders whether this rare or minority male phenomenon could operate in the field when high R:W ratios prevailed. If it could then an apparent decrease in competitiveness of the released males would be found.

We frequently monitor the progress of a medfly SIT program on the basis of the R:W ratio of males caught in trimedlure traps, and on the number of wild males so caught. With Dacus oleae (Gmelin) Kapatos and Fletcher (1983) found that the seasonal efficiency of protein hydrolysate baited McPhail traps varied more than 100 fold with much of the variation being attributable to temperature. Similarly with medfly there was evidence of a negative temperature effect on trimedlure trap captures (Wong et al. 1982). These workers also found that the recovery of released wild flies in traps was lower than that of untreated laboratory flies. Conversely, Prokopy and

Economopoulos (1975) found substantially more released wild D. oleae flies were caught in McPhail traps than released laboratory flies. Cunningham et al. (1978) found that hydrolysed protein traps were much more efficient in trapping medfly, D. cucurbitae, and Dacus dorsalis Hendel in areas of low rainfall compared with areas of high rainfall.

Fitt (1983) found that the response of male Dacus opiliae Drew and Hardy to methyl eugenol traps varied with the seasonal pattern of humidity associated with "wet" and "dry" seasons. Adult males which had experienced field conditions during the dry season responded minimally to methyl eugenol in the laboratory (25°C, 75% RH) while males of the same batch held in the laboratory responded maximally and showed no temporal variation. Fitt pointed out that if lure responsiveness varies seasonally then conclusions about the population density based on trap data must be made with great caution. With respect to the SIT one further point is valid. Without relevant information it would seem dangerous to assume that the response of released males to a male attractant is similar to that of wild males at all times of the year. If the responsiveness of the two types of males differed, this would be reflected in the R:W ratios and false conclusions about the progress of any SIT program, and about the competitiveness of the released males would then be made.

The above rather fragmentary data from work with several tephritid species suggests that we should evaluate the seasonal response of both released and wild males to trimedlure in the various climatic zones where suppression/eradication programs are being, or might be, pursued.

Competitiveness of released males will also depend on how well they are accepted as mates by the wild females. With D. dorsalis mating was optimal when both partners were of equal size; mating was reduced when the male was smaller or larger than the female (Iga 1981). Large Anastrepha suspensa (Loew) females preferred large rather than small males (Burk and Webb 1983). Limited work by Boller et al. (1977) with Rhagoletis cerasi L. indicated that when large males were combined with small females mating was significantly reduced. In the screwworm Cochliomyia hominivorax (Coquerel) large males mated more than small males with large females (Alley and Hightower 1966). On the other hand no effect of fly size on mating was evident with the onion fly, Delia antiqua Meigen (Robinson and Zurlini 1981).

With medfly, recent work by Churchill-Stanland (in press) in California indicates that mating success declined when females were larger than males, and large males were no more successful in mating than males of a size equal to their female partners.

Thus the weight of evidence is that male size is important in determining mating success, and this may be more important in lek-mating species such as the medfly. This aspect is of considerable importance to the manager of a mass-production facility since it sets one of the specifications for production of an optimal fly which can then be monitored by appropriate quality control procedures.

REFERENCES

Alley, D.A. and Hightower, B.G. 1966. Mating behavior of the screw-worm fly as affected by differences in strain and size. J. Econ. Entomol. 59: 1499-1502.

Boller, E.F., Remund, U., Katsoyannos, B.I. and Berchtold, W. 1977. Quality control in European cherry fruit fly: evaluation of mating activity in laboratory and field cage tests. Z. Angew. Entomol. 83: 183-201.

Burk, T. and Webb, J.C. 1983. Effect of male size on calling propensity, song parameters, and mating success in Caribbean fruit flies, Anastrepha suspensa (Loew) (Diptera: Tephtritidae). Ann. Entomol. Soc. Am. 76: 678-682.

Churchill-Stanland, C., Stanland, R., Wong, T.T.Y., Tanaka, N., McInnis, D.O. and Dowell, R.V. (in press). Size as a factor in the mating propensity of Mediterranean fruit flies, Ceratitis capitata (Wiedemann), in the laboratory. J. Econ. Entomol.

Cunningham, R.T., Nakagawa, S., Suda, D.Y. and Urago, T. 1978. Tephritid fruit fly trapping: liquid food baits in high and low rainfall climates. J. Econ. Entomol. 71: 762-763.

Ehrman, L. 1966. Mating success and genotype frequency in Drosophila. Anim. Behav. 14: 332-339.

Fitt, G.P. 1983. The influence of seasonal climatic factors on the development of the methyl eugenol response in male Dacus opiliae. Entomol. Exp. Appl. 33: 171-178.

Fried, M. 1971. Determination of sterile-insect competitiveness. J. Econ. Entomol. 64: 869-872.

Haisch, A. 1970. Some observations on decreased vitality of irradiated Mediterranean fruit flies. In Sterile-Male Technique for Control of Fruit Flies. IAEA, Vienna, pp 71-75.

Hooper, G.H.S. 1970. Use of carbon dioxide, nitrogen and cold to immobilize adults of the Mediterranean fruit fly. J. Econ. Entomol. 63: 1962-1963.

Hooper, G.H.S. 1971a. Gamma sterilization of the Mediterranean fruit fly. In Sterility Principle for Insect Control or Eradication. IAEA, Vienna, pp 87-95.

Hooper, G.H.S. 1971b. Competitiveness of gamma-sterilised males of the Mediterranean fruit fly: effect of irradiating pupal or adult stage and of irradiating pupae in nitrogen. J. Econ. Entomol. 64: 1364-1368.

Hooper, G.H.S. 1972. Sterilization of the Mediterranean fruit fly with gamma radiation: effect on male competitiveness and change in fertility of females alternately mated with irradiated and untreated males. J. Econ. Entomol. 65: 1-6.

Hooper, G.H.S. 1982. The sterile insect release method for suppression or eradication of fruit fly populations. In Drew, R.A.I., Hooper, G.H.S. and Bateman, M.A. Economic Fruit Flies of the South Pacific Region. Qd Dep. Primary Industries, Brisbane, pp 98-114.

Hooper, G.H.S. and Katiyar, K.P. 1971. Competitiveness of gamma-sterilized males of the Mediterranean fruit fly. J. Econ. Entomol. 64: 1068-1071.

Iga, M. 1981. Mating preference of Dacus dorsalis Hendel (Diptera: Tephritidae) with reference to individual variations in size. Japan. J. Appl. Entomol. Zool. 25: 292-294.

Itô, Y. and Koyama, J. 1982. Eradication of the melon fly: role of population ecology in the successful implementation of the sterile insect release method. Prot. Ecol. 4: 1-28.

Iwahashi, O., Itô, Y. and Shiyomi, M. 1983. A field evaluation of the sexual competitiveness of sterile melon flies, Dacus (Zeugodacus) cucurbitae. Ecol. Entomol. 8: 43-48.

Kahinohana, H. 1980. Qualitative change in the mass reared melon fly, Dacus cucurbitae Coq. In Proc. Symp. on Fruit Fly Problems, Kyoto and Naha, Nat. Inst. Agric. Sci., Yatabe, pp 27-36.

Kapatos, E. and Fletcher, B.S. 1983. Seasonal changes in the efficiency of McPhail traps and a model for estimating olive fly densities from trap catches using temperature data. Entomol. Exp. Appl. 33: 20-26.

Meats, A. 1983. Strategies for maximizing the potential of the sterile insect release method: experiments with Dacus tryoni. In Cavalloro, R. (ed.). Fruit Flies of Economic Importance. Balkema, Rotterdam, pp 371-377.

Ohinata, K., Ashraf, M. and Harris, E.J. 1977. Mediterranean fruit flies: sterility and sexual competitiveness in the laboratory after treatment with gamma radiation in air, carbon dioxide, helium, nitrogen or partial vacuum. J. Econ. Entomol. 70: 165-168.

Ohinata, K., Fujimoto, M., Higa, H., Tanaka, N. and Harris, E.J. 1978. Mediterranean fruit fly: gamma-radiation in nitrogen and packaging for sterile-insect release program in Los Angeles. J. Econ. Entomol. 71: 610-612.

Prokopy, R.J. and Economopoulos, A.P. 1975. Attraction of laboratory-cultured and wild Dacus oleae flies to sticky-coated McPhail traps of different colors and odors. Environ. Entomol. 4: 187-192.

Robinson, A.S. and Zurlini, G. 1981. Mating success of differently sized onion flies, Delia antiqua. Entomol. Exp. Appl. 30: 101-105.

Soemori, H., Tsukaguchi, S. and Nakamori, H. 1980. Comparison of mating ability and mating competitiveness between mass-reared and wild strains of the melon fly, Dacus cucurbitae Coquillet (Diptera: Tephritidae). Japan. J. Appl. Entomol. Zool. 24: 246-250.

Wong, T.T.Y., Whitehand, L.C., Kobayashi, R.M., Ohinata, K., Tanaka, T. and Harris, E.J. 1982. Mediterranean fruit fly: dispersal of wild and irradiated and untreated laboratory-reared males. Environ. Entomol. 11: 339-343.

PHEROMONE COMMUNICATION IN THE MEDITERRANEAN FRUIT FLY (*Ceratitis capitata* Weid.)

P.E.Howse and M.E.Foda
Chemical Entomology Unit,
Department of Biology,
University of Southampton, U.K.

Although trimedlure is a relatively cheap attractant for medfly, it has the disadvantages that it usually attracts only males, is not sufficiently active to use in male annihilation techniques, and when used as a lure in sticky traps gives rise to a high variance in catches. For example, P.S. Baker (unpublished results) showed that Jackson traps in Mexican orchards caught between 0.007 and 105 flies per trap per week following release of 500-1000 sterile flies between each trap at a trap spacing of 20-50m. Pheromones offer a possible solution to the problem of increasing the efficiency of monitoring and detection traps. A number of pheromone components have recently been isolated and identified from male flies at the University of Southampton (R.Baker *et al*, 1985). The role of these components in communication has subsequently been investigated in wind-tunnel tests, and field evaluation of various blends has begun in Spain (Howse & Foda, unpublished).

Pheromones extracted from mature male medflies or trapped by aeration techniques are of remarkable complexity, and upwards of twenty components have been identified to date. The elucidation of a sexual attractant from such a mixture presents an almost insuperable problem, because it is not known a) whether all components come from the same glandular source or are liberated together, b) which combination of components may control attraction, c) what their concentration ratios and release rates should be, and d) what are the upper and lower threshold concentrations for any particular blend.

Tests on individual components show that the major ones tend to have an activation effect on virgin females and stimulate upwind flight. By chance, one component alone was found to induce male aggregation. In subsequent tests, this component was found to induce zig-zag approach flights, and landing and arrestance at the source in both males and virgin females. It is suggested that this component is multifunctional in nature, controlling attraction of males to other males (in lek formation), attraction of virgin females to calling males, territorial behaviour of males on leaves, and sexual recognition processes on the leaf territory. These hypotheses are now being investigated in field conditions. In some Spanish citrus orchards catches of males have been obtained which are very similar in number to those obtained with trimedlure, and in others catches of females only have been obtained. The factors affecting such differences may include trap colour, presence of volatiles from ripe fruit, as well as the particular flight stimulants and synergists used in the attractant blend. It is hoped that further research will make possible a male attractant superior to trimedlure, and an attractant for virgin females.

REFERENCE: Baker,R.,Herbert,R.H.& Grant,G.G., J.Chem.Soc.Chem.Commun.No.12 824 (1985)

COMPETITION OF UROPHORA STYLATA F. AND TERELLIA SERRATULAE L. (DIPT., TEPHRITIDAE) IN THE FLOWERHEADS OF CIRSIUM VULGARE

H. Michaelis
LS Tieroekologie 1
University of Bayreuth
8580 Bayreuth, W. Germany

Introduction

Fruit flies infesting the flowerheads of thistles are not generally regarded as being of any economic importance. There are two reasons, however, to include some notes on their biology in this place. There have been attempts to use them as biocontrol agents against thistles that pose a weed problem (Peschken 1979, Schröder 1980). The use of Urophora stylata, especially, against its host-plant Cirsium vulgare in Canada has lead to encouraging results. More important, the "ecosystem in a nutshell" of insects in the flowerhaeds of thistles lends itself well to the study of small, compartmented food-webs. This type of trophic systems has been estimated to contain roughly half of the terrestrial species on earth (Price 1980). A close comparative study of some compartmented systems could well lead to a better understanding of the basic principles underlying the population dynamics of species that are part of plant-phytophage-entomophage-type food-webs. For an example, the review of Prokopy (1983) and the theoretical considerations of Pimm (1982) are mentioned. These more generalizing considerations could inspire the more practical and methodological discussions of people concerned with weed or pest control.

The biology of Urophora stylata and Terellia serratulae has been described by Redfern (1968, 1983). Both species are univoltine and overwinter in the larval stage in the flowerheads of thistles. U. stylata is restricted to Cirsium vulgare, whereas T. serratulae has also been reared from some

NATO ASI Series, Vol. G11
Pest Control: Operations and Systems Analysis
in Fruit Fly Management. Edited by M. Mangel et al.

related plants (Redfern 1983, Baloch et al. 1973). U.stylata oviposits into young, closed flowerheads. The young larvae, that start feeding on the achenes, induce the forming of a large, woody, multiocular gall that encases them until pupation and emergence in next may. T. serratulae oviposits into opened flowerheads. The larvae feed on the achenes all the time and overwinter in a sort of cocoon made of pappus. They do not induce gall-forming.

The investigations were carried out in Upper Franconia W. Germany, where C. vulgare is an abundant ruderal plant on roadsides and disturbed pastures. Flowerheads were sampled in fall of 1979 to 1983 and were dissected in winter. All the experiments with adults were carried out with laboratory-raised individuals from samples of the previous year.

Competition

The larvae of both tephritid flies do not show any aggressive behaviour towards other larvae. A considerable number of them can be kept together in gelatine capsules without harm. Any intraspecific competition should therefore be a competition for space and food and be detectable by their mean accumulation of biomass in flowerheads with different phytophage pressure. To test this a number of third instar larvae samples from the field in autumn when they had stopped eating were weighed. As dry weight turned out to be very strictly proportional to fresh weight in both species ($r=0.99$ and 0.97, respectively, $p<0.001$), the latter was used for convenience. Mean fresh weight was plotted against the initial number of eggs in the flowerheads. For U. stylata this means the number of gall chambers found, for T. serratulae the number of larvae in a head, dead or alive.

Neither U. stylata nor T. serratulae show a marked dependency of mean larval fresh weight and their numbers in the flowerheads (fig. 1a). At first glance this is astonishing, but this mode of correlation used does not heed the individual larvaes' space in the flowerheads. In each class of numbers there are relatively large and small flowerheads.

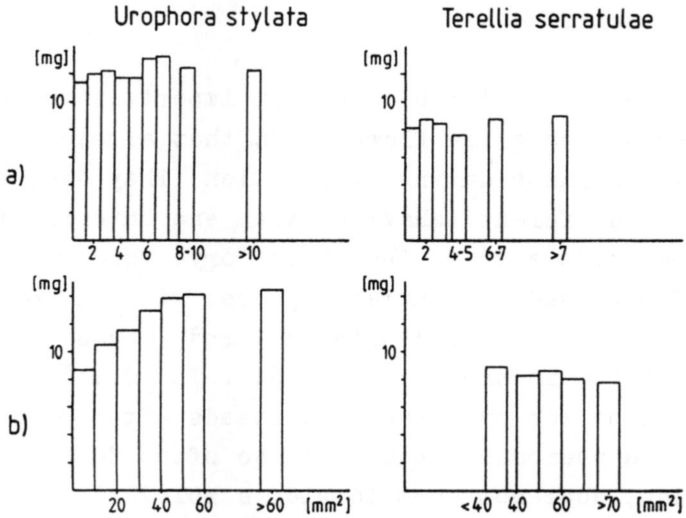

Fig. 1 : Dependency of the mean larval fresh weight of U. stylata (left column) and T. serratulae (right) on their numbers in the heads (a) and the individual larva's mean space (b) (Urophora : n=441, Terellia : n= 152. The classes were formed to contain roughly equal numbers.)

To account for this the same data were classified differently, using a relative value for the spacing of the larvae inside the gall or flowerhead. For U. stylata, the mean gall diameter at its thickest point was used to compute the area of a cross section, this was divided through the number of gall chambers. For T. serratulae, the mean diameter of the flowerhead was used respectively. The resulting values are roughly proportional to the individual larvae's space inside the gall or flowerhead.

The obtained classification shows a marked difference (fig. 1b). Closely spaced U. stylata-larvae remain much smaller than loosely spaced ones. On the other hand, closely spaced larvae of T. serratulae are not smaller than loosely spaced ones, they even appear a little bit heavier.

As an analyses of the frequency distribution of larval weight of U. stylata in densely packed, intermediate and loosely packed galls showed, this reduction in weight is not the result of a few underdevelopped individuals but of an overall shift of the distribution. So at least for U. stylata there is a sign of intraspecific competition for space.

Evidences for discrimination

To avoid competition it would be very important for U. stylata females to recognize flowerheads that already contain eggs and to discriminate during oviposition. They did, however, not show a different behaviour when encountering flowerheads with or without eggs in the laboratory. They moved around on the small flowerheads, eventially probed with the ovipositor, and as well as not laid between 1 and 7 eggs (mean number of eggs found in the field : 3.64). The females did not drag the ovipositor over the flowerheads after oviposition, so that a pheromone deterrent, so often found in other tephritids, does not seem to play a role.

To test, whether or not U. stylata in one way or another discriminates during oviposition, the following experimental setup was used. 100 flowerheads suitable for oviposition (closed buds, ca. 9 mm in diameter) were collected from a fly-free site and offered for oviposition under a cage. Ten females were admitted. After four days half of the flowerheads, chosen in a random manner, were exchanged for new heads sampled at the same time and kept under equal conditions. Another four days later, all the flowerheads were dissected. The comparison of the number of eggs in the heads that were in the experiment all the time and the sum of eggs in the heads before and after the exchange shows that U. stylata does in fact discriminate to some extent (table 1). The "new" heads had received disproportionally more eggs.

table 1 :

1) eggs in heads in the experiment for 8 days	: 305
2) eggs in heads in the experiment the 1st 4 days	: 174
3) eggs in heads in the experiment the 2nd 4 days	: 199
4) sum of 2) and 3)	: 375
expectancy : 1) = 4) = 339	

A reduced oviposition in flowerheads that already contain eggs should also become visible, if the mean number of larvae per infested head in a sample is plotted against the percentage of infested heads. Discrimination should result in a

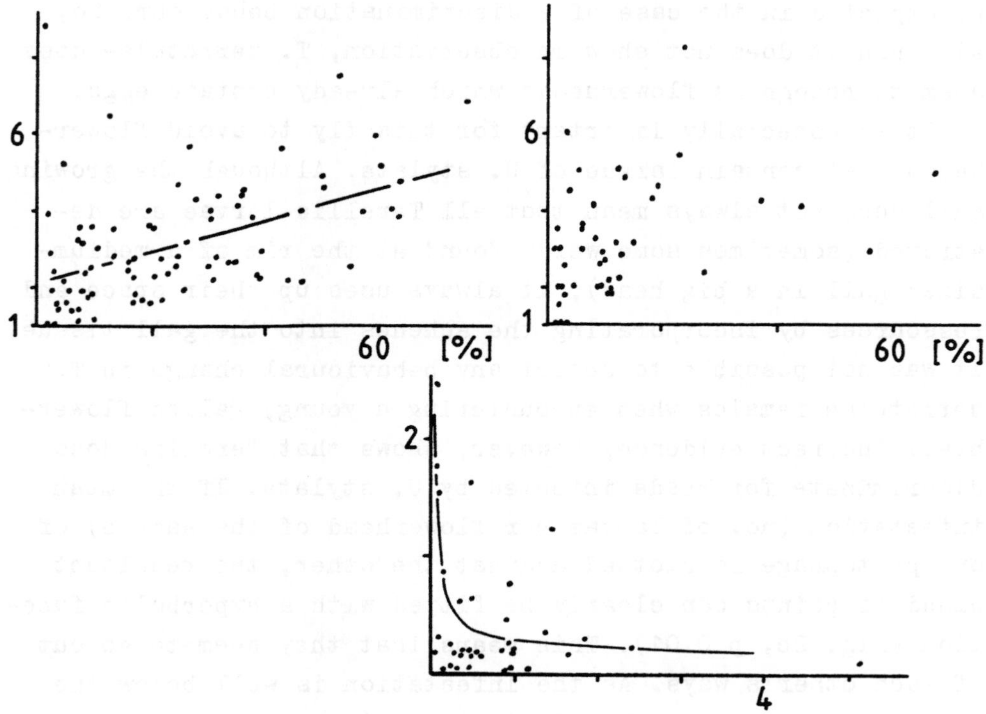

Fig. 2 : Mean number of larvae per head in the sample plotted against percentage of heads infested by Urophora (a) and Terellia (b) and mean numbers of larvae per head in samples with both phytophages, plotted one against the other (x-axis: Urophora, y-axis : Terellia)

saturation of the curve at higher percentages of infestation. This cannot be shown clearly. The correlation between mean numbers in the heads and percentage of infested heads is quite obvious and a linear regression can be fitted with sufficient significance ($p<0.001$) (fig. 2a). The variance is, however, too high to detect any saturation

Terellia just like Urophora does not show any different behaviour when exposed to infested heads. Unfortunately, not enough adults could be reared at the right time to repeat the discrimination test for this species, so only indirect evidence for a discrimination could be found. The plot shown in fig. 2b shows more variance than that of Urophora (fig. 2a) and no significant regression could be fitted. The best regression, however, is that of a saturated curve, which would be expected in the case of a discrimination behaviour. So, although it does not show in observation, T. serratulae does seem to recognize flowerheads which already contain eggs.

It is especially important for this fly to avoid flowerheads that contain larvae of U. stylata. Although the growing gall does not always mean that all Terellia-larvae are destroyed (sometimes some werd found at the rim of a medium-sized gall in a big head), it always uses up their space and ressources by incorporating the achenes into the gall tissue. It was not possible to detect any behavioural change in T. serratulae females when encountering a young, galled flowerhead. Indirect evidence, however, shows that Terellia does discriminate for heads infested by U. stylata. If the mean infestation (no. of larvae per flowerhead of the sample) of one phytophage is plotted against the other, the resultant cloud of points can clearly be fitted with a hyperbolic function (fig. 2c, p 0.01). This means that they seem to go out of each other's ways. As the infestation is well below the carrying capacity of the plant stands (mean percentage of infested heads in samples that contain the flies : 25 % for Urophora, 14 % for Terellia), this can not only be the result of a suppression of Terellia by Urophora, but does indicate a certain amount of discrimination by the latter.

Discussion

Before considering the extent of intra- and interspecific competition found in Urophora stylata and Terellia serratulae, it might be useful to recall an unsolvable, somewhat academic problem. The failure to detect competition or any strong mechanism of avoidance could as well as not mean, that once a strong competitive evolutionary pressure favoured the selection of a way to avoid it, thus lessening its impact. So, theoretically, competition is only detectable if it is uncompletely avoided or counterbalanced, or else there would be no competition at all.

Neither for U. stylata nor for T. serratulae does competition seem to be of any great importance. The larvae are not aggressive and while Terellia is not affected by its spacing in the flowerheads at all, Urophora larvae are affected only as far as larval weight is concerned. The smaller larvaes' chance to emerge are identical to that of big ones. As the smaller larvae emerge as smaller adults which produce fewer eggs, the effect of competition is only felt in the next generation. Compared to mortality due to various parasitoids and predators (Michaelis, in press) this effect can be regarded as being of no great importance. Consequently discrimination during oviposition is also not obvious and only detectable by a closer study. Only interspecific competition may be important, as T. serratulae is outcompeted by the growing Urophora-gall in the same head. There seems to be some discrimination to avoid oviposition in heads that already contain the early-coming gall-former. For phytophage insects that spend most of their lifetimes confined inside a flowerhead such a small importance of competition seems unlikely and it is tempting to speculate upon possible reasons.

One reason is certainly the fact the forming of a gall induces the plant to divert extra assimilates into flowerheads infested by U. stylata. This can be clearly shown by comparing the biomass of infested and uninfested heads. By use of this physiological process, the phytophage indirectly procures its own energy source and is not as dependent on the flowerhead alone. The negative effect of close packing in the gall more than of high numbers means that it is much more important for the ovipositing female to select flowerheads which are liable to get a good supply of assimilates than to avoid its own kind. These are the apical heads at the end of the stems. Although all flowerheads are only suitable for oviposition for a few days and the selection of the best ones poses the problem of synchronisation, it can be clearly shown that Urophora does oviposit much more frequently in apical heads than in heads of secondary or tertiary position. This may be facilitated by their mating behaviour. The early-emerging males defend territories on the apical heads on which the sexes often mate. This may lead to oviposition on the same spot.

For Terellia serratulae the "oviposition window" is much

wider. As it oviposits into the bigger, fully opened flowerheads, synchronisation with the host plant must be easier. Infested flowerheads cannot be shown to be "assimilate sinks" as has been demonstrated by some knapweed-infesting tephritids (Harris 1980), they are not significantly heavier than uninfested ones. The reason that Terellia does not seem to suffer from crowding is unclear. Probably the number of larvae is still lower than the head's "carrying capacity", they do not form an interactive grazing system in the sense of Monro (1975).

Phytophage species confined in flowerheads have generally two options : to minimize competition or to maximize their competitive capacity (Zwölfer 1979). In the case of an intrinsic inferiority (i.e. in the larval stage), extrinsic advantages (i.e. in the searching or oviposition abilities of the adults) can balance the scales. Many examples for this kind of combination of advantages and disadvantages have been demonstrated for thistle-infesting insects (Zwölfer 1980, 1983). In the case of the species-pair Urophora stylata / Terellia serratulae, competition, which is not very severe on the first hand because of infestation rates and other factors, seems to be further reduced by a certain discriminating behaviour of T. serratulae.

This relationship between insects and the host plant seems to be another example stressing Pimm's notion (1982), that generally interrelationships between different steps of the trophic chain, in our case between the insects and the plant, are more important than competition between species of comparable trophic position.

References

Baloch, G.M., A.G.Khan, 1973 : Biological control of Carduus species. 2: Phenology, biology and host specifity of Terellia serratulae L. (Dipt., Tephritidae). CIBC techn. bull. 16, 11-22

Harris, P., 1980 : Effects of Urophora affinis Frfld. and U. quadrifasciata (Meig.) (Dipt., Tephritidae) on Centaurea diffusa Lam. and C. maculosa Lam.(Compositae). Z. angew. Ent. 90, 190-201

Michaelis, H., 1985 : Nahrungsnetzuntersuchungen in den Blütenköpfen von Cirsium vulgare. Verh. Ges. Oekologie, Bremen 1983, in press

Monro, J., 1975 : The exploitation and conservation of resources by populations of insects. J. Anim. Ecol. 36, 531-547

Peschken, D.P., 1979 : Biologische Unkrautbekämpfung in Kanada mit Hilfe von Insekten und Nematoden. Z. angew. Entom. 88, 1-6

Pimm, S.L., 1982 : Food webs. London (Chapman & Hall), 1-219

Price, P., 1980 : Evolutionary biology of parasites. Princeton (Pr. Univ. Pr.), 1-237

Redfern, M., 1968 : The natural history of spear thistle heads. Field studies 2, 669-717

Redfern, M., 1983 : Insects and thistles. Naturalist's handbook 4, Cambridge (C. Univ. Pr.), 1-65

Schröder, D., 1980 : The biological control of thistles. Biocontrol news and information 1, 9-26

Zwölfer, H., 1979 : Strategies and counterstrategies in insect population systems copeting for space and food in flower heads and plant galls. Fortschr. Zool. 25, 2/3, 331-353, Stuttgart, N.York (G.Fischer)

Zwölfer, H., 1980 : Distelblütenköpfe als ökologische Kleinsysteme: Konkurrenz und Koexistenz in Phytophagenkomplexen. Mitt.dtsch.Ges.allg.angew.Ent. 2, 21-37

Zwölfer, H., 1983 : Life systems and strategies of resource exploitation in Tephritids. In : Fruit flies of economic importance (R.Cavalloro, ed.), Proc. CEC/IOBC Int. Symp., Athens 1982, 16-30, Rotterdam (A.A.Balkema)

GENETIC SEXING AND TRANSLOCATIONS IN *CERATITIS CAPITATA*

A.S. Robinson
Insect Genetics Unit
Research Institute ITAL
P.O. Box 48
6700 AA Wageningen, The Netherlands

ABSTRACT

Genetic sexing can be defined as the use of genetic mechanisms to produce unisexual male progenies for release in an insect control programme. The advantages of an all male release can include economic savings and better biological efficiency of the released males. Various genetic mechanisms have been investigated to accomplish this end, the most successful being the use of chromosomal translocations.

Translocations involve the exchange of genetic material between two different chromosomes and a special type of translocation i.e. male-linked translocation will be defined and their characteristics as relevant to genetic sexing will be described. Translocations are often associated with reduced fertility which can be a disadvantage. The fundamental principle exploited in the use of translocations is the creation of a new male-linked genetic entity.

The gene systems used in combination with the translocations can be roughly divided into lethal and visible mutant types. The lethal system includes for example insecticide resistance/susceptibility genes and the visible systems can be characterized by pupal colour mutants. By combining the translocation with the gene system it is possible to link a particular gene with the male sex.

The developments of genetic sexing systems in *C. capitata* will be briefly reviewed. The systems include pupal colour, sensitivity to ethanol and purine. Data is presented on the induction and recovery of translocations and the stability of the translocation complexes.

The paper will conclude with a discussion on the practical application of genetic sexing systems in mass rearing facilities and the difficulties which are likely to be experienced. It is concluded that integration of such systems into a sterile insect release programme would greatly increase the economic feasibility and lead to a more widespread acceptance for this type of insect control.

NATO ASI Series, Vol. G11
Pest Control: Operations and Systems Analysis
in Fruit Fly Management. Edited by M. Mangel et al.

1. INTRODUCTION

Genetic sexing is the use of genetic mechanisms for the production of unisexual male progenies for release in an SIT control programme. In this type of control technique the crucial aspect is the frequency of mating between the released sterile males and the wild females. The released females are therefore redundant. This, together with other considerations, prompted Whitten (1969) to suggest the use of males only for the application of the SIT. Since then several different approaches in many insect species have been attempted in order to produce all male progenies (for review see Robinson 1983). The motivation to develop genetic sexing in the Mediterranean fruitfly, *Ceratitis capitata* has been generated by the following arguments.

1.1 Economic

During mass rearing, irradiation and release the females demand "equal time" as compared to their male counterparts. Therefore if the females could be removed from the population at an early developmental stage, large savings would be possible, in theory costs could be reduced by half. For various on-going control programmes LaChance (1979) has calculated the savings which would accrue if an efficient male producing strain was incorporated into the mass rearing facility.

1.2 Assortative Mating

When large numbers of sterile males and females are released into the wild population, the possibility exists that there will be preferential within population mating and relatively very few wild females will be mated with sterile males. This lack of penetration of sterility into the wild population results in a very low competitiveness of the released males. The causes of assortative mating under these conditions will be a result of behavioural and genetic changes selected for during laboratory colonization. The removal of females from the release will severely reduce the affects of assortative mating.

1.3 Sterile Stings

Sterilized Mediterranean fruit fly females when released are still capable of oviposition on the host, thus leading to bacterial and fungal attack with the consequent lowering of the crop value. If sterile insects are being released at large overflooding ratios this problem can be very serious. The problem of sterile stings prevents the application of the SIT in many countries in S. Europe and North Africa.

1.4 Biological Efficiency

Following a male only release in the consequent absence of laboratory females there could be a strong dispersive tendency in the males. This could mean that lower release ratios could be used and that release strategies could be made more economical. Together with the reduction in assortative matings (1.2), the increased biological efficiency would lead to a much higher "competitiveness" of the released males.

2. CONSTRUCTION OF A GENETIC SEXING SYSTEM

The technique at present demonstrating the most perspective for the successful development of genetic sexing in Medfly rely on the use of chromosomal translocations in combination with different gene systems. This is a technique which has been used with success in several insect species (Whitten et al. 19075, Curtis et al. 1976, Baker et al. 1981 and Seawright et al. 1978).

2.1 Translocation

A translocation is defined as the reciprocal exchange of chromosomal portions between two non-homologous chromosomes (Robinson 1976). This type of rearrangement occurs naturally but the frequency can be considerably increased following irradiation. The type of translocation produced and the position of the chromosomal breaks is a random process therefore following irradiation the translocations are screened to select out the required lines. Fig. I.A shows a schematic representation of a translocation. Translocations have many consequences both at the individual and population levels, however, three characteristics are of relevance to the present discussion.

2.1.1 Semi-sterility

Following meiosis in a translocation individual where homologous chromosomes pair with each other various gamete types are formed. A proportion of these carry duplication/deficiency chromosomes which lead to lethality when these gametes are included in the zygote. The proportion of these "lethal" gametes varies with different translocations but in general corresponds to

50% of the total. Translocations are therefore said to be characterized by semi-sterility. Fig. I.B depicts how the translocated chromosomes pair during meiosis.

2.1.2 Mode of inheritance

Among the balanced gametes produced by a translocation carrying individual, can be found normal gametes and translocation gametes with, in general, equal frequencies. Translocations behave therefore as Mendelian characters and can be followed from generation to generation. Using a combination of semi-sterility and inheritance data, translocations can be easily identified an isolated in the progeny of irradiated individuals.

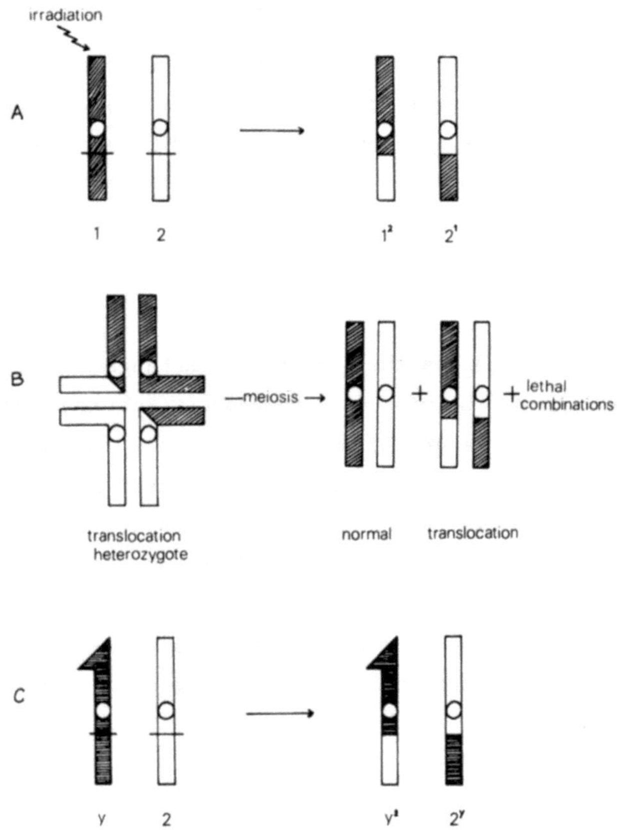

Fig. I. Translocation schematically represented. I.A. Reciprocal translocation, I.B. Meiosis in a translocation heterozygote and I.C. Male-linked translocation.

2.1.3 New linkage groups

Following a translocation the position of genes is changed relative to each other. For example in Fig. I.A a gene normally associated with chromosome 2 becomes associated with chromosome 1 as a consequence of the translocation. A translocation therefore results in the creation of new linkage groups i.e. groups of genes which tend to be inherited together.

2.2 Male-linked Translocations

These represent a special class of translocations with the three characteristics listed above plus a significant "extra" of particular relevance to genetic sexing. In these translocations the male-determining chromosome (Y) is involved in the translocation (Fig. I.C.). This means that the genes on chromosome 2 which originally segregated independently of sex become sex-linked. Therefore using these types of translocation it is possible to shift genes to the male determining chromosome. The consequence is that such genes become male-linked and pass from father to son and so on. It is this particular type of translocation which is exploited in the development of genetic sexing systems. In a population containing a male-linked translocation the males are semi-sterile, pass the translocation on to their sons and demonstrate sex-linkage of genes whereas the females are fully fertile and show normal segregation.

2.3 Gene-Systems

These can be broadly grouped into two classes, firstly those relying on differential lethality effects and secondly those relying on visible differences between the sexes. Although the principles involved are identical the type of gene-system chosen has significant consequences for the practical application of the technique.

2.3.1 Conditional-lethal systems

The principle here is to manipulate the males so that they become resistant to a particular environmental constraint, be it physical or chemical. The types of system include temperature and insecticide resistance/susceptibility. For example the gene responsible for insecticide resistance would be translocated to the male-determining chromosome and a strain can then be constructed in which the females are susceptible and the males resistant. By treating this population at a certain developmental stage with a discriminating dose of insecticide the females will be killed and the males survive. The advantage of these systems is that they can be applied at the population level and are therefore ideally suited for integration into a mass rearing system. They are, however, destructive, the females cannot be used.

2.3.2 Visible systems

Here an easily recognizable phenotype is made sex-linked so that individuals can be sexed on this basis. A good example of this can be found in the silk worm, Bombyx mori where using a translocation an egg colour gene was made sex-linked. Females hatched from black eggs and males from white eggs and using a photoelectric cell large populations could be sexed (Tazima et al. 1951). These systems have the disadvantage that they have to be applied on an individual basis, however, both sexes can be retained and utilized (see below 3.2.2).

3. GENETIC SEXING SYSTEMS IN MEDFLY

There are several facets of the genetics and biology of the Medfly which make it an attractive candidate for the development of a genetic system. Firstly, it belongs to the cyclorohaphid dipteras where crossing-over does not occur in males (LaChance et al. 1966). This has important consequences in that genes which are translocated to the male determining chromosme, irrespective of their position on the chromosome, will remain linked. Secondly single pair mating and general ease of rearing make the isolation of lines and mutants relatively simple. Thirdly cytological techniques have been developed as well as detailed linkage maps (Radu et al. 1975, and Saul and Rössler 1984).

3.1 Induction of translocations

This aspect necessary for the development of a genetic system has presented few problems and many translocations including male-linked have been routinely isolated (Steffans 1983, Robinson and van Heemert 1982 and Robinson and Zapater unpublished results). If a good gene system is available then the screening of a few hundred progenies following irradiation will isolate the required translocation.

3.2 Gene systems

3.2.1 Conditional lethality

In five mosquito species genetic sexing systems have been developed based on insecticide resistance/susceptibility. The insecticide resistance genes used were isolated from field populations. Unfortunately, or fortunately depending on ones point of view, resistance has not yet appeared in field populations of medflies (see Wood, these proceedings). However, laboratory selection has been carried out for dieldrin resistance. A dieldrin resistant line has been produced in the Medfly, however, no genetic sexing system using this line could be produced and the work has been stopped (Wood, pers. comm.).

Saul (1982) has been developing a system based on resistance/susceptibility to purine. In Drosophila, individuals lacking the enzyme xanthine dehydrogenase (XDH) are sensitive to purine in the larval medium (Gelbart et al. 1976), such individuals also exhibit a mutant eye phenotype rosy (ro). A similar phenotype was found in the Medfly and proved to be sensitive to purine. Following irradiation four lines were isolated in which the active XDH allele was translocated to the male determining chromosome. Following the addition of purine to the larval diet virtually all male progenies could be produced (Saul 1984). However, Saul (1984) concluded that none of the lines were ideal for use in an SIT programme.

Another enzyme system, alcohol dehydrogenase (ADH) can be exploited in a similar way. Individuals with no active ADH (nulls) are poisoned by environmental alcohol whereas normal individuals metabolize the alcohol to the aldehyde (Grell et al. 1968). In Drosophila a genetic sexing system based on ADH and male-linked translocations has been demonstrated (Robinson and van Heemert 1980). A series of null alleles at an ADH locus have been produced following irradiation (Riva and Robinson in prep.), however, none of them could be made homozygous. However, one of them can be used in an efficient mating scheme to isolate a new series of null mutations (Robinson and Riva 1984) and it has been translocated to the male determining chromosome. This line does under certain experimental conditions exhibit genetic sexing properties, it is, however, of little practical significance (Robinson, Riva and Zapater 1985 in prep.). The confirmation of a second ADH locus in the Medfly (Milani et al. 1985) complicates substantially the use of ADH nulls to produce a genetic sexing system.

3.2.2 Visible systems

This was first developed by Rössler (1979 and 1980) and he linked the wild type allele for a black pupal mutant to the male determining chromosome and was able to sex individuals using pupal colour. However, as there was some overlap between the phenotypes a second gene system was developed using a white pupal mutant (Robinson and van Heemert 1982). Several translocations were isolated and following maintenance without selection it appeared that

the lines were stable, females emerged from white pupae and males from brown pupae. Using an adapted rice sorting machine up to 10 kg of pupae could be sexed per hour; this is about one million individuals.
It was decided to test one of the lines under field conditions in Procida, Italy, under cage and open field situations. The results in the field cage (Robinson et al. in press) confirmed the increased biological efficiency of releasing males only. Further results in subsequent years are presented by Cirio (1986 these proceedings).

4. CONCLUSIONS

4.1 In retrospect

The most simple system to develop i.e. the use of pupal colour mutants, is probably the most promising for the rapid introduction into a current SIT programme and given certain conditions it could function extremely efficiently. These conditions will be discussed below (4.2). The other systems based on conditional lethalilty are some way away from practical application. Nevertheless considering the economic advantage of a "population treatment" (conditioned lethals) over "individual treatment" (visibles), the former must be actively researched. The results achieved with the two enzyme systems already studied should encourage the search for other systems.
There is a clear lack of field data on the manifold effects, disadvantages and advantages, of an all male sterile release. These effects will be apparent in the released population and in the response of the wild population to the excess of males in the environment. Preliminary results are, however, encouraging. With handsight these field experiments should have had a much higher priority in the early development of genetic sexing systems.

4.2 Problem areas

Two problem areas, both genetic, will be identified assuming that biological and economical equations relating to the integration of genetic sexing into an SIT programme have been solved with a positive balance. Firstly the problem of stability. By stability is meant the tight linkage of sex with the gene system under study. As already reported crossing-over is in general, limited to the females in Medflies, however, there is some evidence (Rössler 1982) that crossing-over at a low frequency can occur in translocation carrying males. In small laboratory populations sampling effects will reduce the significance of such events, however, in mass rearing plants dealing with extremely large populations the occurrence of an extremely low recombination factor would lead to de-stabilization of the genetic sexing line. There is some indirect evidence that this occurred with a genetic sexing line mass reared at the IAEA Seibersdorf laboratory. This problem can be solved by selecting from a series of different translocation lines the one showing the highest degree of stability under some sort of mass rearing conditions. The speed at which de-stabilization occurs is dependent on the frequency of the rare cross-over event. A very low frequency could be tolerated for several generations and still produce an efficient sexing system.

The second problem involves contamination, a much more serious threat than de-stabilization. The accidental introduction of a single normal male, i.e. non translocated, into a mass rearing facility would lead to an extremely rapid deterioration of the sexing system for the following reason. The introduced male has full fertility and rapidly displaces the semi-sterile genetic sexing males. Once contamination has occurred no remedies can be taken and the population has to be disposed of and the genetic sexing strain re-established. Strict quarantine regulations have therefore to be enforced to protect the population from immigration.

It is important that these two aspects receive critical attention during the planning for the integration of genetic sexing systems in mass rearing facilities.

5. ACKNOWLEDGEMENTS

M.E. Riva and M. Zapater are thanked for their contribution to this work. The research was partially supported by IAEA grant 2972/RB.

6. REFERENCES

BAKER, R.H., SAKAI, R.K. RAANA, K. 1981. Genetic sexing for mosquito sterile-male release. J. Hered. 72: 216-218.

CURTIS, C.F., AKIYAMA, J., DAVIDSON, G. 1976. A genetic sexing system in Anopehles gambiae species. A. Mosq. news. 34: 492-498.

GELBART, W., McCARRON, M., CHOVNICK, A. 1976. Extension of the limits of the XDM structural element in Drosophila melanogaster. Genetics 84: 211-232.

GRELL, E.H., JACOBSON, K.B., MURPHY, J.B. 1968. Alterations of genetic material for analysis of alcohol dehydrogenase isozymes of Drosophila melanogaster. Ann. N.Y. Acad. Sci. 151: 441-455.

LACHANCE, L.E. 1979. Genetic strategies affecting the success and economy of the sterile insect release method. In "Genetics in Relation to Insect Management" Eds. M.A. Hoy and J.J. McKelvey Jr. pp. 8-19. The Rockefeller Found. 179 pp.

LACHANCE, L.E., DAWKINS, C., HOPKINS, D.E. 1966. Mutants and linkage groups

of the screw work fly. J. Econ. Entomol. 59: 1493-1496.
MILANI, R., GASPERI, G., MALACRIDA, A.R. 1985. Recent progress and perspectives in the genetical studies on Ceratitis capitata. IAEA Consultants Meeting on Genetic Sexing Mechanim for the Mediterranean Fruit Fly, Ceratitis capitata. Vienna 1985.
RADU, M. RØSSLER, U., KOLTIN, Y. 1975. The chromosomes of the Mediterranean fruit fly, Ceratitis capitata: Karyotype and chromosomal organization. Cytologia 40: 823-828.
ROBINSON, A.S. 1976. Progress in the use of chromosomal translocations for the control of insect pests. Biol. Rev. 51: 1-24.
ROBINSON, A.S. 1983. Sex-ratio manipulation in relation to insect pest control. Ann. Rev. Genet. 17: 191-214.
ROBINSON, A.S., CIRIO, U., HOOPER, G.H., CAPARELLA, M. Field experiments with a genetic sexing strain in the Medfly, Ceratitis capitata I. Field cage studies. submitted to Ent. Exp. and Appl.
ROBINSON, A.S., HEEMERT, C. VAN. 1980. Genetic sexing in Drosophila melanogaster using the alcohol dehydrogenase locus and a Y-linked translocation. Theor. Appl. Genet. 59: 23-24.
ROBINSON, A.S., HEEMERT, C. VAN. 1982. Ceratitis capitata, a suitable case for genetic sexing. Genetica 58: 229-237.
ROBINSON, A.S., RIVA, M.E. 1984. A simple method for the isolation of an allelic series using male-linked translocations. Theor. Appl. Genet. 67: 305-306.
RØSSLER, Y. 1979. Automated sexing of Ceratatis capitata: The development of strains with inherited sex-limited pupal colour dimorphism. Entomophaga 24: 11-16.
RØSSLER, Y. 1980. "69-apricot" A synthetic strain of the Mediterranean fruit fly Ceratitis capitata with sex-linked pupla colour and an eye colour marker. Entomophaga 25: 275-281.
RØSSLER, Y. 1982. Genetic recombination in males of the Mediterranean fruit fly (Diptera: Tephritidae), and its potential for use in a genetic sexing program. Ann. Entomol. Soc. Am. 75: 28-31.
SAUL, S.H. 1982. Rosy-like mutant of the Mediterranean fruitfly Ceratitis capitata (Diptera: Tephritidae) and its relation to automated sexing methods. Ann. Entomol. Soc. Amer. 75: 480-483.
SAUL, S.H. 1984. Genetic sexing in the Mediterranean fruit fly, Ceratitis capitata (Wiedeman) (Diptera: Tephritidae); Conditional lethal translocations that preferentially eliminate females. Ann. Entomol. Soc. Amer. 77: 280-283.
SAUL, S.H. and RØSSLER, Y. 1984. Genetic markers of the autosomal linkage groups of the Mediterranean fruitfly, Ceratitis capitata (Diptera: Tephritidae). Ann. Entomol. Soc. Am. 77: 323-327.
SEAWRIGHT, J.A., KAISER, P.E., DAME, D.A., LOFGREN, C.S. 1978. Genetic method for the preferential elimination of females of Anopheles albimanus. Science 220: 1303-1304.
STEFFANS, R.J. 1983. Methodology of translocation production and stability of translocations in the Mediterranean fruit fly, Ceratitis capitata Wied. (Dipt.: Tephritidae). Z. ang. Ent.. 95: 181-188.
TAZIMA, Y., HARADA, C., OHATA, N. 1951. On the sex discriminating method by colouring genes of silkworm eggs. I. Induction of translocation between the W and the tenth chromosomes. Jap. J. Breed. 1: 47-50.
WHITTEN, M.J. 1969. Automated sexing of pupae and its usefulness in control by sterile insects. J. Econ. Entomol. 62: 272-273.
WHITTEN, M.J., FOSTER, G.G., ARNOLD, J.T., KONOVALOV, C. 1975. The Australian sheep blowfly, Lucilia cuprovina. In "Handbook of Genetics", Ed. R.G. King 3: 401-418. New York Plenum 874 pp.

TRIALS TO ESTABLISH QUALITY CONTROL METHOD OF MASS-REARED AND IRRADIATED FLIES IN THE FIELD

Tadashi Teruya
Fruit Flies Laboratory, Okinawa Prefectural Agricultural Experiment Station, Sakiyama-cho, Naha 903, Japan

ABSTRACT

To establish quality control method of the mass-reared and irradiated melon flies, Dacus cucurbitae, in the field, two preliminary tests have been done with mark-release-recapture method. The survival rate, dispersal and orientation ability to the places where wild flies were located, of the mass-reared and irradiated flies were examined in summer, 1985. Thirty Steiner type traps (Steiner 1957) baited with cue-lure and naled were distributed within a radius of 1200 m. The recapture rate and the total abundance of the males irradiated with a dose of 70 Gy at -3 days adult, which were the dose and the day of irradiation used in our eradication project, were fewer than those of the males irradiated with a dose of 70 Gy at -2 days adult and those of males untreated. Total abundance of the males irradiated with a dose of 70 Gy at -3 days adult in the two tests were almost same. The dispersal pattern and the decline of survival rate of the flies in Test 1 were different from those in Test 2. It did not always follow that the catches of released males depend on the distance of the trap from the release site within the same annulus, although the average catches of annulus decreased as the distance between the trap and the release site increased. The tendencies that more released males were caught in the traps which graded high in wild catches than that graded low, were observed within the same annulus.

NATO ASI Series, Vol. G11
Pest Control: Operations and Systems Analysis in Fruit Fly Management. Edited by M. Mangel et al.

INTRODUCTION

Differences on several performance traits between mass-reared flies and wild flies have been reported by many authors recently. However, it seems no comprehensive studies as to whether or not these changed traits lower the fitness of mass-reared flies as a soldier in Sterile Insect Technique (SIT). Iwahashi et al. (1983) studied the mating competitiveness of the melon fly, Dacus cucurbitae, in the field in the course of the eradication program in Kume Is., Okinawa, Japan. Subsequently, they reported that the mating competitiveness of the males became drastically lowered during a few generations. No such sudden change of the traits concerned with mating competitiveness of insects which have been reared artificially under constant conditions for many generations has been found in laboratory tests. It has not been identified yetdeteriorated characters of the factory produced melon flies as a soldier in SIT. Is such deterioration actualized only in the field? To confirm if such change occurs naturally in mass-reared insects is important for judging whether or not new strains must be introduced periodically to the factory for SIT.

The RAPID test recommended by Boller et al. (1981) is a very useful quality control system for early warning in producing flies in the factory. The RAPID has been carried out in our melon fly mass-rearing factory. The monitoring has actualized some troubles in the production process of the flies. Overcomingthetroubles, percent emergence and fliers have become more than 90% and 95%, respectively. However, as described above, they do not inform sufficiently about the performance of flies in the field. The purpose of this report was to provide preliminary data for establishing a quality control method of the factory produced flies in the field.

MATERIALS AND METHODS

Flies were reared and irradiated in the Melon Fly Mass-rearing and Sterilization Factory of the Fruit Fly Eradication Project Office of the Okinawa Prefectural Government, Naha, Japan. After irradiation with gamma-ray, pupae were marked with fluorescent dyes of different colors to identify each

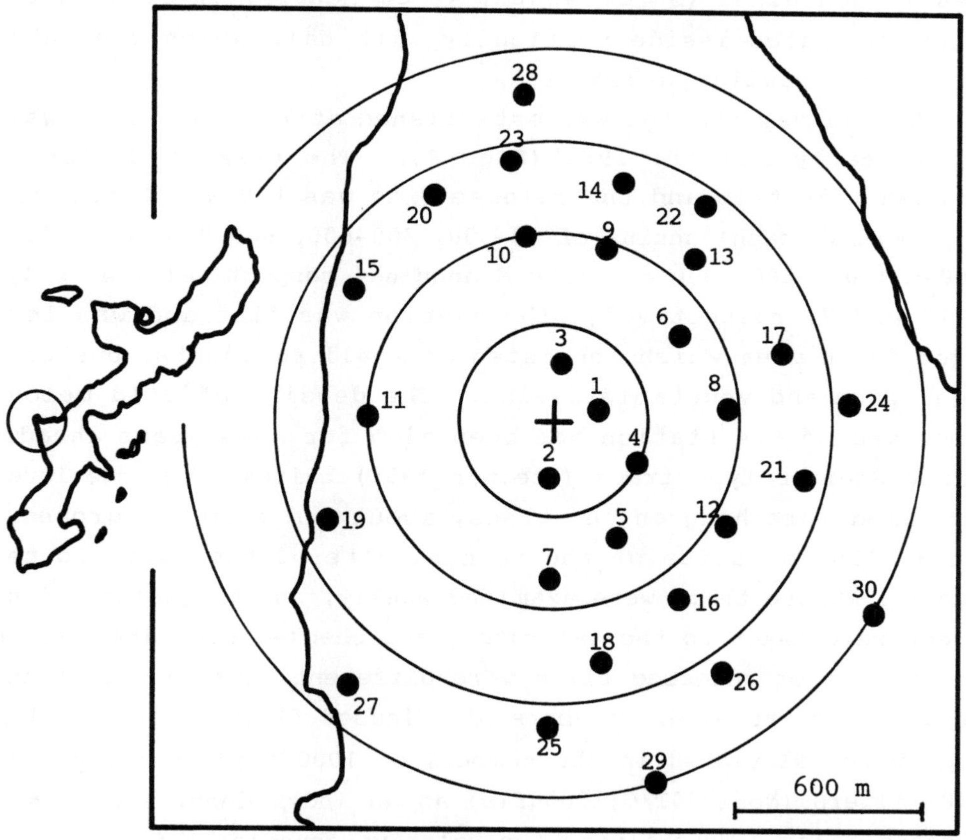

Fig. 1. Survey station at Yomitan. Serial numbers show the traps according to the distance from the release site. The annuli which were used to group the traps for dispersal measurements were 300 meters apart.

category of the flies. Then they were put in the release bags (11 cm width, 18 cm length and 16 cm height) (Kuba 1980) and kept under 25-27 C until the release day. Number of pupae per bag was ca. 4000-5000. Adults emerged were fed with sugar in the bag and were released 1 or 2 days after peak emergence. To determine percentage of emergence and rate of flier, 3 bags were taken from each treatment as samples. Flier is the fly which has some flight ability examined with following method. The flies in the bags sampled were left ca. 4 hours in the plastic container (30 cm width, 40 cm length, 16 cm height) which is coated inside vertically with talc which prevents flies from crawling up the wall.

The survey station was established at Yomitan, Okinawa, Japan, early June in 1985 (Fig. 1). The nearest distance between the trap and the release site was 150 m. Number of traps within each annulus of 0-300, 300-600, 600-900 and 900-1200 m (300, 600, 900 and 1200 M-annulus, respectively) were 4, 7, 12 and 7, respectively. The station was flat and was located in an area which consisted of small residents, bushes, sugar cane and vegetable fields. The density of wild melon flies around the station has been high for some years ahead. Thirty Steiner type traps (Steiner 1957) baited with cue-lure and naled were hung on the trees, about 1 m from the ground, within 1200 m radius of the release site of the flies to be studied. These traps were examined weekly, and any melon flies caught were taken to the laboratory and checked for mark. Since the numbers of released flies were different in each treatment and in each release, numbers of released flies caught in the traps were calculated in the numbers of 1000 traps per day per 10000 fliers (No/1000T/D/10000fls) as an index density of the released males. To evaluate the relative abundance of released males at each annulus, the catches/1000T/D/10000fls/area of 300 m-radius (3.14×300^2 m^2) were used. Thus, on the assumption with that the flies distributed uniformly within each annulus, the numbers 1, 3, 5 and 7, in proportion to each area, were multiplied to the catches of each 300, 600, 900 and 1200 M-annulus, respectively (Tables 4 and 8).

In Test 1, the flies irradiated with a dose of 70 Gy at -2 days adult and -3 days adult were examined. The pupae were separated from sawdust with sifting machine (Taniguchi and Iwaizumi 1984) at 60% of the pupal period. The flies were released 2 days after peak emergence. Numbers of fliers used in this test are shown in Table 1. During the test period average temperature were 26.5 C (25.0-28.0 C) and the precipitations of every 10 days were 77, 33 and 6 mm (116 mm/month). This test was conducted from June through July, 1985.

Table 1. Number of pupae and fliers used in Test 1.

Treatment	Number of pupae	Percent emergence	Percent flier	Number of flier	Factor of 1000T /D/10000 flier
T-1*	43445	85.9	57.1	21309	67.1
T-2	39413	86.1	77.4	29265	54.4

* See text

In Test 2, the flies sifted by hand and unirradiated, and those sifted with machine and irradiated with a dose of 70 Gy were examined. Half of pupae obtained from the same lot were separated from sawdust with sieve by hand and the other half were separated with sifting machine at 60% of the pupal period. The latter were then irradiated with a dose of 70 Gy. Flies emerged were released 1 day after peak emergence. Numbers of fliers released are shown in Table 2. During this test period average temperature was 28.1 C (27.8-28.4 C), and the precipitations of every 10 days were 35, 20 and 16 mm (71 mm/month). This test was conducted from July through August, 1985.

Table 2. Number of pupae and fliers used in Test 2.

Treatment	Number of pupae	Percent emergence	Percent flier	Number of flier	Factor of 1000T /D/10000 flier
T-1*	18992	94.6	70.6	12684	112.7
T-2	61624	91.4	70.0	39427	36.2

* See text.

Catches of wild males of each trap at each week were calculated in No/1000T/D. Each trap were graded for wild catches. The calculation for grading was as follows: the difference between log. maximum catches minus log. minimum catches in each week was divided by 5. Average catches calculated in logarithm were used as an index to evaluate the wild catches of each trap (Tables 13 and 14).

RESULTS AND DISCUSSION

Comparison in Test 1. The accumulated relative density of the released males irradiated at -2 days adult (T-1-1) and that at -3 days adult (T-1-2) were 10325 and 6578, respectively (Table 3). Totals of the relative abundance of them were 16616 and 9596, respectively (Table 4). The recapture rates of T-1-1

Table 3. Average catches of released males* in each annulus at each week in Test 1.

Annulus	6/14		6/22		6/28		7/5		Sum of row	
	T-1	T-2**	T-1	T-2	T-1	T-2	T-1	T-2	T-1	T-2
300M	2012	1727	5633	3169	604	517	235	150	8483	5563
600M	156	154	525	390	134	78	19	8	835	630
900M	95	59	414	159	168	45	34	14	710	277
1200M	86	8	134	62	67	23	10	16	297	109
Sum of column	2349	1948	6706	3780	972	663	297	187	10325	6578

* Numbers are expressed as No/1000T/D/10000fls. ** See text.

Table 4. Average population of released males* in each annulus at each week in Test 1.

Annulus	6/14		6/22		6/28		7/5		Sum of row	
	T-1	T-2**	T-1	T-2	T-1	T-2	T-1	T-2	T-1	T-2
300M	2012	1727	5633	3169	604	517	235	150	8483	5563
600M	469	462	1576	1170	402	233	57	23	2505	1889
900M	475	295	2068	793	838	227	168	68	3549	1383
1200M	604	54	939	435	469	163	67	109	2079	762
Sum of column	3560	2539	10215	5568	2314	1140	527	350	16616	9596

* Numbers are expressed as No/1000T/D/10000fls. ** See text.

and T-1-2 were 0.0695 (741/21309/0.5) and 0.0423 (555/26265/0.5), respectively, when the sex ratio of female to male of the released flies is unity. No noticeable difference were found in the dispersal pattern and the decline in survival rate between T-1-1 and T-1-2 (Tables 5 and 6).

Table 5. Relative density of released males* in each annulus at each week in Test 1.

Annulus	6/14 T-1	6/14 T-2**	6/22 T-1	6/22 T-2	6/28 T-1	6/28 T-2	7/5 T-1	7/5 T-2	Average T-1	Average T-2
300M	86	89	84	84	62	78	79	80	82	85
600M	7	8	8	10	14	12	6	4	8	10
900M	4	3	6	4	17	7	11	7	7	4
1200M	4	0	2	2	7	4	3	8	3	2

* Numbers are expressed as % of total density at each week. ** See text.

Table 6. Relative abundance of released males * in each annulus at each week in Test 1.

Annulus	6/14 T-1	6/14 T-2**	6/22 T-1	6/22 T-2	6/28 T-1	6/28 T-2	7/5 T-1	7/5 T-2	Average T-1	Average T-2
300M	57	68	55	57	26	45	45	43	51	58
600M	13	18	15	21	17	20	11	7	15	20
900M	13	12	20	14	36	20	32	19	21	14
1200M	17	2	9	8	20	14	13	31	13	8

* Numbers are expressed as % of total population at each week. **See text.

Comparison in Test 2. The accumulated relative density of the released males sifted by hand and unirradiated (T-2-1), and those sifted by machine and 70 Gy-irradiated at -3 days adult (T-2-2) were 15673 and 7457, respectively (Table 7). Total of the relative abundance of T-2-1 and T-2-2 were 19686 and 9559, respectively (Table 8). The recapture rate of T-2-1 and T-2-2 were 0.0956 (606/12684/0.5) and 0.0460 (907/39427/0.5), respec-

Table 7. Average catches of released males* in each annulus at each week in Test 2.

Annulus	7/13 T-1	7/13 T-2**	7/20 T-1	7/20 T-2	7/27 T-1	7/27 T-2	8/3 T-1	8/3 T-2	Sum of row T-1	Sum of row T-2
300M	3690	1658	7717	3851	2169	970	676	236	14251	6714
600M	113	47	209	145	467	238	209	78	998	507
900M	56	24	141	82	47	45	19	12	263	163
1200M	64	5	64	62	32	5	0	0	161	72
Sum of column	3923	1734	8132	4140	2715	1258	904	325	15673	7457

* Numbers are expressed as No/1000T/D/10000fls. ** See text.

Table 8. Average population of released males* in each annulus at each week in Test 2.

Annulus	7/13 T-1	7/13 T-2**	7/20 T-1	7/20 T-2	7/27 T-1	7/27 T-2	8/3 T-1	8/3 T-2	Sum of row T-1	Sum of row T-2
300M	3690	1658	7717	3851	2169	970	676	236	14251	6714
600M	338	140	628	435	1400	715	628	233	2994	1522
900M	282	121	704	408	235	227	94	60	1314	815
1200M	448	36	451	435	225	36	0	0	1124	507
Sum of column	4757	1955	9500	5129	4029	1947	1398	529	19683	9559

* Numbers are expressed as No/1000T/D/10000fls. ** See text.

tively. The dispersal pattern and the decline in survival rate of T-2-1 were not different remarkably from those of T-2-2 (Tables 9 and 10).

Table 9. Relative density of released males* in each annulus at each week in Test 2.

Annulus	7/13 T-1	7/13 T-2**	7/20 T-1	7/20 T-2	7/27 T-1	7/27 T-2	8/3 T-1	8/3 T-2	Average T-1	Average T-2
300M	94	96	95	93	80	77	75	72	91	90
600M	3	3	3	4	17	19	23	24	6	7
900M	1	1	2	2	2	4	2	4	2	2
1200M	2	0	1	2	1	0	0	0	1	1

* Numbers are expressed as % of total density at each week. ** See text.

Table 10. Relative abundance of released males* in each annulus at each week in Test 2.

Annulus	7/13		7/20		7/27		8/3		Average	
	T-1	T-2**	T-1	T-2	T-1	T-2	T-1	T-2	T-1	T-2
300M	78	85	81	75	54	50	48	45	72	70
600M	7	7	7	8	35	37	45	44	15	16
900M	6	6	7	8	6	12	7	11	7	9
1200M	9	2	5	8	6	2	0	0	6	5

* Numbers are expressed as % of total population at each week. ** See text.

Table 11. Number of released males* caught in each trap at each week in Test 1.

Annulus	Date / Trap no.	6/14 T-1	6/14 T-2**	6/21 T-1	6/21 T-2	6/28 T-1	6/28 T-2	7/5 T-1	7/5 T-2
300 M	1	67	54	335	218	201	54	335	218
	2	1207	490	15759	7236	1006	1034	268	163
	3	1006	1088	1954	2122	402	272	67	54
	4	5767	5277	4493	3101	805	707	268	163
600 M	5	201	109	201	109	268	54	67	0
	6	268	218	-	-	134	54	0	0
	7	-	-	1006	1251	67	272	67	0
	8	0	0	67	0	0	0	0	54
	9	134	326	738	381	268	109	0	0
	10	201	218	939	218	134	0	0	0
	11	134	54	201	381	67	54	0	0
900 M	12	0	0	0	0	0	0	0	0
	13	335	163	1475	272	201	109	201	0
	14	268	0	604	381	201	0	0	0
	15	0	109	67	163	134	0	0	0
	16	0	54	0	0	67	54	0	0
	17	0	0	536	109	67	0	0	0
	18	134	109	671	272	469	109	0	0
	19	0	0	67	54	469	218	134	163
	20	201	54	604	163	67	54	0	0
	21	0	0	0	0	67	0	0	0
	22	0	218	671	326	201	0	0	0
	23	201	0	268	163	67	0	0	0
1200 M	24	67	0	0	109	67	0	0	54
	25	268	0	536	163	67	54	67	54
	26	134	54	0	0	67	0	0	0
	27	0	0	0	54	201	54	0	0
	28	134	0	201	54	67	54	0	0
	29	0	0	201	54	0	0	0	0
	30	0	0	0	0	0	0	0	0

* Numbers are expressed as No/1000T/D/10000fls. ** See text.

Comparison between Test 1 and Test 2. Tables 11 and 12 show the number of released males caught in each trap in each annulus at each week in the Tests 1 and 2, respectively. The numbers were calculated in the No/1000T/D/10000fls. In the tables order of the traps from top to bottom shows nearness of the trap to the release site (Fig. 1). It did not always follow that within the same radius the catches of the released males

Table 12. Number of released males* caught in each trap at each week in Test 2.

Annulus	Date	7/13		7/20		7/27		8/3	
	Trap no.	T-1	T-2**	T-1	T-2	T-1	T-2	T-1	T-2
300 M	1	1127	435	1352	761	2816	870	1239	326
	2	11379	5038	25348	12432	4844	2247	1014	326
	3	901	507	2478	1124	338	471	338	109
	4	1352	652	1690	1087	676	290	113	181
600 M	5	225	36	338	145	338	326	113	0
	6	0	36	0	0	113	109	0	36
	7	225	72	225	507	1465	834	1014	399
	8	0	0	0	0	113	36	0	0
	9	225	72	113	72	563	181	113	72
	10	0	36	0	36	113	0	113	36
	11	113	72	789	254	563	181	113	0
900 M	12	0	0	0	0	0	36	0	0
	13	113	72	113	109	113	72	0	36
	14	113	36	0	72	113	0	0	36
	15	0	36	225	36	0	217	113	0
	16	0	0	0	0	0	0	0	0
	17	113	36	113	36	113	36	113	0
	18	113	36	338	36	225	0	0	0
	19	113	36	563	580	0	36	0	36
	20	0	0	0	0	0	145	0	36
	21	0	0	0	0	0	0	0	0
	22	113	36	225	109	0	o	0	0
	23	0	0	113	0	0	0	0	0
1200 M	24	0	0	0	0	0	0	0	0
	25	0	0	225	254	113	36	0	0
	26	113	0	0	36	0	0	0	0
	27	225	36	113	109	113	0	0	0
	28	113	0	113	36	0	0	0	0
	29	0	0	0	0	0	0	0	0
	30	0	0	0	0	0	0	0	0

* Numbers are expressed as No/1000T/D/10000fls. ** See text.

depend on nearness of the trap to the release site. On the other hand, the average catches (Tables 3 and 7) and the relative abundances (Tables 4 and 8) of each annulus of the flies decreased as the distance between the trap and the release site increased.

Tables 13 and 14 show the relationship between the released male catches and the wild male catches. Within the same annulus, the tendencies that more released males were caught in

Table 13. Total number* of released males caught in each trap and average grade of male catches of each trap in Test 1.

Annulus	Trap no.	Total no. of males caught		Average grade of wild male catch
		T-1**	T-2	
300 M	1	939	544	1.8
	2	18240	8923	4.3
	3	3420	3536	4.0
	4	11333	9249	4.8
600 M	5	738	272	3.0
	6	402	272	3.0
	7	1140	1523	4.0
	8	67	54	1.0
	9	1140	816	4.0
	10	1274	435	5.0
	11	402	490	4.5
900 M	12	0	0	2.3
	13	2213	544	3.8
	14	1073	381	4.3
	15	201	272	4.8
	16	67	109	2.3
	17	604	109	3.8
	18	1274	490	4.0
	19	671	435	4.0
	20	872	272	5.0
	21	67	0	1.5
	22	872	544	4.8
	23	604	163	4.0
1200 M	24	134	163	2.5
	25	939	272	4.3
	26	201	54	3.8
	27	201	109	3.7
	28	402	109	4.8
	29	201	54	4.5
	30	0	0	1.0

* Numbers are expressed as No/1000T/D/10000fls. ** See text.

the trap which graded high in wild catches than that graded low were observed (Tables 13 and 14).

Table 14. Total number of released males* caught in each trap and average grade of male catches of each trap in Test 2.

Annulus	Trap no.	Total no. of males caught		Average grade of wild male catch
		T-1**	T-2	
300 M	1	6534	2392	4.0
	2	42585	20043	5.0
	3	4056	2211	4.8
	4	3830	2211	4.5
600 M	5	1014	507	2.3
	6	113	181	3.5
	7	2929	1812	3.3
	8	113	36	1.8
	9	1014	399	3.8
	10	225	109	4.0
	11	1577	507	4.5
900 M	12	0	36	1.3
	13	338	290	4.8
	14	225	145	4.0
	15	338	290	4.3
	16	0	0	2.8
	17	451	109	3.5
	18	676	72	4.8
	19	676	689	4.8
	20	0	181	5.0
	21	0	0	2.7
	22	338	145	4.7
	23	113	0	4.5
1200 M	24	0	0	3.0
	25	338	290	4.0
	26	113	36	4.0
	27	451	145	3.5
	28	225	36	4.8
	29	0	0	5.0
	30	0	0	1.3

* Numbers are expressed as No/1000T/D/10000fls. ** See text.

Percentages relative abundance of the 300 M-annulus and 1200 M-annulus of T-1-1 and T-1-2 to the total of the relative abundance were 51-58 and 13-8, respectively (Table 6). Those of T-2-1 and T-2-2 were 72.4-70.2 and 5.7-5.3, respectively (Table 10). Percentages sum of average catches of the first half of the experiment period of T-1-1 and T-1-2 to the total catches were 82 and 85, respectively (Table 5). Those of T-2-1 and T-2-2 were 72 and 74, respectively (Table 9). Total of the average catches of each treatment in the Tables 3 and 7 can be used roughly as an index of accumulated density of the released males in the 300 M-annulus under the conditions where the flies were released in a grid arrangement, 600 m apart, and were released weekly when environmental conditions were identical during the period. The accumulated density would be a standard of the number of release males per unit area per unit time. No/1000T/D/10000fls/4ws/area of 1200 m-radius in the Tables 4 and 8 were used as an index of total catches of released males, on assumption as the number of released flies dispersed beyond 1200 m-radius from the release site and that of flies lived longer than 4 weeks were negligible. The test period should be changed with the effective accumulative temperature.

Here I listed two examples of the data obtained from the mark-release-recapture method and made brief comments. After analyzing the data more detail, the method will be improved to take data more efficiently. The data provided us the informations on the survival rate, dispersal and orientation ability to the places where wild flies were located, of factory produced flies in the field. Those data on the characters of the flies may be changed with environmental conditions even the flies have constant quality. Thus, repetition of the test throughout the year is requisite for bring into relief the performance of the flies in the field. That may be the way to construct the standard in the field to judge the ability of the mass-reared, irradiated, transported and released flies. The labors required for this work, about one assistant-160 hours per one experiment, is not large cost in the eradication project with SIT.

REFERENCES

Boller, E. F., Katsoyannos, B. I., Remund, O., and Chambers, D. L. 1981. Measuring, monitoring, and improving the quality of mass-reared Mediterranean fruit flies, Ceratitis capitata Wied. I. The RAPID quality control system for early warning. Zug. ang. Ent. 92: 67-83.

Iwahashi, O., Ito, Y., and Shiyomi, M. 1983. A field evaluation of the sexual competitiveness of sterile melon flies, Dacus (Zeugodacus) cucurbitae. Ecol. Entomol. 8: 43-48.

Kuba, H. 1980. Release of sterile flies to Kume Is. and its adjacent isles. Bull. Okinawa Pref. Special Pest Control Project. 5: 119-126. In Japanese.

Taniguchi, M. and Iwaizumi, R. 1984. Influence of sifting of pupae on adult eclosion and flight capability of the melon fly, Dacus cucurbitae Coq. (Diptera: Tephritidae) using a newly developed sifter. Bull. Okinawa Agric. Exp. Sta. 9: 107-111. In Japanese with English summary.

IV. MODELING AND ANALYSIS

INTERRELATIONS AND APPLICATIONS OF MATHEMATICAL DEMOGRAPHY TO SELECTED PROBLEMS IN FRUIT FLY MANAGEMENT

James R. Carey
Department of Entomology
University of California
Davis, CA 95616

ABSTRACT

I address 2 sets of demographic problems in this paper. The first group concerns the application of life table methods to statistical and arithmetic problems in tephritid ecology and management. Here I introduce the notion of the life table as a statistical tool and the concept of multiple decrement tables. I then establish connections between formal demography and Abbot's Correction, Key Factor Analysis and Probit Analysis. The second group of problems concerns estimating population growth rate, r, from population age structure or sex ratio if relative survival and relative trappability of 2 age groups or of each sex is known. Formulae are given for determining r using either the nulliparous:parous ratio or the sex ratio of captured adults. Eleven sample problems are interspersed in the text to clarify most of the main techniques or concepts. A general discussion of life tables and stable population theory is included at the end.

INTRODUCTION

Many insect ecologists and tephritid biologists view the field of demography as having only 2 uses--1)the life table as a format for collecting and summarizing age-specific life history information; and 2)the stable population model (Lotka's equation) for determining the intrinsic rate of population increase and other population parameters. This perspective of insect demography was set by 2 seminal papers published nearly 4 decades ago. The first of these was by Deevey (1947) on life tables of animal populations and the second was by Birch (1948) on the rate of increase in insect populations. These papers established the foundation of contemporary insect demography.

My objectives in this paper are to demonstrate that the scope of demography extends beyond that outlined by Deevey and Birch, that interconnections exist between demographic methods and those used in applications both inside and outside of ecology and that the field has practical and not just conceptual relevance. I will attempt to do this by

NATO ASI Series, Vol. G11
Pest Control: Operations and Systems Analysis
in Fruit Fly Management. Edited by M. Mangel et al.

addressing 2 different sets of problems. The first set is concerned with 2 forms of life tables--single and multiple decrement tables. Here I introduce the concept of life table variance in single decrement (conventional) life tables as given in Chiang (1960a,b; 1984) and provide an analytical framework for the multiple decrement table as modified from Keyfitz (1977) and Land and Rogers (1982). I then establish connections among these classical forms of tables and 3 analytical techniques currently viewed as standard tools in the field--Abbot's Correction, Key Factor Analysis and Probit Analysis. The second set of problems is concerned with estimating population growth rate from single age- or sex-structured field samples using stable population theory. These notions represent extensions of techniques derived for application to human populations which are given in Keyfitz (1977) and Keyfitz and Beekman (1984). Example problems are given to illustrate the major concepts.

LIFE TABLES AND RELATED TOOLS

Single Decrement Tables

The standard form of the life table as given in Deevey (1947) and elsewhere (eg. Keyfitz 1977; Pollard 1973; Chiang 1984) is called a single decrement table since it lumps all causes of death into one decrement. The parameters are typically organized in columns by age x. The major life table function is termed the l_x schedule which gives the probability of a newborn individual surviving to age x. Functions usually derived from this schedule include p_x which denotes the probability of surviving from age x to x+1 and is referred to as period survival, q_x which is the complement of p_x and denotes the probability of dying in the age interval x to x+1, d_x which gives the fraction of a cohort dying in the age interval x to x+1 and e_x which is the expectation of life at age x. The appropriate formulae for the functions are as follows:

$$p_x = l_{x+1}/l_x \quad (1a)$$

$$q_x = 1 - p_x \quad (1b)$$

$$= 1 - (l_{x+1}/l_x) \quad (1b')$$

$$= (l_x - l_{x+1})/l_x \quad (1b'')$$

$$d_x = l_x - l_{x+1} \quad (1c)$$

$$e_x = \sum_{y=x}^{\omega} l_y/l_x \quad (1d)$$

An hypothetical, single decrement life table example is given in Table 1.

Table 1. Hypothetical, single decrement life table showing the relationship among various parameters.

Age	Life Table Function				
x	l_x	p_x	q_x	d_x	e_x
0	1.0	.8	.2	.2	1.5
1	.8	.6	.4	.3	0.9
2	.5	.4	.6	.3	0.4
3	.2	.0	1.0	.2	0.0
4	.0	--	--	.0	--
	2.5			1.0	

This table helps to illustrate 2 additional life table relationships, both of which are important for understanding subsequent parts of this paper. First, the survival schedule, l_x, can be reconstructed from the p_x schedule using the following equation:

$$l_x = \prod_{y=0}^{x} p_y \tag{2}$$

Thus the probability of attaining age x (l_x) is equal to the product of the independent probabilities of surviving all previous age intervals. One of the practical reasons for understanding this relationship is concerned with constructing an l_x schedule without the need for observing a single cohort from age 0 through the age at which its last member dies. For example, every parameter given in Table 1 for the hypothetical cohort could have been obtained from 3 cohorts age 0, 1 and 2 by allowing each to 'age' over 1 time step. The number in each of the 3 cohorts at the end of this time step divided by the initial number yields the probability of surviving age x to x+1 (i.e. the parameter p_x). This approach is often referred to as the time-specific method (Southwood, 1971).

A second important relationship among life table functions involves the expectation of life at birth (i.e. at age=0). Note that the formula for expectation of life at age x (Eqn 1d) becomes:

$$e_o = (\sum_{x=0}^{\omega} l_x) - 1 \tag{3a}$$

$$= \sum_{x=1}^{\omega} l_x \tag{3b}$$

when x=0. Since survivorship at age 0 always equals unity (i.e. $l_o=1$) by definition, an alternative and more intuitive expression for e_o is:

$$e_o = \sum_{x=0}^{\omega} x d_x \qquad (3c)$$

where x is weighted by the fraction dying in the interval x to x+1, d_x. The equivalence of (3b) and (3c) can be shown as follows (Pollard, 1973):

$$e_o = \sum_{x=0}^{\omega} x d_x \qquad (4a)$$

$$= 0(l_o-l_1)+1(l_1-l_2)+2(l_2-l_3)+\ldots \qquad (4b)$$

$$= l_1-l_2+2l_2-2l_3+3l_3- \ . \ . \ . \qquad (4c)$$

$$= l_1+l_2+l_3+ \qquad (4d)$$

$$= \sum_{x=1}^{\omega} l_x \qquad (4e)$$

Therefore, the mean age of death in a cohort as given in Eqn(3c) is conceptually and computationally equivalent to the expectation of life at age x, e_x, and can be expressed as either Eqn (3a) or (3b). The reason for pointing out these relationships is to underscore the idea that a life table is a method for gathering statistical information on cohort mortality. It is felt that this aspect has been largely overlooked by ecologists and pest researchers possibly for the simply reason that the notation used in life tables is different than that used in conventional statistics. A comparison of notation and formulae for the mean and variance used the life table and those used in conventional statistics is given in Table 2.

Table 2. Notation and formulae for mean and variance in conventional statistics and life table statistics (f_i denotes normalized frequency in conventional and d_x denotes the normalized frequency at which death occurs at age x in life table).

Statistic	Conventional		Life Table	
	Symbol	Formula	Symbol	Formula
MEAN	$\bar{y}$	$\Sigma y_i f_i$	e_o	$\Sigma x d_x$
VARIANCE	σ^2	$\Sigma f_i(y_i-\bar{y})^2$	s^2	$\Sigma d_x(x-e_o)^2$
		or		or
		$\Sigma f_i(y_i)^2-(\bar{y})^2$		$\Sigma d_x(x)^2-(e_o)^2$

It may be noted that notational differences exist but that computational ones do not. The mean (expectation of life at x=0) of the hypothetical life table given in Table 1 can be computed using the formula given in Table 2 as

$$e_o = (.8)+(.5)+(.2) \quad \text{[from Eqn(3b)]}$$
$$= 1.5$$

or

$$e_o = .2(0)+.3(1)+.3(2)+.2(3) \quad \text{[from Eqn(4a)]}$$
$$= 1.5$$

and the variance of this mean computed as

$$= .2(0-1.5)^2+.3(1-1.5)^2+.3(2-1.5)^2+.2(3-1.5)^2$$
$$= 1.05$$

or

$$=[.2(0)^2+.3(1)^2+.3(2)^2+.2(3)^2] - (1.5)^2$$
$$= 1.05$$

This second method for computing variance uses the second equation for variance given in Table 2 and is often referred to as the computational form.

A statistical perspective of the life table is useful in several respects. This utility will be illustrated with 3 examples using data from previous tephritid life history studies.

Example 1. In medfly studies conducted in Greece at 25°C (Carey, 1984), I found that the proportion of 35 females that survived for 3 months was 0.06 and the proportion of 25 males that survived this same period was 0.35. Question: Was female survival over this period significantly different than male survival?

This question can be addressed by first noting that the proportion of females and males surviving each constitute probabilities (see Chiang, 1984), denoted p_f(=.06) and p_m(=.35), respectively, with complements of q_f(=.94) and q_m(=.65). Therefore, the respective variances for the female and male cohorts are $p_f q_f/n_f$ and $p_m q_m/n_m$, where n_f(=35) and n_m(=25) denote the initial numbers. The null(H_o) and alternative(H_1) hypotheses are:

H_o: $p_m = p_f$ (no survival differences)

H_1: $p_m > p_f$ (male survival greater)

The statistical significance of the sample difference $(p_m - p_f)$ is determined by computing the probability

$$Pr\{Z > (p_m - p_f)/S.D.(p_m - p_f)\}$$

where Z is the normal distribution for a dichotomous random variable with mean p and variance pq/n where q=(1-p). The denominator in this equation is given as

$$S(p_m - p_f) = [(p_m q_m / n_m) + (p_f q_f / n_f)]^{1/2}$$

which is the standard deviation of the difference of two variables. Therefore

$$S(p_m - p_f) = [((.65)(.35)/25)+((.94)(.06)/35))]^{1/2}$$
$$= 0.103$$

Thus

$$Pr\{Z > (.35 - .06)/.103\}$$
$$Pr\{Z > 2.816\} = 0.01$$

from the normal table. Therefore, we can reject the null hypothesis and state that male medfly survival to 90 days is significantly greater than female survival over this same period at the 99% level of significance.

Example 2. Expectation of life at eclosion (e_x) of a wild strain of 35 female medflies studied by Carey (1984) was 65.1 days and of a laboratory strain of 25 females studied by Shoukry and Hafez (1979) was found by Carey (1982) to be 32.9 days. Both investigations were conducted at 25°C. Questions: i)What are the 95% confidence intervals of e_x for the 2 strains? ii) What fraction of all deaths for each strain occurs before 10 days? Before 50 days?

By re-analyzing the data presented in the 2 studies and using the life table variance equation given in Table 2, I found that the variance of the wild strain was 102.4 (S=10.1) and of the laboratory strain was 134.6 (S=11.6). The problems can thus be addressed as follows:

i) The 95% confidence intervals (CI_{95}) of the expectation of life at eclosion for each strain is given by

$$CI_{95} = e_o \pm 1.96S/(n)^{1/2}$$

where n denotes the number of trials (individuals). Therefore the confidence intervals for the wild strain are

$$CI_{95}(\text{wild}) = 65.1 \pm 1.96(10.1)/(35)^{1/2}$$
$$= 65.1 \pm 3.4$$

and for the laboratory strain are

$$CI_{95}(\text{lab}) = 32.9 \pm 1.96(11.6)/(25)^{1/2}$$
$$= 32.9 \pm 4.6$$

Therefore we can state that we are 95% confident that the expectation of life at eclosion of the wild medfly strain falls between the ages of 61.7 and 68.5 days and of the laboratory strain between the ages of 28.3 and 37.5 days.

ii) This question calls for the probabilities

$$\Pr\{x < 10\} \quad \text{and} \quad \Pr\{x < 50\}$$

Converting x to standard normal variables and consulting a normal distribution table yields

$$\Pr\{Z<[(10-65.1)/10.1] \quad \text{and} \quad \Pr\{Z<[(50-32.9)/10.1]$$
$$\Pr\{Z < -5.46\} = 10^{-5} \qquad \Pr\{Z < -1.50\} = .07$$

for the wild strain and

$$\Pr\{Z<[(10-32.9)/11.6] \quad \text{and} \quad \Pr\{Z<[(50-32.9)/11.6]$$
$$\Pr\{Z < -1.97\} = .03 \qquad \Pr\{Z < 1.47\} = .93$$

for the laboratory strain. Therefore if death rates in both cohorts are approximately normally distributed, less that 0.00001 of the wild strain would be expected to die in the first 10 days but about 3% of the laboratory strain would. Also 93% of the wild strain would be expected to die after age 50 but only 7% of the laboratory strain would be expected to survive to this age.

Example 3. Medflies often tend to lay fewer eggs/clutch in small hosts such as coffee than in large ones such as peach (eg. McDonald and McInnis, 1985). Suppose that the average clutch size(s) laid in a small host was s=4 with an egg to adult survival(p) of p=.80 and the average clutch size in a large host was s=13 with an average survival of p=0.65. Questions: i)What is the probability of at least 1 of each sex emerging as adults from the respective hosts? and ii)How much would mortality have to be increased to reduce this probability to 0.001?

i)This demographic question is concerned with 2 different probabilities--the chance of an egg of either sex surviving to adulthood and the chance that an eclosing adult will be either a male or a female. The first part of the problem is basically concerned with determining the probability of obtaining either all males or all females from a sample of n adults. Since the probability of obtaining all females from a randomly-chosen sample of size n is $(.5)^n$ and that this is also equal to the probability of obtaining all males, then the probability of obtaining either all males or all females from n adults is $2(.5)^n$. Therefore, the probability of <u>not</u> obtaining all of either 1 sex or the other, denoted p_{f+m}, is the complement of this term. That is

$$p_{f+m} = 1 - 2(.5)^n$$

The number of adults, n, is the product of eggs/clutch and survival, thus the correct equation is

$$p_{f+m} = 1 - 2(.5)^{(s)(p)}$$

The probabilities of obtaining at least one of each sex from the original question are

$$\begin{aligned} p_{f+m} &= 1 - 2(.5)^{(4)(.8)} \\ &= 1 - 2(.109) \\ &= 0.782 \qquad \text{(small host)} \end{aligned}$$

$$\begin{aligned} p_{f+m} &= 1 - 2(.5)^{(13)(.65)} \\ &= 1 - 2(.003) \\ &= 0.994 \qquad \text{(large host)} \end{aligned}$$

Thus the probability of at least one of each sex eclosing from the small host is slightly over 78% and that for the large host is over 99% (see Landolt et al., 1984 for broader implications).

ii) The answer to this problem is obtained by solving the equation

$$.001 = 1 - 2(.5)^{(13)(.65)x}$$

for x to determine how much the product of clutch size and natural survival must be reduced to insure that only 1 out of 1000 times will both sexes eclose. The value of x can be determined by rearranging and reexpressing the equation as

$$\begin{aligned} 2(.5)^{(8.45)x} &= 0.9990 \\ (.5)^{(8.45)x} &= 0.4995 \\ &= e^{(8.45)(\ln .5)x} \\ \ln(.4995) &= (8.45)(\ln .5)x \\ x &= [\ln(.4995)]/(8.45)(\ln .5) \\ &= 0.119 \end{aligned}$$

Therefore either survivorship or clutch size must be reduced by a factor of 0.119 to insure that only 1 chance out of a thousand will both sexes eclose from the same clutch.

<u>Multiple Decrement Tables</u>

Multiple decrement life table theory is based on the assumption that multiple causes of death act independently and is concerned with the probability that an individual will die of a certain cause in the presence of other causes. The concept itself stems from reliability theory in operations research. The parallel between a physical system and a biological system may be noted by viewing both as consisting of components and that each of these parts possesses a survival schedule. For example, Keyfitz (1977) points out that a watch can operate only as long as all the parts are functioning and that each part has its own life table. The probability that the watch will continue to operate at a given age will be the product of the probabilities that each of the components will 'survive' to that age. Similarly, living vertebrates possess a number of different organs (eg. heart, lungs, etc.) which can also have life tables. The organism dies at the time when any of its component organs dies. The conceptual connection between these internal biological and physical components should be clear.

The same notion of probabilities applied to internal components causing the death of a system can also be applied to external components such as disease and accidents in humans or predation and parasitism in insects. The concept here is that the probability of an insect surviving a certain age (or stage) is the product of all probabilies related to specified causes.

Consider a cohort of individuals subject to i causes (i=1,2,...,k) of death over their lifetime (up to age ω). If $p_{x,i}$ denotes the probability that an individual age x will survive cause i then the probability that an individual will survive to age x+1 subject to all causes will be

$$P_x = \prod_{x=1}^{k} p_i \qquad (5)$$

where P_x denotes this probability. The probability that an individual will attain age x, l_x, can then be computed as

$$l_x = \prod_{y=0}^{x} P_y \qquad (6)$$

This expression corresponds to Eqn. (2) in the single decrement table. The general framework for an age- or stage-specific, multiple decrement life table is given in Table 3.

Table 3. Age-specific, multiple decrement life table model where $p_{x,i}$ denotes the survival probability of an individual from age x to age x+1 subject to cause i (i=1,2,...k) where ω denotes the oldest possible age.

Age(x)	Survival Probability			p_x	l_x
	$p_{x,1}$	$p_{x,2}$	$p_{x,k}$		
0	$p_{0,1}$	$p_{0,2}$	$p_{0,k}$	p_0	l_0
1	$p_{1,1}$	$p_{1,2}$	$p_{1,k}$	p_1	l_1
2	$p_{2,1}$	$p_{2,2}$	$p_{2,k}$	p_2	l_2
.	.	.	.	.	.
.	.	.	.	.	.
ω-2	$p_{\omega-2,1}$	$p_{\omega-2,2}$	$p_{\omega-2,k}$	$p_{\omega-2}$	$l_{\omega-2}$
ω-1	$p_{\omega-1,1}$	$p_{\omega-1,2}$	$p_{\omega-1,k}$	0	$l_{\omega-1}$
ω	--	--	--	--	0

Examples of the use of the multiple decrement table are given in the following two sections.

<u>Abbot's Correction</u>

A problem encountered in determining the effectiveness (eg. kill or sterility rate) of a chemical, radiation or temperature treatment in insect bioassay is separating out treatment-induced effects from effects that would have occurred in the absence of the treatment. For example, a typical insecticide bioassay design involves replicates of cohorts subjected to the test chemical as well as replicates of cohorts not subjected to the chemical (i.e. check or control). After a predetermined period (i.e. 24-48 hrs), the proportion which are dead in each of the 2 sets of cohorts is recorded. The arithmetic problem is concerned with adjusting for the natural mortality which may have occurred in the treatment in order to isolate the "pure" effectiveness of the chemical.

Abbot (1925) derived the original formula for natural mortality adjustment in bioassay by introducing the concept of an "effective kill rate" (EKR) of a chemical. This formula is (notation different than Abbot's):

$$EKR = (P_{nat} - P_{trt})/P_{nat} \qquad (7)$$

where P_{nat} denotes the proportion alive in the check, P_{trt} denotes the proportion alive in the treatment and $(P_{nat}-P_{trt})$ gives the proportion dead due to the treatment alone. This adjustment is known as Abbot's Correction in toxicology and is accepted as convention. It should be noted that the implicit assumption in determining EKR is that the 2 probabilities of dying in the treatment are independent.

My purpose for addressing Abbot's Correction is to demonstrate that the concept is nothing more than that of a multiple (double) decrement life table over a single time (age) interval. The major conceptual difference between Abbot's approach and the demographic approach is that of derivation perspective--Abbot's concerns mortality probabilities and the demographic focuses on survival probabilities.

The starting point for the re-derivation of Abbot's EKR is to view the probability of surviving a treatment (P_{trt}) as conditional upon the product of the probability of surviving natural mortality (P_{nat}) and the probability of surviving the chemically-induced mortality (P_{chem}). These correspond to i causes (i=1,2) in multiple decrement theory (Table 3). Thus

$$P_{trt} = (P_{nat})(P_{chem}) \quad (8a)$$

The proportion remaining alive in the treatment plot or cage equals P_{trt} and the proportion remaining alive in the check equals P_{nat}. Therefore, the probability of surviving the chemical in the absence of natural mortality is found by rearranging Eqn. (8a) as

$$P_{chem} = (P_{trt})/(P_{nat}) \quad (8b)$$

Since P_{chem} denotes the probability of surviving the chemical in the absence of natural mortality, its complement represents the probability of dying due to the chemical alone. This corresponds to Abbot's EKR. Thus

$$EKR = 1 - P_{chem} \quad (9a)$$

$$= 1 - (P_{trt})/(P_{nat}) \quad (9b)$$

$$= (P_{nat}-P_{trt})/P_{nat} \quad (9c)$$

Note that Eqn. (8c) is identical to Abbot's original formula for EKR which is given by Eqn. 6 and that Eqns. 8a, 8b and 8c also correspond with Eqns. 1b, 1b' and 1b'', respectively.

Two points of this re-derivation merit comment. First, by shifting emphasis from mortality rate to survival rate the derivation becomes more mathematically transparent. Second, by establishing the parallel between Abbot's Correction and multiple decrement life table analysis the purpose of both becomes more apparent. Moreover, the multiple time and multiple treatment version of Abbot's Correction becomes the multiple decrement life table.

Example 4. Ebeling (1953; Table 13) reported that of a cohort of 1000 melon flies treated with 0.2 percent DDT, 76, 71 and 68% remained alive after 2, 5 and 15 days, repectively, while 96, 93 and 91% of the checks were alive after these respective periods. Questions: i)What is the effective kill rate of DDT over the entire period?; and ii)What are these rates over each successive period?"

i)This question can be answered by noting that p_{trt}=.68 and p_{nat}=.91. Therefore

$$p_{chem} = p_{trt}/p_{nat}$$
$$= .68/.91$$
$$= .75$$

Thus the effective kill rate (EKR) is

$$EKR = 1 - p_{chem}$$
$$= 1 - .75$$
$$= .25$$

ii)This problem is most easily addressed in the form of a multiple decrement table by computing the p_{chem} (Abbot's Correction) and p_{nat} for each day. The results are

	Probability			
Day(x)	p_{chem}	p_{nat}	p_x	l_x
0	.79	.96	.76	1.00
2	.97	.97	.94	.76
5	.98	.98	.96	.71
15	--	--	--	.68

Therefore, the effective kill rate for the period 0-2 days is (1-.79)=.21, for 3-5 days is (1-.97)=.03 and for 6-15 days is (1-.98)=.02. Additional insight into these data may be gained by converting these effective kill rates over these unequal time periods to average daily effective kill rates. This is done by taking the root of p_{chem} corresponding to the period of observation to obtain average daily survival probabilities and then substracting this value from 1.0 to get daily mortality (kill) rates. For example, the square root (i.e. 2 day period) of p_{chem}(=.79) for the 0-2 day period is 0.89 and its complement (daily effective kill rate) is 0.11; the cube root of p_{chem}(=.97) for the 3-5 day period is 0.99 and its complement is 0.01; and the 9th root of p_{chem}(=.98) for days 6-15 is 0.998 and its complement is 0.002. Hence the average daily EKR for the first 2 days after exposure is around 10-fold greater than the daily EKR for the next 3 days and 50-fold greater than the daily EKR from 6-15 days after initial exposure.

Key Factor Analysis

Varley (1947), Morris (1959) and Varley and Gradwell (1965) all contributed to the notion of partially our mortality factors of an insect population over a single generation. This technique has come to be known as Key Factor Analysis (coined by Morris, 1959). The idea here is that insects experience multiple causes of death both within and between stages and that one of the objectives in ecology is to identify one or a small number of the major (key) determinants (eg. predation, parasitism, etc.) of population mortality.

The basic mathematical framework for Key Factor Analysis as described by Varley and Gradwell (1960; 1963a,b; 1965) involves essentially 3 parts:

1. Mortality induced by the ith factor, denoted k_i, is calculated by subtracting each $\log_{10}$ population from the previous one. That is

$$k_i = \log_{10}N(i) - \log_{10}N(i+1) \qquad (10a)$$

2. The "killing power" of the ith factor is expressed as a factor of 10 as

$$10^{k_i} = \text{killing power} \qquad (10b)$$
$$= N(i)/N(i+1)$$

3. The "total killing power" of all factors over an entire generation is expressed as

$$10^K = \text{"total killing power"} \qquad (10c)$$

where $K = \Sigma k_i$. The factor k_i which has the largest value is deemed the 'Key' factor.

My objective here is to establish the mathematical connections between Key Factor Analysis and conventional life table analysis. This is most easily done simply by reproducing an example of Varley and Gradwell's analysis and establishing the relationship of the conventional parameters l_x and p_x (Table 4).

Table 4. Relationship of Varley & Gradwell's analysis (from Southwood 1971; p302) with conventional life table parameters.

	Varley and Gradwell's			Conventional	
STAGE and k-factor	N	$\log_{10}N$	k's	l_x	p_x
Potential natality	1500	3.176		1.00	.84
k_o(egg reduction)			.076		
Eggs laid	1260	3.1000		.84	.68
k_1(egg loss)			.171		
Eggs hatching	850	2.929		.57	.32
k_2(predation)			.498		
3rd instars	270	2.431		.18	.61
k_3(parasitism)			.201		
Larv. surv. parasite	170	2.230		.11	.55
k_4(predation+)			.276		
Pupae	90	1.954		.06	.50
k_5(overwintering)			.301		
Adults emerging	45	1.653		.03	.33
k_6(dispersal)			.352		
Adults reproducing	20	1.301		.01	--
			K=1.875		

Three relations should be apparent from this table. First, the number surviving to a given stage can be expressed as a fraction of the original (potential) cohort. This is the l_x column. For example, the fraction surviving to the pupal stage is (90/1500)=.06. Second, the inverse of 10^k equals p_x. That is

$$(N_{i+1}/N_i)=(10^k)^{-1}$$
$$=(l_{x+1}/l_x)$$
$$=p_x$$

For example, the inverse of 10 to the k_1-power is $(10^{.171})^{-1}=0.68$ which is equal to the corresponding p_x-value given in Table 4. Third, the inverse of total generation "killing power", 10^K, equals the total generation survival (i.e. l_x for reproducing adults). That is

$$(10^K)^{-1} = (10^{1.875})^{-1}$$
$$= 0.01$$

which is the fraction of the original cohort that survives to become adults.

An overriding assumption in Key Factor Analysis is that each type of mortality acts in succession (Varley et al. 1973). This assumption is obviously valid if only one mortality factor affects each successive stage

However, it is not valid when 2 or more factors affect the same stage. because they present themselves as simultaneous and not sequencial risks. Therefore, a point that appears not to have been previously brought out in the literature on this method of analysis is that the decision by the analyst on which factor comes first in the k-factor sequence within a stage can drastically affect their values. For example, consider k_3(parasitized larvae) and k_4(larvae surviving parasitization which then died of predation and other causes) in Table 4. These factors together accounted for 180 larval deaths (i.e. 270-90) of which k_3 accounted for 100(i.e. 270-170) and k_4 for 80(i.e. 170-90). If predation instead of parasitization was considered as the first mortality factor affecting larval survival, the respective k-values would be

$$k\ (\text{predation}) = \log_{10}270 - \log_{10}190 = 0.153$$

$$k\ (\text{parasites}) = \log_{10}190 - \log_{10}90 = 0.325$$

Hence the k-value for parasitization increases from .201 to .325 and the k-value for predation decreases from .276 to .153--a complete reversal from the original analysis concerning which factor affected larval survival the most. An alternative approach to Key Factor Analysis is the multiple decrement life table and will be illustrated with the following example.

Example 5. Cameron and Morrison (1977) investigated mortality of *Rhagoletis pomonella* in Quebec apple orchards and analyzed the data using Key Factor Analysis. Questions: i)How would one construct a multiple decrement table from their data?; and ii)What level of generation mortality would exist if various mortality factors were eliminated?

i)Cameron and Morrison analyzed 11 k-factors--k_1 for egg hatch, k_2-k_3 for larval mortality and k_6-k_{11} for pupal and emergence mortality. A multiple decrement table can be made by translating their k-factors into survival probabilities for each stage subject to a given cause. A detailed table would consist of 3 rows (1 for each preadult stage) and 11 columns (1 for each cause). A simplified version is given in Table 5 where certain causes of death are lumped into a general category. For example, nutrition, dessication and persistent diapause are lumped into 'natural causes' by stage and all predator species are viewed as one.

Table 5. Analysis of mortality in *Rhagoletis pomonella* using stage specific, multiple decrement life table model applied to K-factor Analysis. Data from Cameron and Morrison(1977).

	Survival Subject to Specified Cause (i)					
Age(x)	Natural (i=1)	Predation (i=2)	Parasitism (i=3)	Disease (i=4)	p_x	l_x
Egg(x=1)	.98	1.00	1.00	1.00	.98	1.00
Larva(x=2)	.32	.44	.25	1.00	.04	.98
Pupa(x=3)	.96	.49	1.00	.95	.45	.03
	.30	.22	.25	.95	.02	.02

ii)Table 5 shows that .30, .22, .25 and .95 of the total generation would survive if natural, predator-induced, parasite-induced and disease-induced mortality were each isolated causes. That is, if all other causes were eliminated except the one in question. This table also shows that larval survivorship subject to all causes is only .04 but if larval parasitism was eliminated ($p_{2,3}$=.25), larval survival would increase by 3.5-fold to .14 and generation survival would increase from .02 to .06 or by 3-fold. This is in contrast to the interpretation of Key Factor Analysis where the elimination of a mortality factor which allows only a quarter of those entering a stage to survive would imply a 4-fold increase if this cause were eliminated. This discrepancy between multiple decrement theory and Key Factor Analysis is a consequence of the difference in assumptions--sequential vs simultaneous mortality for causes within a stage.

Probit Analysis

The pioneer analyst of quantal response in bioassay was Finney (1964) who is best known for developing the technique known as Probit Analysis. The basic notion for this method can be described as follows. An assumption is made that mortality of organisms increases arithmetically as the dose of a toxicant to which they are subjected increases geometrically. Hence, by transforming arithmetic dose to $\log_e$ dose, the percent mortality per increment of $\log_e$ dose is normally distributed (approximately) and the proportion dying is equal to the area under the normal density curve below a given dose. This is referred to as a "probit" (=probability unit). The $\log_e$ dose at which 50% mortality occurs (eg. LD_{50}) is set at Probit 5. This dose plus 1 standard deviation is Probit 6 (e.g. LD_{68}), plus 2 standard deviations is Probit 7 (eg. LD_{95}), and so forth.

The specific purpose of this section is to demonstrate that the statistics of the life table are identical to those of Probit Analysis if age is simply viewed as a dose of time (see also Brass 1971; Hill and Trussell 1977). The connections between Probit Analysis and life table analysis can be established as follows:

1. Life table and Probit Analysis are both concerned with the exact same quantal response (i.e. 0 and 1 for death and survival, respectively).
2. The d_x schedule of the life table is the exact analog of the normalized distribution of deaths per increment of dose in Probit Analysis. For example, all deaths that occur in a series of bioassays covering a range of toxicant doses are normalized so that their sum equals unity. This establishes the frequency distribution of deaths in a hypothetical cohort. When this distribution is plotted against the $\log_e$-transformed dose, it symmetrically (normally) distributed.
3. The $\log_e$ dose vs mortality curve in bioassay is typically presented as the cumulative normal distribution which represents cumulative mortality with increasing dose. The l_x parameter of the life table is the exact complement of this curve. Thus the curve ($1-l_x$) vs age is conceptually identical with the mortality vs dose relationship in bioassay.
4. It is assumed in Probit Analysis that the $\log_e$ dose at which 50% mortality occurs (LD_{50}) represents both the mean dose and the median dose. The expectation of life at birth (e_o), as noted earlier, can be redefined as the exact mean age of death at birth of the cohort and usually is quite close to the median age of death (age at which l_x=.5). Hence, zero dose corresponds with age zero (birth) in a cohort and the LD_{50} of Probit Analysis corresponds exactly with the expectation of life at birth of the life table. Therefore, the parameter, e_o, can be reinterpretted as the lethal 'dose' of time at which 50% of a cohort is dead.

5. The number of standard deviations (see Table 2) added to e_o in a life table will give the area under the normal curve and the corresponding expected level of cohort mortality. For example, e_o+1SD represents 68% mortality, e_o+2SD represents 95% mortality, and so forth. This is identical with Probit Analysis.

The notion that age and dose are demographically interchangeable suggests the possibility that other life history phenomena could be substituted for age or for the dependent variable, death rate. This is the case if death is viewed in the abstract as simply the exit of an individual from its current (alive) dichotomous state or if age (or dose) is viewed metaphorically as representing a discretization of a continuous life history phenomenon (see Hoem and Jensen, 1982). Consider the following 2 examples. A tephritid larva is subject to 2 competing 'risks' of exiting the larval state. The first is dying which is a literal risk while the second is pupating which is a figurative risk. A pupation schedule by day for all ultimate survivors is identical to a survival (l_x) schedule in which survival is viewed as remaining in the current (alive larval) state. Therefore, the complement of this schedule ($1-l_x$) represents cumulative pupation by age (day) x and the d_x schedule represents the frequency distribution of pupation by day. All other parameters that were given in the section on single decrement life tables apply as well.

A second example among innumerable conceivable ones concerns exchanging age for reproductive level. A tephritid female that lives to, say, 10 days also must obviously have lived to day 9, to day 8 and so forth. Likewise, a female that lives to lay its 100th egg must also have lived to lay its 99th, it 98th, etc. Therefore, it is possible to view death rate as a function of 'egg class' rather than only as a function of age class (or dose). Consider the following examples.

Example 6. Back and Pemberton (1916; Table 1) examined the effect of cold temperatures upon medfly eggs by refrigerating them within a host and removing subsamples on preselected days. The data were presented as the number of eggs remaining alive vs the number under observation for the given period. For example, egg trials at 0^oC for the first 4 out 15 days revealed that all of the 81 eggs under observation for 1 day exposure hatched, 520 out of 528 eggs hatched after a 2-day exposure, 135 out of 150 hatched after a 3-day exposure and 216 out of 336 eggs hatched after a

4-day exposure. Questions: i)How would one construct a life table from these data?; ii)What are Probits 5-9?; iii)What are the 95% confidence limits for the proportion dead after 4 days of cold exposure?

i)These data simply give the probability of an individual surviving to day x at 0^oC. Therefore, l_0=1.00 (by definition), l_1=1.00 (=81/81), l_2=.98 (=520/528), l_3=.90 (=135/150), l_4=.64 (=216/336) and so forth. The d_x schedule can then be constructed from this l_x curve and the mean and standard deviation computed.

ii)The expectation of life and life table variance from Back and Pemberton's data was found to be e_o=3.61 days and S=0.71 using the formulae given in Table 2. Adding 1 standard deviation, S, to e_o for every probit over Probit 5 yields

PROBIT	Days at 0^oC	LD
5	3.61	LD_{50}
6	4.32	LD_{68}
7	5.03	LD_{95}
8	5.74	$LD_{99.9}$
9	6.45	$LD_{99.997}$

iii)The proportion of the 336 eggs that survived the 4 day cold exposure was 0.64(=l_4) thus the proportion which died was 0.36. The formula for a sample variance is

$$S^2 = q(1-q)/n$$

where q is the probability of dying(=.36) and n is the number observed (=336). Therefore the variance and standard deviation for these egg data are

$$S^2 = (.36)(.64)/336$$
$$= 0.00068$$
$$S = .0261$$

The 95% confidence intervals for the proportion dying after the 4-day exposure are

$$CI_{95} = q \pm 1.96S$$
$$= .34 \pm .05$$

Thus we conclude with 95% confidence that the probability of a medfly egg dying after 4 days exposure to 0^oC is between 0.29 and 0.39.

Example 7. Larval developmental rates of a Hawaiian laboratory strain of medflies seeded on unlimited quantities of each of 2 hosts are given in Fig. 1 (D. Krainacker, unpub.). The total number of pupae recovered from the initial cohorts of newly-hatched 1st instars was 162 and 356 for peach and papaya, repectively. Questions: i) What are the equivalents of Probits 5-9 for these data? ii)Is developmental time for larvae reared on peach significantly different than those reared on papaya?

i)The first step required for answering this question is to determine both the expectation of 'life' (mean age of pupation) and the standard deviation (S) of this mean. These were determined from Krainacker's original data using the formulae given in Table 2. The statistics for papaya are: e_o=8.68 days, S=1.16; and for peach are: e_o=10.67 days, S=1.50. Thus the corresponding probits are

PROBIT	HOST		Pupated
	Papaya	Peach	
5	8.68	10.67	50%
6	9.84	12.17	68
7	11.00	13.67	95
8	12.16	15.17	99.9
9	13.32	16.67	99.997

ii)To compare the developmental rates of medflies reared on 2 hosts, we test the null hypothesis that development is the same on peach as on papaya against the alternative that development on peach is longer. The test statistic is

$$Z = [(e_o(\text{peach}) - e_o(\text{papaya})]/S(\text{diff})$$

where S(diff) is the standard deviation of the difference of 2 means (see Chiang 1984; p55). The value of this statistic was determined to be 0.13. Therefore

$$Z = (10.67-8.68)/.13$$
$$= 7.47$$

which exceeds the critical value of $Z_{.99}$=2.33 at the 1% level of significance. We conclude that under laboratory conditions, development time of medflies reared on peach is longer than their development on papaya.

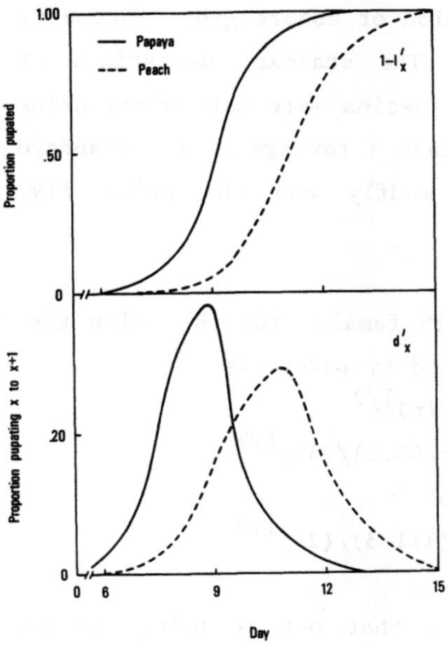

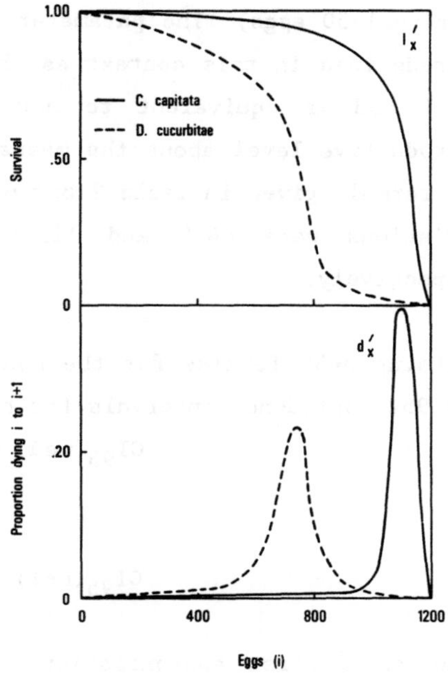

Fig. 1. Pupation rates of Mediterranean fruit flies, *Ceratitis capitata*, reared on 2 hosts given as cumulative ($1-l_x'$) and frequency (d_x') distributions (from D. Krainacker, unpub.).

Fig. 2. Survivorship of the Mediterranean fruit fly, *C. capitata*, and the melon fly, *Dacus cucurbitae*, expressed as the proportion living to 'egg class' i (l_x') and as the frequency distribution of death in the cohort from 'egg class' i to i+1 (d_x') (from Carey 1984 and Carey et al. 1985).

Example 8. Gross fecundity (total eggs laid in lifetime of an adult which lives to the last possible age) of both the medfly and the melon fly is approximately 1200 eggs/female but their net fecundities (reproduction weighted by survival) are different--1070 for the medfly and 710 for the melon fly (Carey 1984; Carey et al. 1985). The lower net fecundity of the melon fly is despite its much greater longevity. Questions: i)How is cohort mortality distributed with respect to cohort gross reproductive levels for the 2 species?; ii)What are the 95% confidence intervals for the net fecundities of both species?; and iii)What fraction of all females would be expected to have died when the cohort gross reproductive level is 1000 or less?

i)The data from the original studies were used to re-plot the relationships given in Fig. 2. Thus l_x' represents the probability of an individual surviving to cohort 'egg class' i (i=x) and d_x' represents the fraction of the cohort dying in the 'egg class' interval i to i+1 (interval=50 eggs). The parameter expectation of life at birth (eclosion) is redefined in this context as the expectation of cohort gross fecundity level and is equivalent to net fecundity. The standard deviations of reproductive level about the means for each species were determined using the formula given in Table 2 but using egg class i for age x. The standard deviations were 44.2 and 111.3 for the medfly and the melon fly, respectively.

ii)Using n=35 females for the medfly and n=25 females for the melon fly, the 95% confidence intervals for the net fecundity rates are

$$CI_{95}(\text{med}) = e_o+1.96S/(n)^{1/2}$$
$$= 1070+1.96(44.2)/(35)^{1/2}$$
$$= 1070+14.6$$
$$CI_{95}(\text{mel}) = 710+1.96(111.3)/(25)^{1/2}$$
$$= 710+43.6$$

Thus as a first approximation we can state that net fecundity of the medfly probably falls between 1055 and 1084 eggs/newly-eclosed female and for the melon fly between 666 and 754 eggs/newly-eclosed female. The intervals for the 2 species do not overlap hence we can be relatively certain that their net fecundities are significantly different.

iii)This question is identical to the one given in the second part of Example 2 and calls for the probability

$$Pr\{x < 1000\}$$

Converting x to standard normal variable and consulting a table of the normal distribution we find that

$$Pr\{Z < (1000-1070)/44.2\}$$
$$Pr\{Z < -1.58\} = .058 \quad \text{(medfly)}$$
$$Pr\{Z < (1000-710)/111.3\}$$
$$Pr\{Z < 2.61\} = .995 \quad \text{(melon fly)}$$

Therefore if the frequency of death is assumed to be normally distributed with respect to cohort gross reproductive level, only 6% of a medfly cohort would be expected to have died before the gross cohort fecundity averaged 1000 eggs but over 99% of a melon fly cohort would be expected to have died.

ESTIMATING r FROM POPULATION STRUCTURE

Concept

Perhaps one of the most misunderstood relationships among population ecologists is that between population age structure and population growth rate. If two populations have identical fixed survival schedules and reproductive patterns but different reproductive levels, the population which has the higher reproduction will eventually possess the higher growth rate and be more youthful. Both of these population characteristics are intuitive but often not fully appreciated. A high growth rate makes a population younger but this youthfulness also contributes to the higher growth rate itself. This is because a greater proportion of the population is skewed into the younger, more fecund age groups. In general, growth rate and age structure are conceptually and computationally inseparable and population growth rate and cohort survival are the sole determinants of age structure in closed populations.

My objective in this section is to show, in theory, how age structure and sex ratio can be used to determine population growth rate if relative survival of two age classes (groups) or of the two sexes is known. The theoretical basis for the age-structured case can be described as follows. The fundamental tenet of stable population theory is that closed populations with fixed birth and death rate will eventually grow at a fixed rate of increase and the fraction in each age class will be constant. Because the fraction of the total population in each age class is constant, the ratio of fractions contained in 2 age classes will also be fixed. This ratio, like its component fractions, will be unique to and a consequence of both the population growth rate and the relative survival between the 2 age groups.

The analytical basis for determining r in the simplest case (2 age classes) is shown as follows (see Keyfitz 1977; Keyfitz and Beekman 1983). If n_x and n_y denote the number in age classes x and y, respectively, c_x and c_y denote the fraction of the total population in age classes x and y, respectively, and N denote the total number in the population (all age classes), then the following relations hold:

$$c_x = n_x/N \qquad (11a)$$

$$c_y = n_y/N \qquad (11b)$$

$$(c_x/c_y) = (n_x/n_y) \qquad (11c)$$

The point here is that the ratio of fractions in each of the 2 age classes is equal to the ratio of the numbers in each as given in Eqn(12c). The analytical formulae for the fractions of the total stable population at ages x and y are

$$c_x = be^{-rx}l_x \quad (12a)$$

$$c_y = be^{-ry}l_y \quad (12b)$$

where r denotes the intrinsic rate of increase, b denotes the intrinsic birth rate and l_x and l_y represent survival to ages x and y, respectively. Therefore the ratio of c_x to c_y is

$$(c_x/c_y) = (be^{-rx}l_x)/(be^{-ry}l_y) \quad (13a)$$

$$= (e^{-rx}l_x)/(e^{-ry}l_y) \quad (13b)$$

Since $(c_x/c_y) = (n_x/n_y)$ from Eqn. (12c) then

$$(n_x/n_y) = (e^{-rx}l_x)/(e^{-ry}l_y) \quad (14a)$$

$$= [e^{r(y-x)}l_x]/l_y \quad (14b)$$

By rearranging Eqn.(14b) and taking logs of both sides yields the solution of r as

$$r = [(y-x)^{-1}]\log_e[n_x l_y/n_y l_x] \quad (15)$$

Note that the term inside the right brackets of Eqn.(16) can also be expressed as the product of 2 ratios (i.e. $\{n_x/n_y\}$ and $\{l_y/l_x\}$). The importance of noting this relationship is that relative, and not just absolute numbers may be used for each ratio. For example, if the number of individuals in the 2 age classes x and y are subsampled by the same fraction, the ratio of the number in each of these subsamples will be identical to the ratio of the total number in each. Likewise, if survivorship in each age class is decreased by a constant fraction, for example by a mortality factor which is indescriminant with respect to age, the ratio of these altered survival fractions will also be identical to the ratio of the unaltered survival fractions.

<u>Example 9</u>. Suppose a large number of a single species of medfly host are collected and all eggs and 1st instar larvae dissected out for counting. A total of 2941 eggs and 1994 1st instars are recovered. It is known from previous studies that egg hatch in the medfly population being considered is 100%, that 17% of all 1st instars die before reaching the 2nd instar and that the duration of both the egg and the 1st instar is exactly 2 days. Question: What is the growth rate of this stable population?

This information is sufficient to determine population growth rate (Eqn(16)) in this hypothetical population by noting that x=2 (egg duration), y=4 (2 days egg + 2 days 1st instar), n_x=2941, n_y=1994, l_x=1.00

(100% egg survival) and l_y=.83 (17% 1st instar mortality). Inserting these values into Eqn(16) yields

$$r = (4-2)^{-1}\log_e[(2941)(.83)/(1994)(1.00)]$$
$$= (1/2)\log_e 1.224$$
$$= 0.1$$

The data for this hypothetical problem were generated from an earlier paper which contains the stable age distribution and l_x-values for laboratory-reared medflies (Carey, 1984; Table II). For example, the fraction of a total medfly population contained in age classes 0-2 days (eggs) and 2-4 days (1st instars) was .294 and .202, respectively for a ratio of 1.475 eggs:1st instars. This is also the ratio of eggs to 1st instars I gave in the hypothetical census (i.e. {2941/1994}=1.475). Two points of this illustration merit comment. First, the correct r-value for the stable population was recovered by using knowledge of the age distribution and survival; and second, only relative survival is necessary. For example, if host-induced mortality decreased both egg and 1st instar survival by half and the same ratio of eggs to 1st instars existed, the same growth rate would be implied. That is, if l_x equaled 0.5 instead of 1.00 and l_y equaled .415 instead of .83, the value of the term inside the brackets in Eqn(17) would still equal 1.224. Therefore the value of r would still equal 0.1.

<u>Estimating r from Nulliparous:Parous Ratios</u>

Adult female fruit flies can be divided into 2 age groupings--nulliparous and parous. The first refers to reproductively immature individuals (preovipositional) and the second to those that are reproductively mature (ovipositional). Although the length of the nulliparous period for tephritids is conditional upon a number of factors such as temperature, food and host availability (Back and Pemberton, 1918; Christenson and Foote, 1960; Bateman, 1972), I will assume in the current discussion that conditions for attaining reproductive maturity are both optimal and constant between generations.

The fraction of a stable female fruit fly population in the nulliparous and parous age groups, denoted c_{nul} and c_{par}, repectively, is simply the sum of the fractions contained in the age classes comprising each age group. That is

$$c_{nul} = b\sum_{x=\epsilon}^{\alpha-1} e^{-rx} l_x \qquad (16a)$$

$$c_{par} = b\sum_{y=\alpha}^{\beta} e^{-ry} l_y \qquad (16b)$$

where ϵ, α and β denote the ages of eclosion, reproductive maturity and last reproduction, respectively.

By analogy with Eqn(12c), the ratio of the fractions of the total population contained in each of 2 age groups is equal to the ratio of the numbers in each. That is

$$(c_{nul})/(c_{par}) = (N_{nul})/(N_{par}) \qquad (17)$$

where N_{nul} and N_{par} denote the number in each of the age groups. Setting the right side of Eqn(18) equal to the ratio of the terms to the right in Eqns(17a-b) yields

$$(N_{nul})/(N_{par}) = (\sum_{x=\epsilon}^{\alpha-1} e^{-rx} l_x)/(\sum_{y=\alpha}^{\beta} e^{-ry} l_y) \qquad (18)$$

which is the formula for the nulliparous:parous ratio in a stable fruit fly population. Unlike the problem with only 2 age classes [Eqn(14b)] which has an analytical solution for r [Eqn(16)], the value of r in Eqn(19) must be found using numerical methods.

Example 10. Suppose that the probability of trapping a parous female relative to a nulliparous female is 0.8 for the medfly and 0.5 for the melon fly and the following numbers were captured over a 3-month period

	Medfly		Melon Fly	
MONTH	nulliparous	parous	nulliparous	parous
June	77	613	1034	517
July	1256	1142	2736	912
August	216	865	489	815

Question: What are the implied population growth rates for both populations if each is considered stable?

Answering this hypothetical question requires 3 steps--i)determining the stable nulliparous:parous ratios of the 2 species over a range of r-values; ii)converting trap catch numbers to ratios corrected for the trappability factors; and iii)assigning estimated r-values to the trap catch data by month.

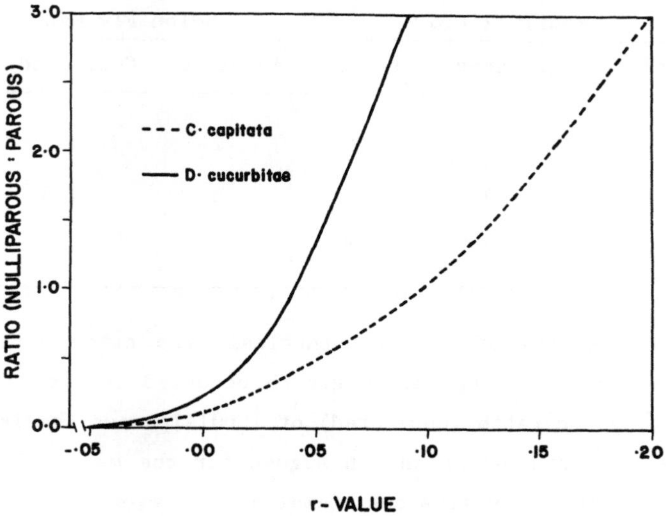

Fig. 3. Nulliparous:parous ratios of 2 tephritids for different r-values (data taken from Carey 1984 and Carey et al. 1985).

Estimates of the nulliparous:parous ratios for the 2 species over a range of r-values was determined using Eqn(19) and the survival and preovipositional data given in Carey (1984), Carey et al. (1985) and Wong et al. (1978). These results are given in Fig. 3. The preovipositional periods were set at 6 and 16 days for the medfly and the melon fly, respectively with no mortality occuring during this period. Expectation of life was approximately 2 months for the medfly and 3 months for the melon fly. The ratios of the numbers in each of the 2 reproductive categories can be corrected for trapping bias by multiplying the number of nulliparous captured by the relative trappability of parous individuals and dividing this number by the number of parous captured to obtain the corrected ratio in the population. For example, 1034 nulliparous and 517 parous melon flies were captured in June. Since nulliparous are, in this hypothetical example, twice as likely to be captured than are parous, the correct ratio in the population is (1034)(.5)=517 nulliparous to 517 parous or a 1:1 ratio. The r-value implied by each ratio can then be taken directly from Fig. 3. A summary of ratios and r-values is given in the following table (ratios given as the number of nulliparous relative to 1 parous):

MONTH	Medfly Ratio Absolute	Corrected	r-est	Melon Fly Ratio Absolute	Corrected	r-est
June	.13	.1	.00	2	1	.05
July	1.1	.9	.10	3	1.5	.07
August	.25	.2	.02	.6	.3	.00

Therefore the results of this hypothetical trap catch data for the two species show that zero population growth occurred in June for the medfly as indicated by the ratio (corrected) of 1 nulliparous female for every 10 parous females (ratio =0.1) and in August for the melon fly as indicated by the ratio of 3 nulliparous females for every 10 parous females (ratio=0.3). It can also be noted that similar nulliparous:parous ratios between the species can imply quite different r-values. For example, a 1:1 ratio for the melon fly indicates an r-value of r=.05 while a ratio of .9:1 for the medfly, which is approximately the same, indicates an r-value which is 2-fold greater (r=.1).

<u>Estimating r from Population Sex Ratio</u>

Population sex ratio is a form of internal structure and has direct parallels with population age structure in the 1-sex stable model. If a 2-sex population contains fixed sex-specific survival rates and a fixed maternity schedule for the production of male and female offspring, it will eventually contain fixed fractions in each age class by sex and grow at a fixed rate of increase. Because each age class contains a fraction of the total 2-sex population and this fraction is always constant, the sex ratio at age x (males age x to females age x) and the overall sex ratio (all males to all females) will also be fixed.

If primary sex ratio (sex ratio at birth) is unity and survival of the sexes is equal, a 2-sex model is not needed since both age-specific and overall sex ratio will equal 1. However, when the primary sex ratio does not equal 1 or when one sex lives longer than the other, a 2-sex model is required in order to fully understand the sex ratio implications of the life history traits for each sex. The reason for this is simple in concept. Because growing populations are more youthful than are declining ones, the sex with the highest survival of young individuals will skew the sex ratio toward that sex in expanding populations and the longer-lived sex will skew the sex ratio in the other direction in declining populations.

By analogy with Eqn(13), the fractions of a 2-sex population age x in the male and female populations are given by

$$c_x(f) = [e^{-rx} l_x(f)]/s \qquad (19a)$$

$$c_x(m) = [e^{-rx} l_x(m)]/s \qquad (19b)$$

where

$$s = \sum_{x=\epsilon}^{\omega} e^{-rx} l_x(f) + \sum_{x=\epsilon'}^{\omega'} e^{-rx} l_x(m) \qquad (19c)$$

and r denotes the intrinsic rate of increase for the female population, $l_x(f)$ denotes female survival to age x, $l_x(m)$ denotes male survival to age x, ϵ and ϵ' denote age of eclosion for females and males, respectively, and ω and ω' denote the last possible ages for females and males, respectively. Therefore, the ratio of all adult males to all adult females (denoted SR) is simply the ratio of sums of all fraction in each age class by sex. That is

$$SR = \sum_{x=\epsilon'}^{\omega'} e^{-rx} l_x(m) / \sum_{x=\epsilon}^{\omega} e^{-rx} l_x(f) \qquad (20)$$

The s-terms cancel when Eqns(19a) and (19b) are divided.

<u>Example 11</u>. The expectation of life at eclosion for both male and female medflies was found by Carey (1984) to be approximately 2 months. However, when 2 months of age the average male lived another 1 1/2 months while the average female only lived another week. Question: What are the sex ratio implications for a stable, 2-sex population if the population were growing at different r-values and sex-specific survival was similar to that found by Carey?

Using Eqn(20) and the survivorship schedules given in Carey (1984) for males and female medflies, I found that the sex ratio of adults would be approximately 1:1 over all positive r-values but would be skewed towards males at r-values less than zero (Fig. 4). The reason for this is that old individuals (>60 days) are in the extreme minority in growing populations. Therefore, the greater longevity of males that of females is swamped by the influence of growth rate on age structure. However, as population growth rate decreases, population age structure shifts to older age groups and the greater male survivorship at old ages skews the sex ratio toward males. The general implication here is that sex-specific survival may play a key role in establishing sex ratio in tephritid populations and that the

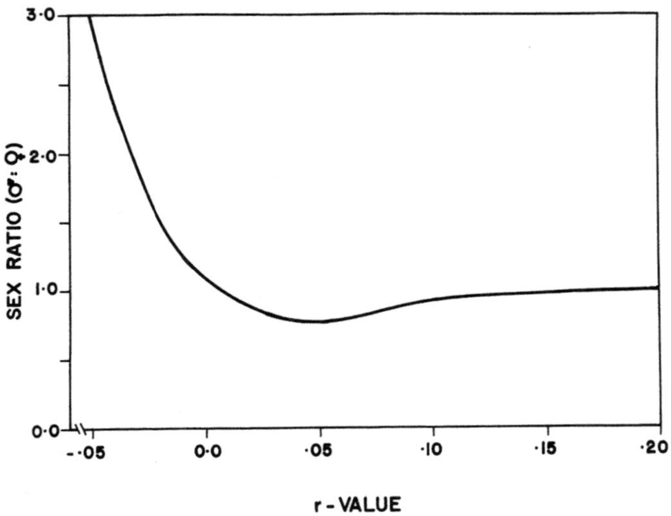

Fig. 4. Adult sex ratio of the Mediterranean fruit fly for different r-values using sex-specific survival schedules from Carey (1984).

reason sex ratio may shift (eg. Houston, 1984) may have nothing to do with changes in sex specific survival but rather may be due to changes in the population growth rate.

GENERAL DISCUSSION

Life Tables

One of the obvious advantages of establishing statistical, computational and conceptual interconnections among some of the more commonly-used tools in pest management including conventional statistics, is that it greatly simplifies the framework for analysis. A small number of statistical and computational methods can be brought to bear on a wide variety of problems, a simple dicotomy of alive or dead can be used figuratively to characterize a broad range of phenomena and age class or dose may serve as abstractions for any discretized series of life history events.

For example, degree-day (phenology) models (eg. Tassen et al., 1983) effectively predict the frequency distribution for exiting certain developmental stages (including diapause) as a function of temperature 'dose'; radiation dose for sterilizing fruit fly pupae (eg. Hafez and Shoukry, 1972) is a linear function of time of exposure thus time and dose are interchangeable; mass-trapping or male annihilation methods used in fruit fly control simply represent an additional mortality cause and can be put into a multiple decrement life table context, mean dispersal distance in mark-recapture dispersal experiments is the exact equivalent of expectation of life at birth if the point of release is viewed as birth (=0), distance as age and death as the cessation of dispersal (i.e. recaptured); and the mating of virgin females can be viewed as the death of their virginity and may be conditional upon the 'dose' of males to which they are exposed in terms of either time or number. In short, establishing a unity among analytical techniques and concepts in tephritid management may provide the first step toward development of general management principles

Stable Population Theory

The origin of the stable population model is often viewed by insect ecologists as being singular and that the model bears little resemblence in either concept or structure to other models in ecology and pest management. These perceptions are unfortunate since the reality is quite different. Euler (1760) was the one of the first to advance the notion that a population has the potential to increase geometrically but Lotka (1907,1922) and Sharpe and Lotka (1911) established the critical link between stable theory and real populations (Smith and Keyfitz, 1977). Fisher (1930) derived the renewal equation in discrete form and Leslie (1945) developed the use of matrices in population projection and showed the relationship with Lotka's work. These collective developments laid the foundation for the papers by Birch (1948) and Cole (1954) who were among the first to fully recognize the importance of stable theory in ecology.

Many of the major concepts in population biology and pest management stem from these fundamental demographic models. For example, the intrinsic rate of increase is frequently used as a measure of fitness in evolutionary biology, the Lotka-Volterra predator/prey and competition equations each contain r, the logistic equation and, in turn the concepts of r and K selection rely on r and assume a stable age distribution, and the parameters used in extinction theory in island biogeography represent the sole components of the intrinsic rate of increase--intrinsic birth and death rates.

Stable theory enters into fruit fly management models in ways which appear not to have been fully acknowledged. For example, the mathematical description of the sterile male principle as given in Berryman (1967) assumes that in the absence of sterile males a wild population will grow exponentially and that adding steriles will modify their population growth rate, r, Ito (1977) and Ito and Koyama (1982) assume that melon fly population size will change according to a logistic equation which necessarily incorporates their intrinsic rate of increase. Although never stated explicitely, all of these models of sterile release implicitely assume that the populations are at the stable age distribution. Similarly, models concerned with the effects of intervention techniques on fruit fly populations such as one developed by Argentesi et al. (1983), are often structured in such a way that if the mortality factor in question (in their case traps) is eliminated, the population would increase exponentially and eventually attain a stable age distribution. The model of fruit fly mass rearing developed by Carey and Vargas (1985) is a direct extension of stable theory which could not have been developed had the original basic demographic work not been done by Vargas et al. (1984). To dismiss stable theory as useless is to necessarily discard a large number of fundamental concepts in ecology and pest management and to reject it due to its 'lack of realism' is to completely misinterpret its purpose and strengths.

ACKNOWLEDGMENTS

I gratefully acknowledge Carroll Calkins, Melvin Couey, Roy Cunningham, Bob Dowell, Marc Mangel and Dick Plant for discussion, the California Department of Food and Agriculture for financial support and Nic Liquido and Harold Gordon for comments on the manuscript.

REFERENCES CITED

Abbot, W.S. (1925) A method of computing the effectiveness of an insecticide. J. Econ. Entomol., 18, 265-267.

Argentesi, R., Cavalloro, R. and Di Coli, G. (1983) Populations' dynamics parameter estimation in fruit-flies pest systems. In: Fruit Flies of Economic Importance (ed. R. Cavalloro). A. A. Balkema, Rotterdam. pp 307-314.

Berryman, A.A. (1967) Mathematical description of the sterile male principle. Can. Entomol., 99,858-865.

Back, E.A. and Pemberton, C.E. (1916) Effect of cold-storage temperatures upon the Mediterranean fruit fly. J. Agr. Res., 5, 657-666.

Back, E.A. and Pemberton, C.E. (1918) The Mediterranean fruit fly in Hawaii. USDA Bull., 536:119pp.

Bateman, M.A. (1972) Ecology of fruit flies. Ann. Rev. Entomol., 17, 493-518.

Birch, L.C. (1948) The intrinsic rate of natural increase of an insect population. J. Anim. Ecol., 17,15-26.

Brass, W. (1971) On the scale of mortality. In: Biological Aspects of Demography (ed. W. Brass), Taylor and Francis, London. pp.69-110.

Cameron, P.J. and Morrison, F.G. (1977) Analysis of mortality in the apple maggot, *Rhagoletis pomonella* (Diptera: Tephritidae) in Quebec. Can. Entomol., 109, 769-788.

Carey, J.R. (1982) Demography and population dynamics of the Mediterranean fruit fly. Ecol. Modelling, 16, 125-150.

Carey, J.R. (1983) A life table examination of growth rate and age structure trade-offs in Mediterranean fruit fly populations. In: Fruit Flies of Economic Importance (ed. R. Cavalloro), A. A. Balkema, Rotterdam. pp 315-320.

Carey, J.R. (1984) Host-specific demographic studies of the Mediterranean fruit fly, *Ceratitis capitata*. Ecol. Entomol., 9, 261-270.

Carey, J.R. (1986) Demographic analysis of fruit flies. In: World Crop Pests. Fruit Flies: Their Biology, Natural Enemies and Control. (eds. G. Hooper and A. Robinson), Elsevier Science Publishers, Amsterdam. (in press).

Carey, J.R. and Vargas, R.I. (1985) Demographic analysis of insect mass rearing: a case study of three tephritids. J. Econ. Entomol., 78, 523-527.

Carey, J.R., Harris, E.J. and McInnis, D.O. (1985) Demography of a native strain of the melon fly, *Dacus cucurbitae*, in Hawaii. Entomol. exp. appl., 38, 195-199.

Chiang, C.L. (1960a) A stochastic study of the life table and its applications: I: Probability distributions of the biometric functions. Biometrics, 16, 618-635.

Chiang, C.L. (1960b) A stochastic study of the life table and its applications: II: Sample variance of the observed expectation of life and other biometric functions. Human Biology, 32,221-238.

Chiang, C.L. (1984) The Life Table and its Applications. R. E. Krieger Publ. Co., Malabar, Florida. 316pp.

Christenson, L.D. and Foote, R.H. (1960) Biology of fruit flies. Ann. Rev. Entomol., 5, 171-192.

Coale, A.J. (1972) The Growth and Structure of Human Populations. Princeton University Press, Princeton. 227pp.

Cole, L.C. (1954) The population consequences of life history phenomena. Quart. Rev. Biol., 29, 103-137.

Deevey, E.S. (1947) Life tables for natural populations of animals. Quart. Rev. Biol., 22, 283-314.

Ebeling, W. (1953) Laboratory experiments on the control of three species of fruit flies (Tephritidae). Hilgardia, 21, 515-562.

Finney, D.J. (1964) Probit Analysis. Cambridge University Press, Cambridge. 318pp.

Fisher, R.A. (1930) The Genetical Theory of Natural Selection. Dover, New York.

Hafez, M. and Shoukry, A. (1972) Effect of irradiation on adult fecundity and longevity of the Mediterranean fruit fly, *Ceratitis capitata* Wied. in Egypt (Diptera: Tephritidae). Z. Angew. Entomol., 72,59-66.

Hill, K. and Trussell, J. (1977) Further developments in indirect mortality estimation. Pop. Studies, 31, 313-334.

Hoem, J.M. and Jensen, U.F. (1982) Multistate life table methodology: a probabilist critique. In: Multidimensional Mathematical Demography. K.C. Land and A. Rogers (eds). Academic Press, New York. pp.155-264.

Houston, W.W.K. (1981) Fluctuations in numbers and the significance of the sex ratio of the Mexican fruit fly, *Anastrepha ludens* caught in McPhail traps. Ent. Exp. & Appl., 30, 140-150.

Ito, Y. (1977) A model of sterile insect release for eradication of the melon fly, *Dacus cucurbitae* Coquillett. Appl. Ent. Zool., 12,303-312.

Ito, Y. and Koyama, J. (1982) Eradication of the melon fly: role of population ecology in the successful implementation of the sterile insect release method. Prot. Ecol.,4,1-28.

Keyfitz, N. (1977) Applied Mathematical Demography. Wiley-Interscience, New York. 388pp.

Keyfitz, N. and J.A. Beekman (1984) Demography Through Problems. Springer-Verlag, New York. 141pp.

Land, K.C. and A. Rogers (eds.) (1982) Multidimensional Mathematical Demography. Academic Press, New York. 604pp.

Landolt, P.J., Chambers, D.L. and Chew, V. (1984) Alternative to the use of Probit 9 mortality as a criterion for quarantine treatments of fruit fly (Diptera: Tephritidae)-infested fruit. J. Econ. Entomol., 77, 285-287.

Leslie, P.H. (1945) On the use of matrices in certain population mathematics. Biometrika 33,183-212.

Lotka, A.J. (1907) Relation between birth rates and death rates. Science, 26,21-22.

Lotka, A.J. (1922) The stability of the normal age distribution. Proc. Nat. Acad. Sci., 8,339-345.

McDonald, P.T. and McInnis, D.O. (1985) *Ceratitis capitata*: Effect of host fruit size on the number of eggs per clutch. Entomol. exp. appl. 37, 207-211.

Morris, R.F. (1959) Single-factor analysis in population dynamics. Ecology, 40, 580-588.

Pollard, J.H. (1973) Mathematical Models for the Growth of Human Populations. Cambridge University Press, Cambridge. 186pp.

Sharpe, F.R. and Lotka, A.J. (1911) A problem in age-distribution. Phil. Mag.,21,435-438.

Shoukry, A. and Hafez, M. (1979) Studies on the biology of the Mediterranean fruit fly, *Ceratitis capitata*. Entomol. Exp. Appl., 26, 33-39.

Smith, D. and Keyfitz, N. (1977) Mathematical Demography: Selected Papers. Springer-Verlag, Berlin. 514pp.

Southwood, T.R.E. (1971) Ecological Methods. Chapman and Hall, London. 391pp.

Tassen, R., Palmer, T., Hagen, K., Cheng, A., Feliciano, G. and Blough, T. (1983) Mediterranean frut fly life cycle estimations for the California eradication program. In: Fruit Flies of Economic Importance. R. Cavalloro (ed). A.A. Balkema, Rotterdam. pp.564-579.

Vargas, R.E., Meyashita, D., and Nishida, T. (1984) Life history and demographic parameters of three laboratory-reared tephritids. Ann. Entomol. Soc. Amer., 77, 651-656.

Varley, G.C. (1947) The natural control of population balance in the knapweed gall-fly (*Urophora jaceana*). J. Anim. Ecol.,16, 139-187.

Varley, G.C. and Gradwell, G.R.,(1965) Interpreting winter moth population changes. Proc. XXI Int. Congr. Ent., 377-378.

Varley, G.C., Gradwell, G.R. and Hassell, M.P. (1973) Insect Population Ecology. Univ. California Press, Berkeley. 212pp.

Wong, T.T.Y. and Nakahara, L.M. (1978) Sexual development and mating response of laboratory-reared and native Mediterranean fruit flies. Ann. Entomol. Soc. Amer., 71, 592-596.

STATISTICS TO FIND SPATIAL AND TEMPORAL STRUCTURES IN POPULATIONS

D. Debouzie and J. Thioulouse
Laboratoire de Biométrie
Université Claude Bernard
69622 Villeurbanne. France

Spatial distribution is one of the most characteristic properties of insect populations ; in most cases it allows us to define them (Legay and Debouzie, 1985). Spatial distributions of insect populations are not fixed but dynamic. Thus variations in space and in time of the pattern of individuals occur not only at the scale of the species, but also at the population one. For example, differences between the dynamics of two populations of the butterfly _Euphydryas editha_ may be as great as those between _E. editha_ and _E. chalcedona_ (Ehrlich et al., 1975).

It is important to find spatial and temporal structures in populations for applied and fundamental reasons. No field sampling can be efficient without understanding the underlying spatial distribution (Taylor, 1984). In pest control, we need to know how the insects are dispersed in the region under study ; we have to define the "hot spots" of the infestation so as to modulate the dose of pesticides, or at the orchard scale to take profit of variations in the infestation rate with host varieties ; at last we have to study adult flow from backyard orchards to commercial ones.

Spatial structures **also give** information on theoretical population biology (Legay and Debouzie, 1985). Within a population, groups of individuals may have their own dynamics and contribute differentially to the global population ; then the effective population size will be greatly reduced. The mechanisms that influence or regulate the population dynamics usually differ if, for a given density, individuals are dispersed at random or in clumps. Moreover key-factors implied in the dynamics of a local population, like that in an orchard, may have no influence at a regional scale ; climatic factors or competition for food illustrate such variations. So we need to separate the different levels of heterogeneity when searching the factors that regulate a population.

NATO ASI Series, Vol. G11
Pest Control: Operations and Systems Analysis
in Fruit Fly Management. Edited by M. Mangel et al.

In this paper we propose methods to collect and analyze basic data so as to find the spatial distribution of the individuals inside the universe sampled. Our proposals do not apply to the case where only the mean and the variance of the data are measured with confidence, as did Taylor (1984). We propose to replace random sampling by systematic one. In each sample the spatial positions of the points collected give a relevant information to the structures of the population ; so we must find adequate statistics which take into account such data and give a biologist readable picture of the structures within a sample or between samples when data are collected more than once.

Three main questions will be tackled :

1. What is a population of insects ?
2. How can we sample such a population ?
3. Which statistics can we use to analyze data obtained by systematic sampling ?

1. THE NOTION OF POPULATION

There is an intermediate level between the species and the individuals : the population. We do not consider it as a theoretical object like the mendelian population. We propose to study the populations such as they really live in the field and their characteristics must be found. A population can be defined as one or several ensembles of individuals of the same species that can live together at one time of their life history and then can interact on the mechanisms of their reproduction.

This definition does not imply that populations are fixed in space and in time ; a peculiar population may disappear at one time from a given place or many populations, well-defined in space at one time, may mix up altogether in other environmental conditions.

With insects, we postulate that populations also exist, that they are structured in space and in time and also genetically. So we must find their spatial limits, how these limits vary during the life history of the insects, what are the phenotypic and in some favourable cases the genotypic compositions of the population.

The population approach may have several practical applications, for instance, for fruit fly control. In an heterogeneous area, we must sample all the different types of habitats : large-scale orchards, small orchards and backyard gardens. But we must also sample the classical larval hosts and the alternate host plants which may be attacked by fruit flies under unfavourable conditions. The flow of adults between these different habitats must also be studied and modelled, specially for multivoltine fruit flies. Recently this approach has been successfully developed in the agronomic study of the cabbage stem flea beetle infesting rapeseed fields (Thioulouse, 1985).

In a similar way a fruit fly population cannot be correctly described by counting only one stage of develoment, like adults alone or larvae alone. Indeed demographic studies, such as Carey's one (1982), clearly show that adults represent only 4 % of the total population size. So all the stages of development must be studied and their dynamics quantified.

When population structures are known in a location, the sampling intensity can be reduced and pest control improved. Successful practical results have been obtained with rapeseed insects (Kühne, 1977) and cockchafer larvae (Chessel et al., 1984).

To find spatial and temporal structures we need adequate sampling designs and correlatively statistical methods fitted to the data obtained. The concept of population, the sampling design and the statistics used or available are three closely tied questions. An experimental design cannot be conceived if the universe to be studied is not correctly defined, if adequate statistics do not exist, but also if the existence of structures is denied or unknown.

2. THE SAMPLING DESIGN

Four classical sampling designs will be briefly discussed : random, sequential, stratified and systematic. I want to show the weakness or the advantage of each method when spatial structures are searched for.

Random sampling, but it is also true for sequential sampling, is based implicitely or explicitly on the lack of spatial structures ; the universe sampled is assumed to be homogeneous and in most examples the population from which samples are drawn, is not referred to. For our point of view that insect populations are structured in space, random sampling must be abandoned, unless previous experiments have shown that individuals disperse randomly in their habitat.

The major difficulty with random sampling comes from the statistical method used. In most papers, dispersion is measured by several indices computed from the average value of the number of individuals per unit of sampling and from the variance of this variable. For example, when the index of dispersion is greater than one, we classically conclude that the distribution is aggregative, that individuals are clumped at the scale of the elementary unit of sampling and biological hypotheses are then put forward.

However, this conclusion is only true when no variation in density occurs in the universe sampled. But we know that natural populations are rarely homogeneous ; several works on fruit flies (Fletcher, 1974 for the Queensland fruit fly ; Neuenschwander and Michelakis, 1979 for the olive fruit fly) lead to the same conclusion.

A distribution of frequency, which is the main statistical product of random sampling, may be generated by many patterns of distributions. For instance, the negative binomial distribution can be generated by more than fifteen different models. We must separate what Feller called in 1943 the <u>true aggregation</u> from the false one ; in true aggregation, the fact that a sample already contains one individual, increases the probability of choice of this sample by the individuals. False aggregation is obtained by numerous ways such as all the kinds of variation in density. When we adjust an observed distribution of frequency to a theoretical law, the true pattern of dispersion within the sample cannot be found. An illustration is given in Figure 1 which represents two distributions corresponding to the same mean and variance of X, the number of individuals per unit. The same value of the index of dispersion (variance/mean) masks two different patterns of dispersion within the area sampled.

Simple graphical representations immediately give a good picture of the true distribution. But they cannot be used alone, without statistics, because graphical tools are never neutral (Auda, 1983).

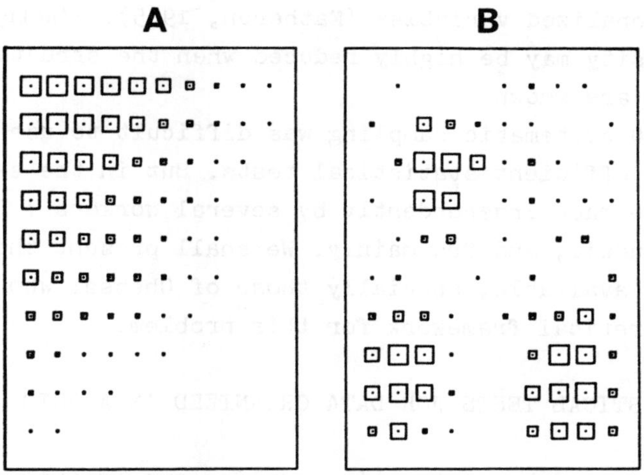

Figure 1 : Two theoretical distributions corresponding to the same mean and variance of individuals per unit of sampling. The index of dispersion equals 3.52. The size of each square is proportional to the number of individuals.
A : Individuals are distributed along a gradient.
B : Three clumps of mean size averaging about 9 units are visible ; elsewhere the distribution is random.

What is important for the population biologist is to find how the individuals disperse in the field, how they occupy their habitat. In 1985 we must go further than the ancient trilogy based on the three classical types of distribution : random, contagious, and regular. Statistical methods should be able to describe the true dispersion, that is to find the size of the aggregates or the presence of a gradient and of any variation in density.

Stratified sampling has no ability to describe spatial structures since on the contrary it assumes that the universe studied is already stratified. Then it is very efficient to estimate the population size and the accuracy of this size.

For many years systematic sampling is well-known to be able to detect spatial structures, specially in phyto-sociology. It offers several practical and theoretical advantages. Usually it is rather easy to be set up in the field ; the starting point of the sampling may be a difficulty but periodic distributions of individuals are rather rare in insect populations. Systematic sampling allows more statistical computations than random sampling. Several scales of heterogeneity can be found in an habitat. Accuracy of the population size estimation can be computed from the theory of regionalized variables (Matheron, 1965). **Finally,** the sampling intensity may be highly reduced when the structures of the population are known.

Till 1970 systematic sampling was difficult to use because of the lack of efficient statistical tests. But in the seventies progresses were made independently by several workers : Chessel, Cliff, Hill, Ludwig, and Ord mainly. We shall present the most powerful tests available, specially those of Chessel who gives a general theoretical framework for this problem.

3. STATISTICAL TESTS FOR DATA ORGANIZED IN A GRID

Data collected by systematic sampling are organized along a line when one dimension is sampled, or in a rectangular grid when two dimensions are explored. The case of grids will only be presented since transects are just a peculiar case.

Field distributions of individuals in their habitat are complex and several scales of heterogeneity are often mixed. Intuitively we can reasonably assume that one statistical test cannot detect these different scales ; a set of tests is needed. Chessel (1978) proposed non parametric statistics, the efficiency of which varies according to the hypotheses tested. A test must be preferentially used against only one precise alternative hypothesis ; a test can be very powerful against a hypothesis but weak or bad fitted to another.

The principle of the statistical analyses can be summarized as follows. The elementary units are summed up into blocks ; the size of the blocks varies from one unit to the whole grid. For each block, three measures of heterogeneity are proposed : global heterogeneity where variation is calculated between the blocks ; intra-block local heterogeneity, and heterogeneity at one scale calculated between two contiguous blocks.

For all the tests used, the value of the statistics is plotted against that of k, the size of the block. The shape of the different curves is used to detect the model of distribution ; reference curves have been obtained after analyzing more than a hundred of grids either in vegetation problems or in animal ones (Chessel, op. cit.).

We have selected the more efficient tests available for analyzing data that are counts, which are the most frequent measures in fruit fly studies, but other tests exist for binary data.

The systematic sampling of cockchafer larvae in the soil will be treated as example ; one sample of one sixth meter square was collected per hectare in a region of meadows in the North Jura, in France. The goal is to find how larvae disperse in this region ; is there some meadows more infested than others ? Is there a general trend in the infestation rate ?

Basic data are presented in Figure 2.

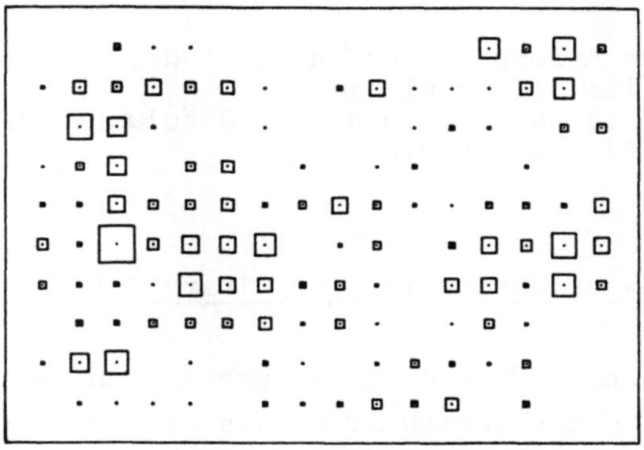

Figure 2 : Distribution of cockchafer larvae (stage L2) in the soil in Nancray (France, 1976). A sample unit 1/6 m2 large is collected per hectare ; about 160 hectares are sampled. The size of the largest square represents 24 larvae.

3.1 Global heterogeneity

The index of dispersion

The different curves giving the variations of the index according to the size of the block k are rather classical (Figure 3). We observe an increase when blocks are made of two columns but with blocks of five or ten columns variations are cahotic. In all the cases the values of the index are highly significant which means that at least one scale of heterogeneity exists in the data.

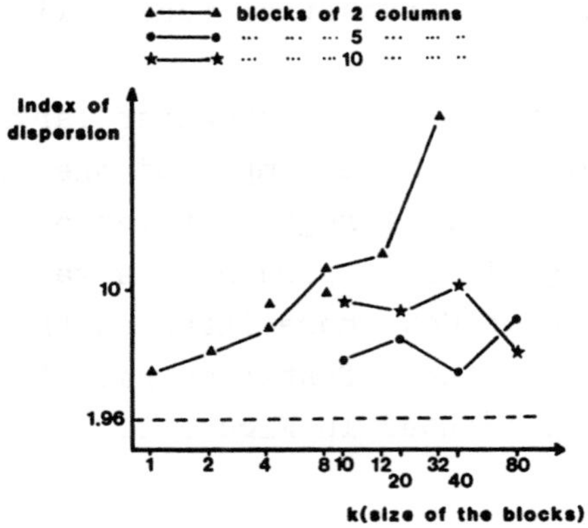

Figure 3 : Variations of the index of dispersion (variance/ mean) against the size of the blocks.
The initial grid has 16 lines and 10 columns (see Fig. 2). The abscissa scale is logarithmic.

The index of dispersion for binary data

To eliminate the effect of true aggregation, due for instance with insects to the laying of a mass of eggs, we propose to converse the raw data into binary ones. When a count is greater or equal than the median value, it is converted into 1, and into 0 if lower. The distribution of the ones in the grid allows us to calculate the index of dispersion (see Appendix).

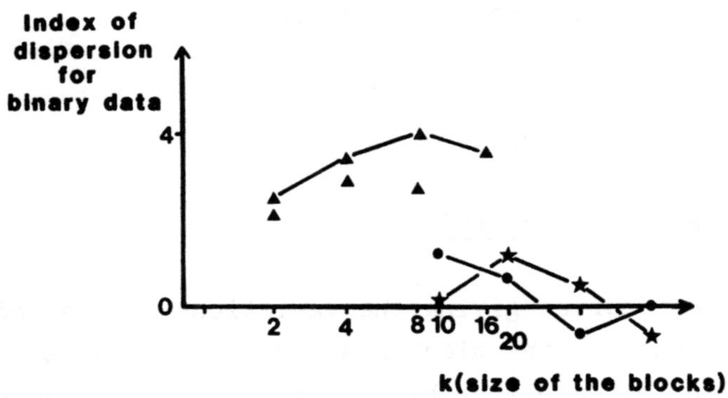

Figure 4 : Variations of the index of dispersion when data are conversed into binary ones.
For a complete legend, see Fig. 3.

The index has a maximum and significant value for blocks

made of 4 lines and 2 columns (Figure 4). But, for blocks of 5 and 10 columns, the dispersion of the local aggregates (corresponding to the ones) is random.

The spatial auto-correlation matrix

The degree of correlation between values recorded at two neighbouring points can be tested by a statistic due to Geary in 1954 and later to Cliff and Ord in 1973. Chessel (1981) applies this general test to two contiguous blocks in a grid : if individuals disperse randomly over the grid, the values measured at two any points are not correlated. The principle of the statistics is to test the variability between two contiguous blocks against the total variability of the measures.

The test applies to quantitative or qualitative data and also to non-rectangular or incomplete grids ; moreover it can be used between two points separated by a distance greater than one.

The test applied to the cockchafer larvae data shows a maximal variability at two sizes of blocks : 8 and 32 (Table 1).

Columns		1	2	5	10
	1	-	2.4	1.0	2.7
	2	2.1	3.0	0.5	1.5
Lines	4	4.3	4.4	0.9	1.4
	8	3.3	3.4	0.0	0.2
	16	4.5	4.8	0.5	-

Table 1 : Values of the spatial auto-correlation test according to the size of the blocks.
Values higher than 1.96 are significant. The test detects two scales of heterogeneity : 4 lin. x 2 col. and 16 lin. x 2 col.

No structure can be detected when all the columns or half the columns are summed up ; so the different lines seem to be homogeneous.

The three tests used to study global heterogeneity do not give contradictory conclusions ; of course they are linked by the sole fact that they use the same data. Nevertheless they are more or less expressive and difficult to interpret according to the examples.

3.2 Heterogeneity at fixed scale

Specific tests were developed to compare the number of individuals in two contiguous blocks (Chessel and DeBelair, 1973). Two blocks are defined as contiguous if their union constitutes a block at a higher scale. So blocks can be compared horizontally on the same line of the grid, or vertically on the same column.

The principle of the test lies on the fact that the difference between two blocks is compared to that calculated under the hypothesis of a random distribution of all the individuals present in the two blocks. Under this hypothesis, the statistic follows a gaussian distribution (see Appendix).

Two major points emerge from Figure 5 which shows the variations of the index of true aggregation.

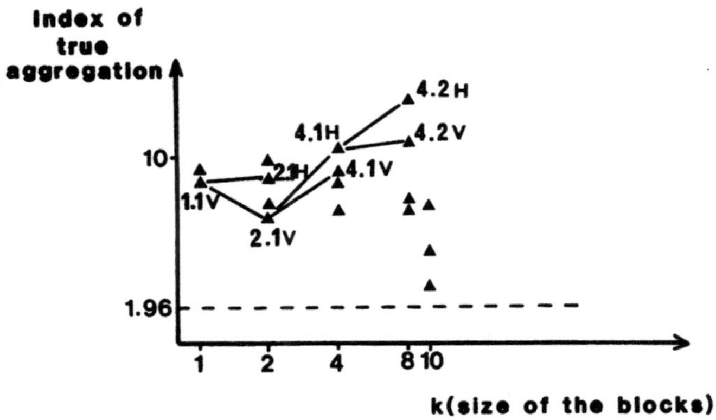

Figure 5 : Variations of the index of true aggregation. Blocks can be compared horizontally (H) or vertically (V). A block formed by 2 lines and 1 column is noted : 2.1 ; only the most interesting combinations of lines and columns are noted.

The counts of two elementary samples are not randomly distributed ; the very high value of the index, between 9 and 10, higher than the threshold which is 1.96, means that many larvae are found in an elementary sample while in the other contiguous sample they are few. This is a true aggregation which is usually observed with insects when many eggs are laid together and when just-born larvae do not migrate very far. In other studies, we observed that the intensity of the true aggregation decreases when cockchafer larvae become older, owing to density-dependent mortality and dispersion ; larvae in the third stage are only slightly contagiously dispersed.

The maximum value of the index corresponds to a block of 4 lines and 2 columns. This scale of heterogeneity has already been found by other statistics and thus is confirmed by the index of true aggregation.

3.3 Local heterogeneity

The local heterogeneity measures the within-block variation calculated between the units of a block. Two statistics are commonly used : the index of local dispersion and the index of the number of empty cases (see Appendix). These statistics are well-fitted to detect locally regular dispersion without any interaction with other heterogeneity structures. The index of the number of empty cases is powerful against the presence of empty units which have a great biological interest, specially in pest control.

No application of the tests to the cockchafer larvae will be given since no peculiar scale of variation was found inside the blocks.

3.4 Conclusions

Two scales of heterogeneity are found : the elementary sample where true aggregation is observed, and a block of eight elementary units, representing an area of about 8 hectares. It means that some meadows, or groups of meadows, are heavily infested, about 45 millions of second stage larvae per km square. At the opposite, other meadows are slightly infested with a rate near 15 millions per km square. The initial map (Fig. 2) could be drawn again, by representing the number of larvae within each block of 8 hectares.

It is worth noting that the two intermixing scales of heterogeneity imply <u>different biological factors</u> : egg-laying and density-dependent mortality and dispersion for the first, ecological characteristics of the meadows and memory of the adult females for the second (see for instance Chessel et al., 1984).

4. STATISTICS FOR DATA ORGANIZED IN SPACE AND IN TIME

Usually population biologists do not collect grid-organized data only once ; they must take into account the temporal structures which are always observed with temperate insects. Then two major questions are asked to the statistician :

- describe the spatial structures at each date,
- test the stability of these structures during the period of survey.

The first question can be answered by using the statistics described in the previous section. But for the second, till now we do not know any test able to study intra-grid dispersion over several times. The approach of Taylor and Iwao, who modelled the relation between the variance and the mean, either for different locations of samples or for different dates, is not supported by the same philosophy. These authors do not aim to describing the distribution within the basic samples ; they compare samples on the basis on the relations between variance and mean. In ecological words, their approach refers more to the species, while ours concerns more the population.

Always with this population view, we propose to describe temporal variations in spatial structures by multivariate analyses. We choose an example from the agronomic field to present the statistical method.

The data correspond to captures of cabbage stem flea beetle adults in 33 traps regularly spaced in a 1.24 hectare rapeseed field ; the traps are examined 64 times from the beginning of the infestations in September 1983 to the last emergence of the new generation in June and July 1984.

Data can be presented as a collection of 64 grids, each of the same size, or as a collection of 33 curves, each curve corresponding to a sampling point (Figure 6).

We propose to use the factorial analysis of correspondences (F.A.C.) to treat the data. The latter are reorganized in a table with two dimensions, instead of three as in the raw data (two for

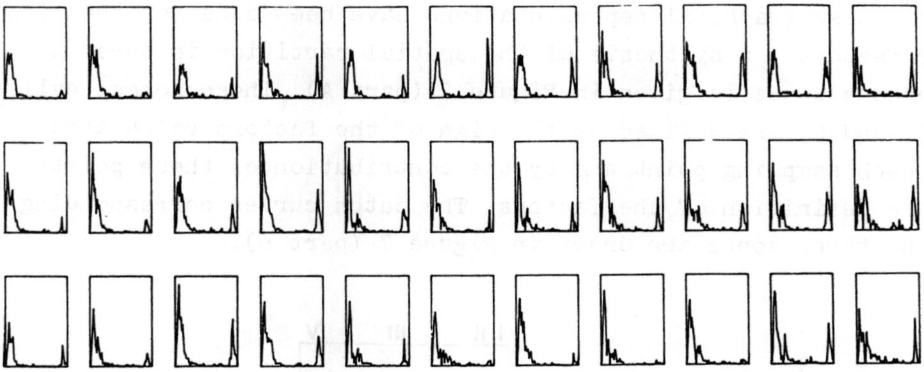

Figure 6 : Curves giving the numbers of cabbage stem flea beetle adults captured in water-traps regularly spaced in a rape field.

Each curve corresponds to a sampling point ; traps are collected 64 times since the invasion of the field in September 1983 to the emergence of the new generation in June and July 1984. For the ordinate axis, the maximum value corresponds to 125 adults captured in a trap (for the curve at the second line and the fourth column).

space, one for time). The lines of the table correspond to the sampling points and the columns to the dates (or inversely since in the F.A.C. model chosen here, lines and columns play a symmetrical role).

The goals are twofold : i) compare the temporal curves of catches per location, that is compare the distribution of each line of the table ; ii) compare the spatial distribution of catches per date, which is the distribution of each column.

The factorial analysis of correspondences is well-aimed at responding to these goals. It gives factors of decreasing magnitude according to the eigen values. Coordinates of the lines and of the columns can be computed for each factor. In the inertia model of the F.A.C. which is used here, contributions of lines and columns to the factors and inversely descriptions of a line or a column by a factor, help the interpretation.

Since in F.A.C. the factor-line maximizes the variance of the conditionnal means of the columns and reciprocally, we can represent graphically the factors-lines and the factors-columns in space or in time when data are organized in space or in time, as it is the case in our problem.

These graphical representations have been done for the first four factors ; a synthesis of the spatial partition in three homogeneous zones is given in Figure 7 (part A). These zones, called A, B, and C, are defined by the sign of the factors calculated for each sampling point and by the contribution of these points to the definition of the factors. The catch curves corresponding to the three zones are drawn in Figure 7 (part B).

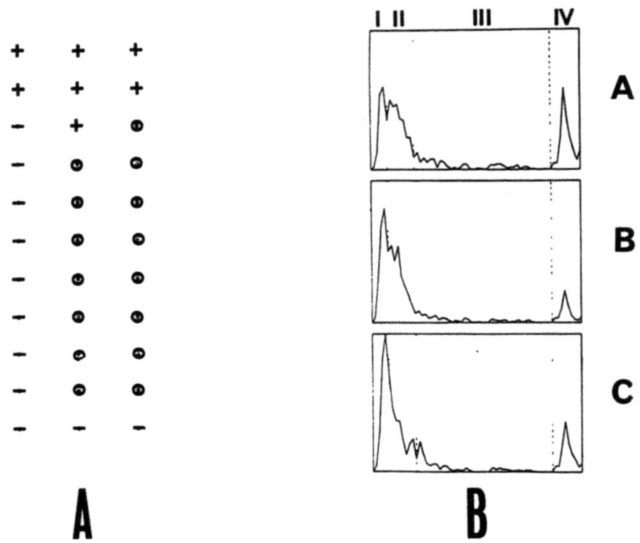

Figure 7 : Part A : Spatial partition of the rape field.
Part B : Curves of captures of adult insects relative to the three zones defined within the field (part A).
Zone A (symbol +) : 7 points. Factors 1 and 2 positive, factor 4 negative.
" B (" -) : 11 " . Factor 1 negative.
" C (" o) : 15 " . Factors 1 and 4 positive.
Period I : 6 dates from 12/09 to 29/09/83.
" II: 8 " " 30/09 to 27/10.
" III: 41 " " 28/10 to 04/06/84.
" IV: 9 " " 05/06 to 11/07/84.
In the ordinate axis, the maximum value represents 93 adults captured per trap.

In a symmetrical way, dates may be partionned in four periods ; of course the spatial partition is linked to the temporal one owing to the reciprocal averaging of the factors-lines and the factors-columns. It is worth noting that the spatial partition and the temporal one have biological and ecological interpretations, since zones are formed by contiguous points, located at the North of the field (zone B) or at the East (zone C), and since periods correspond to different phases in the life history of the insect (main invasion, late invasion and autumn survival, winter survival, and emergence).

At last, data can be reorganized by summing up all the captures for a given period and a given zone (Table 2).

Zone		A	B	C
Period	I	152	218	265
	II	249	232	194
	III	101	67	97
	IV	167	61	112

Table 2 : Mean number of adult insects captured per trap for a given period and zone.
Periods and zones are defined in Figure 7. The numbers correspond to the total captures during a period but periods have variable lengths.

We have shown that spatial and temporal structures may exist within a rape field : in the example, the emergence rate may be three times greater in a zone than in another. The factorial analysis of correspondences describes such structures.

5. CONCLUSIONS

Spatial and temporal structures exist in insect populations even at a fine scale. The existence of local heterogeneities is now well-documented ; for example one fruit tree may be heavily attacked by insects and not its neighbour (Debouzie et al., 1985) ; rape fields may be infested by insects at very variable rates (Thioulouse et al., 1984). These fine structures, which often vary during the life history of the insect, can only be found because we abandon random sampling and adopt systematic survey ; it is also because a set of statistical methods specially fitted to this sampling design is now available. The population concept, the sampling design, and the statistics used remain three closely tied subjects.

Multivariate analyses offer much more possibilities to the

population biologist. For example, using another model of the F.A.C., the factorial analysis of multiple correspondences, the interactions between an insect crop insect and its vegetal host can be studied. The canonic model of the F.A.C. contributes to defining the width of the niches.

One last interesting advantage of the systematic sampling is that the precision of the population size estimate can be obtained. Of course the question of the estimation of the population size remains a difficult one but we must go further in some favourable examples ; in some applied works a confidence interval for this estimation is necessary. The theory of the regionalized variables applied to regular spaced data gives such an interval as a function of the intensity and the nature of the spatial structures found within the grid (see an example for the unidimensionnal case in Thioulouse et al., 1985).

6. LITERATURE CITED

AUDA Y., 1983. Rôle des méthodes graphiques en analyse des données : application au dépouillement des enquêtes écologiques. Thèse Doct. 3° cycle, Lyon.

CAREY J.R., 1982. Demography and population dynamics of the Mediterranean fruit fly. Ecol. Model., 16 : 125-150.

CHESSEL D., 1978. Description non paramétrique de la dispersion spatiale des individus d'une espèce. In : Biométrie et Ecologie. LEGAY J.M. and TOMASSONE R. (Eds.). Soc. Fr. Biom., Jouy (France), pp. 45-135.

CHESSEL D., 1981. The spatial autocorrelation matrix. Vegetatio, 46 : 177-180.

CHESSEL D. and DeBELAIR G., 1973. Mesure de la contagion vraie en échantillonnage par carrés dans l'analyse des populations végétales. C. R. Acad. Sc., sér. D, Paris, 277 : 1483-1486.

CHESSEL D., DEBOUZIE D., ROBERT P. and BLAISINGER P., 1984. L'échantillonnage des larves du hanneton commun Melolontha melolontha L. Acta Oecol., Oecol. Appl., 5 : 173-189.

CLIFF A.D. and ORD J.K., 1973. Spatial autocorrelation. Pion, London.

DEBOUZIE D., LEBRETON J.D., ALLAINE D. and PALLEN C., 1985. Contribution à la notion de groupes et de populations. Exemples de populations d'oiseaux et d'insectes. Rapport A.T.P. C.N.R.S., Biol. des Popul., 73 p.

EHRLICH P.R., WHITE R.R., SINGER M.C., Mc KECHNIE S.W. and GILBERT L.E., 1975. Checkerspot butterflies : a historical perspective. Science, 188 : 221-228.

FELLER W., 1943. On a general class of contagious distributions. Ann. Math. Stat., 14 : 389-400.

FLETCHER B.S., 1974. The ecology of a natural population of the Queensland fruit fly, Dacus tryoni. VI. Seasonal changes in fruit fly numbers in the areas surrounding the orchard. Austral. J.

Zool., 22 : 353-363.
GEARY R.C., 1954. The contiguity ratio and statistical mapping. Incorp. Statist., 5 : 115-145.
KUHNE W., 1977. Untersuchungen zur Befallsverteilung der Rüsselkäfer Ceutorrhynchus napi Gyll., Ceutorrhynchus quadridens Panz. und Ceutorrynchus assimilis Payk. innerhalb grossflächiger Winterrapsbestände. Arch. Phytopatho. Pflanzenschutz, 13 : 109-115.
LEGAY J.M. and DEBOUZIE D., 1985. Introduction à une biologie des populations. Masson, Paris.
MATHERON G., 1965. Les variables régionalisées et leur estimation : une application de la théorie des fonctions aléatoires aux sciences de la nature. Masson, Paris.
NEUENSCHWANDER P. and MICHELAKIS S., 1979. Mc Phail trap captures of Dacus oleae (Gmel.) (Diptera, Tephritidae) in comparison to the fly density and population composition as assessed by sondage technique in Crete, Greece. Bull. Soc. entomol. Suisse, 52 : 343-357.
TAYLOR L.R., 1984. Assessing and interpreting the spatial distributions of insect populations. Ann. Rev. Entomol., 29 : 321-357.
THIOULOUSE J., 1985. Structures spatio-temporelles en biologie des populations d'insectes. Application à l'étude de l'altise du colza (Psylliodes chrysocephala L.) : résultats méthodologiques et biologiques. Thèse Doct. 3° cycle, Lyon.
THIOULOUSE J., DEBOUZIE D. and BALLANGER Y., 1984. Structures spatiales et temporelles des populations d'un ravageur du colza (Psylliodes chrysocephala L., (Col., Chrysomelidae)) dans plusieurs parcelles de culture. Acta Oecol., Oecol. Appl., 5 : 335-353.
THIOULOUSE J., HOUILLER F. and ONILLON J.C., 1985. Variables régionalisées et dénombrement d'insectes : cas unidimensionnel. C. R. Acad. Sc., sér. III, Paris, 301 : 423-428.

APPENDIX

1. The index of dispersion for binary data

We note N = the number of elementary units,
k = the size of a block,
B = the number of blocks (= N/k),
M = the number of values greater (or equal) than the median, that is the number of ones (see text).

Y_i represent the number of ones in the block i.

We calculate : $T = \sum_i Y_i^2$

$$\varepsilon = (T - E(T)) / \sqrt{var(T)}$$

with

$$E(T) = M (1 + (k-1)(M-1)/(N-1))$$

$$var(T) = 2M(M-1)(k-1)(N-k)(N-M)(N-M-1)/(N-1)^2/(N-2)/(N-3)$$

Under the null hypothesis, ε follows a gaussian distribution, slightly skewed at right, since the null hypothesis is rejected at the 5 % level when ε >3.

2. The spatial auto-correlation matrix

Let X = the number of individuals per unit,

$\bar{x}$ = the mean value of X,

σ^2 = the estimated variance of X.

V is a symmetrical matrix defined by :

$V_{ij} = V_{ji} = 1$ if i and j are two neighbouring points,

$V_{ij} = V_{ji} = 0$ otherwise.

Note : $L_i = \sum_j V_{ij}$; $A = \sum_i L_i$; $C = \sum_i L_i(L_i - 1)$.

The total variability of the measures is : $H_T = 2\sigma^2$, and the variability between two neighbouring points is :

$$H_V = (\sum_{i,j} V_{ij}(x_i - x_j)^2)/A.$$

We calculate $Z = H_V / H_T$

and $\varepsilon = (E(Z) - Z) / \sqrt{var(Z)}$

with

$$E(Z) = 1$$
$$var(Z) = (U_1+U_2+U_3)/(A^2.N(N-2)(N-3))$$

and

$$U_1 = ((N^2-3)-(N-1)^2.U_4).A^2$$
$$U_2 = 2A(N-1)(N^2-3N+3-(N-1)U_4)$$
$$U_3 = (N-1)(D+A)((N^2-N+2).U_4 -(N^2+3N-6))$$
$$U_4 = N.\sum_i(x_i-\bar{x})^4/(\sum_i(x_i-\bar{x})^2)^2$$

Under the null hypothesis, ε follows a gaussian distribution.

3. The index of true aggregation

We note : $D = |L - R|$

$P = L + R$

where L and R are the numbers of individuals in two contiguous blocks (left and right).

We calculate : $\varepsilon = \sum_{couples} (D - E(D))/ \sqrt{var(D)} / \sqrt{n}$

with : n = number of couples of blocks where $P > 1$

$E(D;P=2) = 1$

Let P be an even number, so :

$$E(D;P) = E(D;P-1)$$
$$E(D;P+1) = E(D;P) + (P!)/(((P/2)!)^2.2^P)$$

For all values of P :

$$var(D;P) = P - (E(D;P))^2$$

Under the null hypothesis, ε follows a gaussian distribution as soon as $n \geqslant 10$.

4. The index of local dispersion

In the block i made of k elementary units, we note s_{ij} the number of individuals in the unit j.

We calculate : $$LD_i = \sum_{j=1}^{k} s_{ij}^2 \quad ; \quad S_i = \sum_j s_{ij}$$

and $$\varepsilon = (1/\sqrt{B})\left(\sum_{i=1}^{B} (LD_i - E(LD_i))/\sqrt{var(LD_i)}\right)$$

with
$$E(LD_i) = S_i(S_i + k - 1)/k$$
$$var(LD_i) = 2(S_i - 1)S_i(k - 1)/k^2.$$

Under the null hypothesis, ε follows a gaussian distribution as soon as $B \geqslant 10$.

5. The index of the number of empty units

We note ZO_i the number of empty units in the block i ; S_i is defined as in §4.

We calculate : $$\varepsilon = (1/\sqrt{B})\left(\sum_{i=1}^{B} (ZO_i - E(ZO_i))/\sqrt{var(ZO_i)}\right)$$

with $$E(ZO_i) = k((1 - 1/k)^{S_i})$$
$$var(ZO_i) = k(k-1)((1-2/k)^{S_i}) + E(ZO_i) - (E(ZO_i))^2$$

Under the null hypothesis, ε follows a gaussian distribution as soon as $B \geqslant 10$.

ABSTRACT

Spatial structures are one of the most characteristic properties defining insect populations ; they are not fixed but vary during the life history of the insect.

Fitted sampling designs and adequate statistical tests are needed to find spatial and temporal structures. We propose to use systematic sampling and to abandon random designs.

Non parametric statistics (Chessel, 1978) allow us to find how the individuals disperse within the habitat studied. The tests are not based on the mean and the variance of the number of individuals per elementary unit, but on the spatial arrangement of the individuals. Several tests, instead of one index as in the classical studies, are needed to find several scales of heterogeneity. For all the tests, the elementary units are summed up into blocks of growing size and we study the variations of the statistics according to the size of the blocks. We propose statistics adapted to detect three kinds of variation : global (between blocks of units), local (between units within a block), and between two contiguous blocks. An example is given to illustrate the tests.

When grid organized data are collected many times, we propose to use the factorial analysis of correspondences. This method is well aimed to describing temporal variations in spatial structures inside the grid, as seen in an example with insects infesting rape fields.

The Effect of Demographic Factors on the Population Dynamics of the Mediterranean Fruit-Fly

Lloyd Demetrius
Max-Planck-Institut für biophysikalische Chemie
3400 Göttingen, F.R.G.

Abstract

This paper examines the properties of two demographic variables, the Malthusian parameter and population entropy, in the study of the control and eradication of medfly populations. The Malthusian parameter describes the rate of increase of population numbers. Population entropy determines the rate at which fluctuations in population numbers are damped. In stationary populations, entropy predicts the change in magnitude of the population growth rate when the fecundity and mortality rates are subject to perturbations. The parameter entropy thus provides an index for classifying populations according to their capacity for outbreaks and tendency to extinction. These observations are illustrated using data from the medfly population CERATITIS CAPITATA.

Introduction

The control and eradication of insect population involves an understanding of both the dynamics of populations under natural environmental conditions and the effect of eradication strategies on this dynamic.

NATO ASI Series, Vol. G11
Pest Control: Operations and Systems Analysis
in Fruit Fly Management. Edited by M. Mangel et al

Eradication strategies are mainly of two kinds, full coverage sprays and sterile male technique. Both strategies are designed to reduce population numbers but do so in distinct ways. Coverage sprays reduce total population size by increasing death rate. Sterile male techniques control numbers by reducing the birth rate.

Insect populations in general and the Mediterranean fruit fly in particular show great variability in their age-specific fecundity and mortality rates. Hence, in order to understand the effectiveness of eradication strategies, we need to develop models in which the influence of age-specific variability on populations dynamics is studied.

Classical demography theory, as developed by Lotka [1] and Leslie [2], study changes in the population age-distribution due to the age-specific fecundity and mortality rates. The central parameter in these models is the Malthusian parameter or the population growth rate. On the basis of these models, one can predict the sensitivity of the growth rate to changes in the age-specific fecundity and mortality schedule [3]. As a result of this sensitivity analysis, one can assess the relative effect of different eradication strategies on population numbers and determine, for example, the number of sterile males needed for the success of an eradication program.

A problem which is not addressed by the Lotka-Leslie theory

concerns the effect of fluctuations in age-specific birth and death rates. This problem arises not only in eradication strategies but also in mass rearing programs where populations of fixed size are maintained for egg-production. The importance of this problem is derived from the fact that small random fluctuations in the age-specific fecundity and mortality of a stationary population can, in some circumstances, induce population extinction; in other circumstances, exponential growth. Understanding the effect of these fluctuations is therefore crucial in predicting outbreaks and extinctions.

A new class of population models which deal, among other things, with the response of growth rate to fluctuations, was introduced in [4]. These models revolved around a new demographic parameter, called population entropy. This parameter describes the rate at which fluctuations in population numbers are damped. The significance of this parameter for problems of control and eradication is derived from the following property: In stationary populations, the change in the Malthusian parameter due to small perturbations in the age-specific fecundity and mortality rates is determined by entropy. More precisely, the <u>direction</u> of change of growth rate is determined by the nature of the fluctuations. The <u>magnitude</u> of change in the growth rate is determined by entropy. This magnitude measure the capacity of population numbers to increase exponentially or to decay. Hence the entropy parameter emerges as a useful index in classifying populations according to the intensity of possible outbreaks and the rapidity of extinction.

The main aim of this paper is to bring the entropy concept to the attention of scientists involved in the control and eradication of fruit fly populations. This paper gives a brief review of some of the properties of the concept and illustrates its applicability by examining data for the population CERATITIS CAPITATA.

In Section (1), we discuss the relations between the population parameters that occur in demography. We distinguish between the <u>individual parameters</u>, namely; the age-specific fecundity and mortality variables, and the <u>aggregate parameters</u>, namely the growth rate parameters as developed in Lotka and Leslie, and the entropy parameters as introduced in [4]. The relations between the two kinds of aggregate parameters are analysed in Section (2). The application to eradication strategies are based on these relations. In Section (3) we illustrate the meaning of these parameters using data taken from [5], [6].

1. The Population Parameters

1.0 The Individual Parameters

The age-composition of a population that neither gains nor loses by migration and lives in a stable environment is determined by the individual parameters - the age-specific fecundity m(x), and the age-specific mortality μ(x). The dynamics of the age-distribution u(x,t) is given by

$$\frac{\partial u}{\partial x} + \frac{\partial u}{\partial t} = -\mu(x)u(x,t) \qquad (1.0)$$

The boundary conditions are

$$u(o,t) = \int_0^{\infty} m(x)u(x,t)dx \qquad (1.1)$$

$$u(a,o) = f(a)$$

where f(a) denotes the initial age-distribution. The parameter m(x) in insect populations represent the number of eggs laid by a female aged x. The age-specific survivorship l(x) is given by

$$l(x) = \exp[-\int_0^{x} \mu(t)dt] \qquad (1.2)$$

The parameter l(x) represents the probability of an individual attaining age x.

1.1 The Aggregate Parameters

The aggregate parameters are functions of the age-specific fecundity and mortality schedule. We will distinguish between two classes of aggregate parameters. The first class consists of the growth rate and the net-reproduction rate parameter. These variables are derived from the Lotka theory, Lotka [1], based on the dynamics of the age-distribution. The second class consists of the entropy functions which are derived from the dynamics of the distributions of genealogies, Demetrius [4].

1.10 Growth Rate Parameters

The central notion in the Lotka theory is that of a stable age-distribution. This notion describes the state of the population when the relative proportion of individuals in each age class is constant. The Malthusian parameter and net-reproduction rate emerge naturally from this notion of a stable age-distribution.

The Malthusian parameter r is the rate of increase of the population at the stable state. This parameter r can be shown using the dynamical system given by (1.0), to be the unique real root of the equation

$$\int_0^{\infty} \exp(-rx)\,l(x)\,m(x)\,dx = 1 \qquad (1.4)$$

The net-reproduction rate R which is the rate of growth of population numbers per generation in the stable state is given by

$$R = \int_0^{\infty} l(x)\,m(x)\,dx \qquad (1.5)$$

This number is precisely the area under the net-fecundity function $V(x) = l(x)m(x)$, and represent the number of offspring who from birth is subject to the fecundity schedule m(x) and survivorship schedule l(x).

The expressions for the gross reproduction rate and mean

life-expectancy can be derived from R.

The gross reproduction rate is the number of offspring born per individual who passes the entire reproductive span giving birth to offspring according to the schedule m(x). This number $\bar{m}$ is given by

$$\bar{m} = \int_0^\infty m(x)dx \qquad (1.6)$$

The quantity $\bar{m}$ can be considered as the rate of increase per generation of a population with fecundity schedule m(x) and a mortality schedule given by $l(x) = 1$.

The mean life expectancy e_0 is the average expectation of life at birth. This quantity is given by

$$e_0 = \int_0^\infty l(x)dx \qquad (1.7)$$

The number e_0 can be considered as the rate of increase per generation of a population with mortality schedule l(x), and a fecundity schedule $m(x) = 1$.

1.11 The Entropy Parameters

The models which constitute the basis of the entropy parameters revolve around the notion of a genealogy, Demetrius [4], [5]. A genealogy represents the set of descendants and ancestors of a given individual, these ancestors and descendants being indexed by the ages at which they produce offspring. The individuals in the population are considered as generating genealogies. The frequency distribution of the genealogies is determined by the age-specific fecundity and mortality schedule. The models investigate: (a) the distribution of genealogies at equilibrium; (b) the asymptotic rate of increase of the typical genealogies, that is, the genealogies which occur with greatest frequency. The growth rate of the number of typical genealogies are described by entropy functions.

The equilibrium distribution of the genealogies is given by a probability measure, whose marginal distribution expressed in terms of the age of reproducing individuals in the stable population is given by

$$p(x) = \exp(-rx)l(x)m(x) \qquad (1.8)$$

Here r is the Malthusian parameter defined by (1.4).

The asymptotic rate of increase per unit time of the number of typical genealogies is given by the quantity

$$H = -\frac{\int_0^\infty p(x)\log p(x)dx}{\int_0^\infty xp(x)dx} \qquad (1.9)$$

This number is the ratio of the entropy of the probability distribution p(x) and the mean age at reproduction in the stable population. this mean age denoted by T is called the generation time.

When the generation time T is finite, we have

$$H \leq \frac{1 + \log T}{T} \qquad (1.10)$$

Population entropy determines the convergence rate to the equilibrium distribution given by (1.8). It was shown in [8] that at maximal entropy, the birth sequence is immediately stabilized when the stable age-distribution is perturbed, while at zero entropy perturbations give rise to oscillations in the birth sequence. Tuljapurkar [9] exploiting the notion of a genealogy, has shown in the context of the discrete model that H determines the stochastic convergence rate to the equilibrium distribution.

The rate of increase per generation of the number of typical genealogies is the quantity H_R given by

$$H_R = - \int_0^{\infty} q(x) \log q(x) dx \qquad (1.11)$$

where

$$q(x) = \frac{l(x)m(x)}{R} \qquad (1.12)$$

The expression (1.11) is the entropy of the net-fecundity distribution q(x).

The quantity H_R is bounded provided the mean age at reproduction in a cohort μ_1 is finite. We have

$$H_R \leq 1 + \log \mu_1 \qquad (1.13)$$

where

$$\mu_1 = \frac{\int_0^\infty x l(x) m(x) dx}{\int_0^\infty l(x) m(x) dx} \quad (1.14)$$

The expressions for the entropy of the fecundity distribution and the life table can be derived from H_R.

The entropy of the fecundity distribution H_m is given by

$$H_m = - \int_0^\infty \frac{m(x)}{\overline{m}} \log \frac{m(x)}{\overline{m}} dx \quad (1.15)$$

where $\overline{m}$ is given by (1.6).

This quantity describes the dispersion of the fecundity distribution and can be considered as measuring the entropy of the net-fecundity distribution for a population with a fecundity distribution m(x) and a life-table l(x)=1.

The entropy of the life-table is given by

$$H_l = - \int_0^\infty \frac{l(x)}{e_o} \log \frac{l(x)}{e_o} dx \quad (1.17)$$

where e_o is given by (1.7).

This expression measures the heterogeneity in age-specific survivorship curve. The quantity H_l can be considered as describing the entropy of the net-fecundity distribution for a population with a fecundity distribution m(x) = 1 and a life table l(x).

2. Relations between the Aggregate Parameters

There exist simple relations between the growth rate parameters described by the Lotka models and the entropy parameter. The central result asserts that the entropy parameters predict the response of the growth rate parameter to small fluctuations in the age-specific fecundity and mortality schedule.

A complete review of these perturbation results is given in [10]. We will briefly recall some of the main results.

Continuous perturbations in the net-fecundity schedule $V(x)$ are characterized by

$$V^*(x) = V(x)^{1+g(x)} \qquad (2.0)$$

Here $g(x)$ is a bounded function and $V^*(x)$, the perturbed function.

Such changes in the function $V(x)$ will induce changes in the growth rate r, the mean life-expectancy e_0 and the gross-fecundity rate $\bar{m}$. Our analysis in [10] shows that these changes are determined by the entropy functions.

A case of interest arises when $g(x) = \delta$, corresponding to a perturbation uniform over all ages. The parameter r, e_0 and $\bar{m}$ are now functions of δ and we have for the growth rate r

$$\frac{dr}{d\delta} = -\Phi \qquad (2.1)$$

In (2.1), the derivative is evaluated at $\delta = 0$ and the function Φ is given by

$$\Phi = - \frac{\int_0^\infty p(x)\log V(x)dx}{\int_0^\infty xp(x)dx} \qquad (2.2)$$

Writing Δr to denote the change in r, we have from (2.1)

$$\Delta r \simeq -\Phi\delta \qquad (2.3)$$

The equation (2.3) states that the expression Φ, which is called the reproductive potential, predicts the change in growth rate due to small fluctuations in the fecundity-mortality schedule.

Now, from (2.2) and (1.9), we have

$$\Phi = H - r$$

Hence (2.3) can be written

$$\Delta r \simeq (r - H)\delta \qquad (2.4)$$

Consider populations in the stationary growth phase, $(r = 0)$, and write $\Delta r = r^*$, where r^* is the new growth rate. Equation (2.4) becomes

$$r^* \simeq -H\delta \qquad (2.5)$$

Now, the number δ may be positive or negative, depending on the nature of the fluctuations, whereas for insect populations H is strictly positive. Hence, from (2.5) we can conclude the following:

a) The <u>direction</u> of change of the growth rate is determined by the fluctuation δ.

b) The <u>magnitude</u> of change of the growth rate is determined by population entropy H.

The magnitude of change in r determines the intensity of outbreaks and the rapidity of the decay rate to extinction. Hence the entropy parameter provides an index for classifying the different populations according to these characteristics.

Results analogous to (2.5) hold for the net-reproduction rate R_o. We have

$$\frac{\Delta R}{R} \simeq (\log R - H_R)\delta \tag{2.6}$$

This result can be used to derive the effects of changes on the mean life expectancy e_o and the gross fecundity rate $\bar{m}$.

Following (2.6), changes in the mean life-expectancy is given by

$$\begin{aligned}\frac{\Delta e_o}{e_o} &\simeq (\log e_o - H_1)\delta \\ &\simeq -\Phi_1\delta\end{aligned} \tag{2.7}$$

where

$$\Phi_1 = - \frac{\int_0^\infty l(x)\log l(x)dx}{\int_0^\infty l(x)dx} \tag{2.8}$$

The expression Φ_1 is a measure of the convexity of the survivorship curve.

Following (2.6), changes in the gross fecundity is given by

$$\frac{\Delta \bar{m}}{\bar{m}} \simeq (\log \bar{m} - H_m)\delta \qquad (2.9)$$

$$\simeq -\Phi_m \delta$$

where

$$\Phi_m = - \frac{\int_0^\infty m(x) \log m(x) dx}{\int_0^\infty m(x) dx} \qquad (2.10)$$

The expression Φ_m characterizes the convexity of the fecundity distribution.

The relation (2.7), see also Keyfitz [11], Chap. (3), is particularly relevant to mass rearing strategies where the effects of fluctuation on mean longevity is of ecological interest. This relation also enables us to predict the effect of uniform changes on the life-table on the mean life-expectancy.

(3) Analysis of a Med-fly Population

The life-history parameters of the Mediterranean fruit fly, CERATITIS CAPITATA at 25° is given by Table (I). This data is taken from [5]; see also [6] for an analysis of some aspects of this data.

Table I

Life history parameters of the Mediterranean fruit fly, CERATITIS CAPITATA, at 25°C

Stage	x	l_x	m_x
Egg	2	1.00	0.00
Larva	4	0.83	0.00
Larva	6	0.69	0.00
Larva	8	0.58	0.00
Larva	10	0.48	0.00
Pupa	12	0.40	0.00
Pupa	14	0.39	0.00
Pupa	16	0.38	0.00
Pupa	18	0.37	0.00
Pupa	20	0.34	0.00
Pupa	22	0.32	0.00
Adult (pre-oviposition)	24	0.32	0.00
Adult (pre-oviposition)	26	0.31	0.00
Adult (oviposition)	28	0.31	2.41
Adult (oviposition)	30	0.30	7.06
Adult (oviposition)	32	0.29	16.51
Adult (oviposition)	34	0.28	22.36
Adult (oviposition)	36	0.27	25.69
Adult (oviposition)	38	0.26	26.49
Adult (oviposition)	40	0.25	25.96
Adult (oviposition)	42	0.24	25.69
Adult (oviposition)	44	0.23	25.42
Adult (oviposition)	46	0.22	24.63
Adult (oviposition)	48	0.21	23.69
Adult (oviposition)	50	0.20	22.63
Adult (oviposition)	52	0.19	21.56
Adult (oviposition)	54	0.18	20.23
Adult (oviposition)	56	0.17	18.77
Adult (oviposition)	58	0.16	17.44
Adult (oviposition)	60	0.15	15.31
Adult (oviposition)	62	0.14	13.71
Adult (oviposition)	64	0.13	12.11
Adult (oviposition)	66	0.11	10.38
Adult (oviposition)	68	0.10	9.05
Adult (oviposition)	70	0.08	7.46
Adult (oviposition)	72	0.06	6.39
Adult (oviposition)	74	0.04	5.06
Adult (oviposition)	76	0.03	4.00
Adult (oviposition)	78	0.02	2.80
Adult (oviposition)	80	0.02	1.20

The Individual Parameters

Survivorship $l(x)$: This table shows that larval survival is 0.40 and pupal survival is 0.8. Adulthood is described by x = 24. Thus one third of the newborn med flies survive to adulthood, however, adult survival is high.

Fecundity $m(x)$: The pattern of fecundity is in 2-day age intervals. Females begin to lay eggs after a short pre-ovipositional period. At peak reproduction med-flies lay approximately 26 eggs/2 days.

Aggregate Parameters

The Malthusian Parameter r: The growth rate is $r = 0.109$. This indicates that the fly population will increase on average by approximately 10 each day.

The Mean Generation Time T: $T = 40.8$ days. This is the mean age at which individuals reproduce.

Population Entropy H: $H = .0609$. This number should be contrasted with the maximum entropy, denoted H^* (see 1.10) that could be attained for the generation time $T = 40.8$ days. $H^* = .115$. At maximal entropy it is known that no oscillatory components exist when departures from the stable age-distribution occur. The value of H, low when compared with H^*, indicates that in this population large fluctuations in population numbers due to departures from the stable state will occur.

<u>Reproductive Potential</u> Φ: $\Phi = -.0447$. The reproductive potential predicts the response of r to fluctuations. We note the reproductive potential Φ^* corresponding to a population with maximum entropy H* is $\Phi^* = r - H^* = .005$. The value Φ which is much larger in absolute value that Φ^*, indicates that large fluctuations in r will result from perturbations of the life-table - an observation also derived from the comparison of H and H*.

<u>The Function</u> Φ_1: $\Phi_1 = .517$. The parameter Φ_1 measures the convexity of the survivorship curve. $\Phi_1 = 0$ describes a rectangular survivorship curve characterizing the case where all individuals live to some maximum age ω. $\Phi_1 = 1$, describes the exponential survivorship curve representing an age-independent mortality. The number Φ_1 predicts the proportional change in mean life-expectancy due to changes in the shape of the survivorship curve.

<u>The Function</u> Φ_m: $\Phi_m = -.2908$. The absolute value of this function characterizes the convexity of the fecundity distribution. This function is analogous to Φ_1 and measures the proportionate change in gross fecundity due to changes in the shape of the fecundity distribution.

References

1. Lotka, A.J. (1922). The Stability of the age-distribution. Proc. Natl. Acad. Sci. 8, 339-345.

2. Leslie, P.H. (1945). On the use of matrices in

population mathematics. Biometrika 33, 183-212.

3. Demetrius, L. (1969). The sensitivity of population growth rate to perturbations in the life-cycle components. Math. Biosc. 5, 129-136.

4. Demetrius, L. (1974). Demographic parameters and natural selection. Proc. Natl. Acad. Sci. 71, 4645-4647.

5. Carey, J.R. (1982). Demography and population dynamics of the mediterranean fruit fly. Ecological Modelling 16, 125-150.

6. Shoukry, A. and Hafez, M. (1979). Studies on the biology of the Mediterranean fruit fly, Ceratitis capitata. Entomol. Exp. Appl. 26, 33-39.

7. Demetrius, L. (1979). Relations between demographic parameters. Demography 16, 328-340.

8. Demetrius, L. (1977). Measures of fitness and demographic stability. Proc. Natl. Acad. Sci. 74, 384-386.

9. Tuljapurkar, S.D. (1982). Why use population entropy It determines the rate of convergence. J. Math. Biol. 13, 325-337.

10. Demetrius, L. (1985). The meaning of entropy in demographic theory (submitted).

11. Keyfitz, N. (1977). Applied Mathematical Demography. New York. Wiley.

INTERFACING BIOLOGY AND SYSTEMS ANALYSIS IN PEST MANAGEMENT

Wayne M Getz
Division of Biological Control
University of California
Berkeley, CA 94720
USA

ABSTRACT

Mathematical techniques for analysing population processes and designing pest management programs can be loosely divided into two broad categories: agroecosystem analysis, and statistical decision theory. Here the relative strengths and weaknesses of these two approaches are presented. Methodologies are discussed for using systems analysis to design pest management programs and for developing computer software to implement these programs. Although the material presented here deals with the general problem of interfacing biology and mathematics in the context of pest management, the discussions and examples focus to some extent on questions relating to fruit fly management.

INTRODUCTION

In the last 10 to 15 years there has been a strong thrust to apply modeling and systems analysis techniques to problems in insect pest management (Ruesink 1976, Huffaker 1980, Levins and Wilson 1980, Getz and Gutierrez 1982). Progress has been disappointing, perhaps because our expectations of what could be achieved were too high. Initially it was hoped that crop and pest models would contain sufficient realism so that techniques applied to the control of industrial processes could be similarly applied to the control of pest population and pest-crop interaction processes. Several factors, as discussed below, are responsible for thwarting these efforts.

Pests pose a serious problem world wide: they threaten the subsistence of peoples in under-developed countries and are a major economic factor in crop production in developed countries. In California, for example, $100 million was recently spent to eradicate an invasion of the mediterranean fruit fly. Also such pests as mosquitos threaten the health and well being of many human and animal populations, especially in the Third World. Pest management is an ongoing process and we need to bring all our skills to bear on this process. These skills include the application of systems analysis techniques. We also need to learn from past mistakes in our application of systems analysis to pest management problems. Our most serious mistake has been the belief that biological processes can essentially be modeled in the same way as physical and industrial processes.

NATO ASI Series, Vol. G11
Pest Control: Operations and Systems Analysis
in Fruit Fly Management. Edited by M. Mangel et al.

In dealing with these issues I will keep the discussion general, but I will also draw on problems that relate to fruit fly management. The position that I take relates to personal experience in applying systems analysis to pest management. My ideas emerge through mists of frustration in dealing with approaches that others may feel have worked for them. For me, the lack of success is apparent.

A pest management problem has three interlinked but identifiable components: biological, analytical, and socio-economic. At one end are the economic issues of whether control is justified (see Farnsworth, this volume) and the socio-economic issues of implementing environmentally hazardous controls (see Norton, this volume). At the other end we need to understand the biological impact of particular controls and their long term effect on the agroecosystem. Analytical techniques can be used to tie together the biological impact of a program with the cost and logistics of its implementation. Not surprisingly, scientists in different disciplines have differing views on which apsects of the problem should be most emphasized. If a control program is not economically viable, it will not (or should not) be implemented. If essential biological aspects of the crop/pest system are ignored, the program is unlikely to succeed. These statements are obvious. What is not so obvious is whether a particular program is economically viable and what the essential biological aspects of the problem are. A priori we can estimate the costs of a control program and evaluate what level of benefits (pest reduction) are required to at least break even (see Farnsworth, this volume). A posteriori we can evaluate the cost and impact of a program, although we can never be quite sure that observed reductions are due to the control program or to some stochastic or undetected event. The issue of how much biology must be known to design and implement effective programs is a much more difficult question. In the past I believe that we have generally taken the wrong approach to dealing with the biological aspects of the problem.

AGROECOLOGY MODELING PERSPECTIVE

We can either endeavour to eradicate a pest or, if we accept a pest as a biological component of an ecological system, suppress and regulate its population. Systems analysis can be applied to both of these problems and it is commonly held that biologically detailed models of the crop-pest ecosystem will facilitate the application of systems analysis to the pest and crop management problems on hand. While such models are useful in crystalizing our understanding of the biology of a crop-pest system, that is they have a descriptive or qualitative value, they fail to provide the desired quantitative tools for systems analysis in an 'optimization of dynamic systems' setting. Before providing a more detailed discussion of these issues it would be just as well to briefly review the modeling of agroecosystems.

At the center of an agroecosystem model is a sub-model of crop growth and development. This sub-model can either be constructed from a detailed plant physiology model that may include heliotropic movements of leaves and stems or a less detailed but conceptually integrative 'carbohydrate pool' model (Gutierrez et al., 1983). The latter model allocates carbohydrate resources to the stems, leaves, roots and fruits of plants based on supply/demand relationships and an allocation priority scheme. These relationships depend on environmental factors, especially temperature. The impact of pests on the plants is easily integrated into the model. For example, leaf eating insects reduce the plants ability to produce carbohydrates while phloem and xylem feeders tap directly into the carbohydrate pool and reduce the availibility of carbohydrates for plant growth. Plant diseases and root nematodes can also be included through carbohydrate and nutrient uptake considerations. The parameters of these plant models are set to represent average growing conditions for a whole field and, as such, include elements of intraspecies competition for light and nutrients. Further, competition between plants and weeds can also be taken into account, but this requires a considerable amount of additional biological data to assess competitive affects. Although the model simulates biomass growth of leaves, stems, roots and fruit, in non-forage crops, it is the fruit component that is required for economic analysis; and often, both quantity and quality of fruit is important.

Some pests, such as fruit flies, damage fruit directly and the model should be designed to simulate the proportion of damaged fruit and the extent to which it is damaged. For pests such as thrips on oranges, this damage is only cosmetic but it does reduce the value of the fruit. Some plant models are sophisticated enough to include intrinsic fruit quality. In grapes, for example, Williams et al. (1985) simulate the number of 'brix' units in the fruit which is an important consideration for making wine. Since grapes are a perennial vine, the modeling task is complicated by the interannual factors that need to be taken into account. In tree crops such as peaches and almonds, branches do not bear fruit until their second season. Also walnuts, for example, require a certain level of exposure to cold temperatures (modeled by the number of temperature units below a threshold) before beginning a new seasons growth. Thus modeling the growth and development of tree crops requires considerable data relating the state of the trees from one season to the next. No such biologically detailed models currently exist.

Typical insect pest population models partition individuals into several life stages and respectively use phenology and life table methods to model developmental events (eg. adult emergence) and population numbers. Phenology models often use the concept of degree-days above a lower developmental threshold. Refined models may include upper developmental thresholds or the concept of an optimum development temperature (for a review see Wagner et al., 1984) but the application of most phenology model to field situations is frought with difficulties (see Hochberg et al., in press). Phenology models assume that development rates are directly controlled by temparature. Even if this were true (other factors such as nutrition and humidity may also play a role), it is virtually impossible to characterize the full range of micro-climates that each individual is exposed to and therefore accurately assess the temperatures that influence the development of each insect. Furthermore, only the daily minimum and maximum temperatures are usually available at a site which may be several miles removed from the field or orchard in question; and these two measurements are used to reconstruct the required temperature profile (most often a piecewise linear or sinusoidal approximation is used). The problems with life table data are even more tenuous. Firstly, survival rates are determined by many factors, some of which can never be fully characterized. For example, a new strain of disease may arise and decimate a population. The boundary of an ecosystem needs to be drawn somewhere and many species of predators and parasites that prey on the pests in question will either not be included or will be set at constant levels. Secondly, fecundity rates will depend on the health and nutritional status of individuals. Also, the inherent genetic variability of individuals and differential survival rates under different environmental conditions will affect the calculation of average population fecundity rates.

In addition to linking together the various biological components of the ecosystem model, biotic and abiotic inputs need to be linked as well to the model. Abiotic inputs, such as temperature, are usually included through the phenology components of the system model while inputs, such as precipitation, and humidity need to evaluated in conjunction with irrigation practices. The effects of adding fertilizers needs to be integrated into the crop submodel while the action of chemical pesticides and or biological actions such as the release of sterile insects need to be coupled with the appropriate insect submodel (see Plant, this volume). The efficacy of these actions is strongly linked to environmental conditions at the time of their application (eg. temperature, humidity). Futher, in the application of pesticides, the efficacy may be critically dependent on mechanical factors such as height of spray and size of droplets, making it almost impossible to accurately quantify the effects of some management actions.

Once we have what we regard as an acceptable model of our crop-pest system we can use it in conjuction with various mathematical techniques to address several management questions, the following being most typical. If a set of measurements is made on the system at some point in time (eg. stage and density of pest populations) what is the appropriate set of management actions that must be taken to maximize profit at the end of the season. The answer to this question may be in the form of an optimal spray schedule, or combination of spray schedules, cultural practices and biological control actions (i.e. an IPM solution). It also may involve the planning of an optimal harvest schedule. In almonds, for example, research is in progress to determine if the date of hullsplit can be manipulated using irrigation. As the nuts are most vulnerable to damage from an infestation by navel orange worm larvae, manipulation of hullsplit may save the application of one pesticide spray. In theory, existence of appropriate fruit flies models would allow us to compare the efficacy of bait sprays and sterile insect release techniques (SIT) as isolated actions or combined actions which may turn out to be more effective then either action alone. Fletcher (this volume) considers the application of a model to developing a spray/SIT program for the olive fruit fly, Dacus oleae, while Van den Driessche (this volume) presents models that can be used to analyse mating disruption of an SIT or sex pheromone release program. Implementation of an SIT program requires the production of sterile (irradiated) but competitive males, a requirement that presents some biological problems (see Hooper, this volume).

In reality, the application agroecosystems models to pest management problems is thwarted by the following factors:

* **limited precision** — heritable variation in the biology of individuals, local variation in temperature and humidity (micro-climate) and spacial variation in soil conditions places severe restrictions on the precision of the model or requires a model with an almost unachievable level of resolution.
* **limited accuracy** — an inability to include all ecosystem components in the model results in important processes being ignored and others are assumed to take place at a constant rate when they are, in fact, time varying.
* **limited extrapolation** — prediction requires extrapolation into the future under unknown abiotic conditions. The major abiotic driving variables of an agroecosystem, especially precipitation, are usually highly stochastic.
* **limited knowledge of the state and input variables** — it is expensive to intensively monitor crop and pest populations in the field and to measure the micro-climates at a level of resolution where a precise and accurate model (if it were available) could simulate actual changes in populations.

In summary, we are dealing with a biological system, yet we have no reliable way of computing the trajectory of that system into the future: we are barely able to estimate the present state of the system or plot its past course. If we try to pretend that the uncertainty in our knowlege of the state of the system is just a minor nuisance we are perhaps missing the essence of the problem. An approach that takes cognizance of this uncertainty is available and is described in the next section.

STATISTICAL DECISION THEORY AND SAMPLING

The biology of a pest determines which management strategies will be effective in its control and at what point in its life history it is most vulnerable to particular actions. Thus an ongoing study of the biological aspects of a pest management problem is required to develop new and more effective management strategies. For example, the effects of pesticides and pheromones need to be evaluated, traps for monitoring population levels can be improved through the development of better attractants, the life history and migration behaviour of populations needs to be understood, and so on. Since it appears that, even with extensive knowledge of the biology of the pest crop system, biologically detailed predictive models have a very limited use, we may get futher by taking a purely empirical view of the data associated with a particular crop-pest system. A good example of this approach is to use 'time series analysis' to predict future population levels from past data. This approach does not presuppose some deterministic or, for that matter, stochastic mechanistic structure is responsible for the particular values observed. It assumes that the only information available is in the data itself, and uses the correlations of a particular value with previous values, that is time series analysis, as a basis for predicting future values. This approach has been used with some success to model mosquito populations (Hacker et al. 1973). Detailed mechanisitic models have been used to simulate Nicholson's classical blow fly data (Auslander et al. 1974) but a time series model turned out to more closely replicate this data (Brillinger et al. 1980); and this was a situation where the data was collected from a population isolated in the laboratory and reared under controlled conditions.

Time series analysis provides a method of forecasting, where the forecast can then be used to help make a managment decision. On the other hand statistical decision theory provides various techniques for making management decisions based on guessing or estimating the probability distributions for the current state of the crop and/or levels of pests and on estimating the profit or yield that would be realized given that the crop is in a certain state and/or the pest populations are at certain levels.

Statistical decision theory is a well developed branch of mathematics (for example see Berger, 1980) and has been applied to agricultural problems (Mangel, 1985; Mangel et al., 1985). Recently Stefanou and Mangel (in press) used stochastic decision analysis to develop an adaptive scouting program to rationalize the application of pesticides for the control of lygus bug in cotton. Monitoring and sampling is a relatively important component of a pest management program. For some pests, such as mites and aphids, where all life stages of individuals are relatively sessile unless dispersal is taking place, sampling in the field is the primary method for estimating population levels. For pests with a mobile life stage, trapping techniques often exist, but the problem of relating trap counts to absolute population density levels is often very difficult. Usually traps use visual or olfactory cues to attract pests, but the 'attractiveness' of the cues is often a function of environmental conditions due to the existence of competing natural cues. The range and efficiency of a trap are difficult to assess under field conditions, although the release of tagged pests can be used to estimate trap parameters. Many traps use volatile chemical attractants, the strength of which dependends on temperature and wind conditions and decays with time. The cost of sampling or maintaining traps and the information gained through expending each additional unit of effort is an important component of any pest management program. If a highly attractive substance such as a sex pheromone is available, the traps themselves can be used to significantly reduce adult population levels. This type of control appears to be feasible for the olive fruit fly D. oleae (Haniotakis, this volume)

In using traps to estimate population density, there is firstly the question of how intensively an area must be sampled to ensure that the estimate of the actual population level falls within a prespecified confidence interval. This is an important consideration when trying to assess whether the population exceeds a predetermined economic threshold. The problem is dealt with by sampling sequentially until the desired confidence in the population estimate is obtained. These methods have been explored for several pest problems (Taylor, 1971; Wilson et al., 1983; for a review see Plant and Mangel, unpublished manuscript). Next the question is not only one of improving the accuracy of the estimate but the trade-off between the cost of obtaining additional information and the benefit that would accrue from that information if it led to improved management. For example, a risk averse grower is willing to spray if he is uncertain whether a pest population exceeds a threshold level. If increased sampling reveals that the population is indeed above the threshold than the cost of the additional sampling program is not recovered. If, however, increased sampling reveals that the population is below this threshold, then the cost of one spray application can be saved. To evaluate whether additional sampling should take place, one needs to be able to estimate the amount of information gained per unit additional sampling cost, the probability that the outcome will go one way or another, and the expected gains or losses associated with that outcome. Usually the probabilities involved are not known, but educated guesses are made.

A number of interesting questions arise in the placement of traps for the detection of pest infestations and the monitoring of population levels. If one has a limited number of traps and large area in which to assess a population then the optimal distribution of traps depends on whether one expects the population to be clumped or overdispersed and an <u>a priori</u> knowlege (perhaps obtained from the biology of the pest) of where clumps are most likely to be found. In particular, if one is trapping to detect an invasion using a set of regularly maintained traps, and a pest is found, then the question arises as to how many additional traps should be used and where they should be placed in the field. This problem has been looked at by Mangel et al. (1984).

Beyond trap placement, several other challenging sampling problems exist relating to the clumping of individuals in a population with respect to both space and time. Special sampling procedures will be more efficient at estimating the densities of clumped populations than those methods designed for randomly distributed populations (for example see Debouzie and Thioulouse, this volume).

DEVELOPMENT OF A PEST MANAGEMENT PROGRAM

The development of a pest management program is an interdisciplinary activity that will ususally take several years to reach fruition. Even then the development process is never complete, as the acquisition of new information and the evaluation of the performance of the program lead to modification and improvements. A pest management program may alter the production aspects of a particular crop or changes in an agricultural industry (e.g. in marketing or technology) may require modification of the existing pest management program. In general, one can consider the problem of 'development pathways' (Norton, this volume) in implementing a pest management program over the long term. Under a particular set of conditions, however, the process of designing a pest management program can be broken down into a four components (also see Plant and Mangel, 1985):

* **identification of strategies** — identify pesticides, pheromones, biological and cultural control options that can be used to disrupt or retard the development of pest populations. We also need to understand how each of these controls impacts the pest and affects other key components of the agroecosystem.
* **development of tactics** — here both the biology and economics of the problem need to be taken into account. Timing of control actions is important and will be related to the phenology of the crop-pest system.
* **implementation procedures** — once a pest management program has been formulated in theory, the question of implementation needs to be addressed. This may include developing protocols for sampling and/or application of actions. Standardization is important since field measurements and the efficacy of actions respectively depend on exactly how the measurements were taken and how the action was applied.
* **evaluation and adaptation** — inherent in the design of a pest management program must be the ability to evaluate and improve the efficacy of the program.

The first aspect of evaluation is to assess whether the desired levels of control are achieved (eg. can the crop be marketed as desired). If these are not achieved, we then need to be able to assess what went wrong and how the same error can be avoided in the future. For example, perhaps thresholds need to be lowered, cause and effect relationships reexamined, or perhaps problems with particular strategies have arisen that weren't anticipated. Examples of the latter are: unanticipated resurgence of secondary pests before the application of a pesticide that disrupted the trophic ecology of the system; and conflicts between strategies, where the sum of their effects when both are applied is less then the sum of their effects when considered separately.

Note that when developing tactics, it is often difficult to assess at what point a particular control action should be taken. Here the concept of an economic threshold is invoked and the control action is only taken if the pest exceeds a level where the benefits of improving the quality of the yield outweighs the costs of that action (for example see Fletcher, this volume). The economic threshold, however, is a dynamic quantity since a given pest level at one point in the season will not have the same effect as the same level at a different point in the season. Futhermore, thresholds will depend on a mix of management actions if several options are contemplated. For example, to control pink bollworm in cotton, both a mating disruption pheromone and a pesticide can be applied. Whether either or both should be applied depends on initial infestation levels; that is, several different thresholds are involved (Stone 1984). Rules also need to be developed on when to monitor especially as it relates to what measurements were taken in the past, what values were obtained and what control actions were undertaken or are contemplated. In cases where there is zero tolerance for a pest, eg. marketing a fruit in a country that will not permit the importation of infested shipments, the economic threshold is zero and the question then is one of how much effort should be put into monitoring the population so that any low level infestation is detected.

COMPUTER AIDED ANALYSIS

Computer hardware and software technologies have improved rapidily in recent years and are at the point where even the smallest commercial farm operation in a developed country can afford a microcomputer system powerful enough to be used both in the business operation of the farm and in making crop-pest management decisions. Besides storing weather and biological data, or running phenology and even simple pest and crop growth models, these micro computers can be used to plan and schedule irrigation and the application of sprays. For example, Getz et al. (unpublished manuscript) recently designed a package that can be used by mosquito abatement districts to schedule ricefield adulticide sprays. The distribution of rice and human population densities is summerized using county one-square-mile-grid maps. A scheduling algorithm uses this information and uses estimated adult mosquito emergence and dispersal rates in different ricefields to identify those fields that contribute the greatest number of mosquitos to urban areas on a weekly basis. The adult emergence rates can be estimated both through monitoring and the application of an appropriate population model.

The design of a computer package for almond orchard management has also been considered (Getz and McGrail 1981). The central component of this package is an irrigation scheduling routine. Irrigation has an impact on tree phenology, frost control (using overhead irrigation systems in cold weather), application of fertizlizers and the quality of the developing nuts. Since the application of ground sprays requires a firm orchard floor, irrigation schedules also need to be planned around the application of pesticides, fungicides and herbicides, and coordinated with the anticipated date of harvesting.

The elements of a crop management computer package would usually include: a method for predicting crop phenology and, if possible, yield; a method for predicting pest phenology and, if possible, population levels; a method for relating yield quantity and quality to pest populations at certain periods during the season; methods for evaluating the impact of control actions on pests, and hence on yield; algorithms for scheduling management actions, including irrigation, that ensure that logistical conflicts do not arise (eg. having to spray immediately after an irrigation treatment); and an 'accounting' module for calculating costs and estimating returns based on anticipated market prices. The output from an almond orchard irrigation scheduling algorithm is illustrated in the accompanying figure.

A computer package should be designed with the following features in mind:

* **modularization** — if modules are regarded as 'black boxes' with input and output plugs at either end, then they can easily be replaced by improved black boxes with compatible

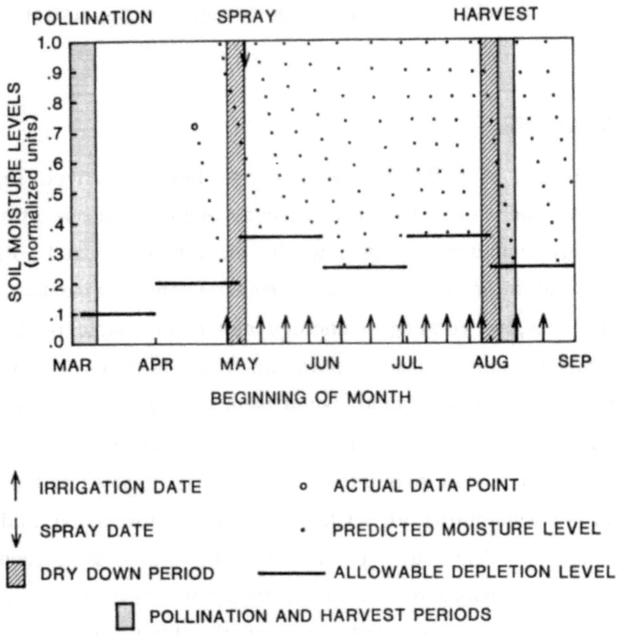

Figure. Allowable depletion levels are a priori determined based on seasonal considerations. Irrigation is scheduled on dates when it is anticipated that these depletion levels will be reached using a suitable evapotranspiration model. Just prior to harvest, however, irrigation is scheduled early to allow for a required 'dry down' period.

plugs. For example, the input into one of the black boxes may be the level of a certain pest at a particular stage of crop growth, and the output may be the estimated level of damage. Initially, the black box may be a simple regression relationship, later to be replaced by a dynamic model of pest plant interactions but supplying the same output that, for example, is required by an accounting module.

* **library of files** -- the library would contain general information relating to the application of pesticides etc. and would be updated as required. It would contain site specific historical records on weather, the yield performance of fields, and other data that is updated on a seasonal basis. Finally, files keeping track of the current status of the crop, current pest levels and the availability and condition of equipment would also be required.

* **decision algorithm** — an algorithm that can use the various modules to evaluate the combined effects of all feasible combination of management actions is required and it would also select those that are 'best' according to a predefined set of measures. These may include maximizing expected net revenue subject to a risk constraint and/or a pesticide usage constraint.

* **management scenarios** — it is desirable to generate several different options, each including information that may be especially noteworthy in implementing that option. For example, it may be likely the with certain programs there is a risk that secondary pests may emerge or that a particular spray schedule is conducive to the development of resistence. In this way the final management decision is left in the hands of the manager.

To include this latter type of information requires that qualitative information be processed along with quantitative information. One way of achieving this is through the implementation of an 'Expert System' as part of the total software package. An Expert System has been developed for the management of diseases in soy beans (Michalski et al., 1983), one is currently being developed for IPM in cotton (N. Stone, personal communication) although expert sytems do exists in other fields including medicine (Shortcliffe 1976), mineral exploration (Gaschnig 1982) and wildlife management (Starfield and Bleloch, 1983 and 1986).

An expert system consists of three components, one of which is a set of questions where each question has two or more answer options available. For example some of the questions asked by a fruit fly eradication expert system might be:

Q1. What species are we concerned with?

A1. Mediterranean fruit fly Ceratitis capitata.

A2. Melon fly Dacus cucurbitae.

A3. Oriental fruit fly Dacus dorsalis.

Q2. Which of the following were found?

A1. One or more adults, but only one gravid female.

A2. At least two gravid females.

A3. At least one adult, but no gravid females.

A4. Life stages other than adult.

Q3. Where were individuals trapped.

A1. Urban area only.

A2. In a commercial crop.

A3. In a rural non-commercial area.

Q4. If two or more individuals were found how far apart were the most distant two?

A1. Less than 100m.

A2. Less than 1km.

A3. 1-5km.

A4. Greater than 5km.

An expert system also has a section containing the set of possible decisions. For fruit fly, these may include:

D1. Continue to monitor at same intensity.

D2. Trap more intensively around the point of detection.

D3. Spray and strip within one kilometer square area of the point of detection.

D4. Apply aerial bait sprays in areas were infestations have been detected.

D5. Organize an eradication program that includes the release of sterile insects.

Based on the knowledge of a group of experts, the third section contains a set of logical decision rules which are devised to process the answers to the questions. For example:

R1. If Q2.A3 and Q3.A1 and Q4.A1 then D2 (i.e., if a localized infestation of adults is found in an urban area then intensify trapping and wait to see what happens before taking any futher steps)

R2. If Q2.A2 and Q3.A2 and (Q4.A3 or Q4.A4) then D4 (i.e., if more than one gravid female is found in a commercial area and the extent of the infestation is more then 1 sq km then organize an aerial spray program).

The ideal situation would be the interface of an expert system with predictive models so that the our 'expert knowledge' base is supplemented by our predictive capabilities. The development of such a package requires the collaboration of a multidisciplinary team of 'experts' and the services of a system modeler. This, of course, takes money and time but appears to be a worthwhile project.

CONCLUSION

Pest management is a pervasive problem in agriculture. Systems analysis can be employed to reduce the costs of managing pests, as well as to reduce the hazards associated with the application of toxic chemicals. Little progress, however, can be made without an adequate understanding of the biology of the systems under consideration. The trap we fall into is the belief that a more detailed understanding of the biology leads to better systems models. In my experience I have not found this to be true. Rather, a more detailed understanding of the biology leads to the development of new control methods and to implementing new and existing control methods more efficiently. We can then use systems analysis to cut through the complexity of the problem and find the best (in an appropriately defined sense) combination of options (i.e., control methods, timing and mode of application, etc.) in implementing a pest management program.

Most of the papers in this volume deal with problems relating to the management of one or more species of fruit flies. From these papers it is clear that models may prove to be useful in streamlining mass rearing techniques for SIT methods of control (Plant, this volume) and predicting phenological events in the field (Fletcher, this volume). However, the application, of systems analysis to designing an integrated control program that includes detection of infestations, the monitoring of population levels, and the application of one or more control techniques, requires a concerted interdisciplinary effort as well as a consonant level funding.

REFERENCES

Auslander, D. M., G. F. Oster and C. B. Huffaker, 1974. Dynamics of interacting populations. J. Franklin Inst. 297: 345-376.

Berger, J. O., 1980. Statistical Decision Theory, Springer-Verlag, New York.

Brillinger, D. R., J. Guckenheimer, P. Guttsop and G. Oster, 1980. Empirical modelling of population time series data: the case of age and density dependent vital rates. In: G. F. Oster (ed.), Some Mathematical Questions in the Life Sciences, Am. Math. Soc., Providence, Rhode Island.

Debouzie, D. and J. Thioulouse, 1986. Statistics to find spatial and temporal structures in populations. This volume.

Fletcher, B. S., 1986. Some issues in fruit fly ecology. This volume.

Farnsworth, R. L., 1986. An economic model to measure costs and benefits of eradication. This volume.

Gashnig, J., 1982. PROSPECTOR: an expert system for mineral exploration. In: D. Michie (ed.), Introductory Readings in Expert Systems, pp. 47-64, Garden Breach, New York.

Getz, W. M. and A. P. Gutierrez, 1982. A perspective on systems analysis in crop production and insect pest management. Ann. Rev. Entomol. 27: 447-466.

Getz, W. M., E. R. Lichtenberg and M. E. Hochberg, unpublished manuscript. Spray schedule decision models and the economics of ricefield mosquito control.

Getz, W. M. and L. McGrail, 1981. Computer Aids for CP/IPM in Almonds. AG Comp Bulletin 2: 53-69.

Gutierrez, A. P., J. E. DeVay, G. S. Pullman and G. E. Friebertshauser, 1983. A model of verticillium wilt in relation to cotton growth and development. Phytopathology 73: 89-95.

Hacker, C. S., D. W. Scott and J. R. Thompson, 1973. Time series analysis of mosquito population data. J. Med. Ent. 10: 533-543.

Haniotakis, G. E., 1986. The role of sex pheromones in the contol of the olive fruit fly, Dacus oleae: present status—prospects. This volume.

Hochberg, M. E., J. Pickering and W. M. Getz, in press. Evaluation of phenology models using field data: case study for the pea aphid Acyrthosiphon pisum Harris, and the blue alfalfa aphid Acyrthosiphon kondoi Shinji (Homoptera: Aphidoidea). J. Environ. Ent.

Hooper, G.M.S., 1986. Some observations on the competitiveness of sterilized mediterranean fruit fly. This volume.

Huffaker, C. B., 1980. New Technology of Pest Management. John Wiley & Sons, New York.

Levins, R. and M. Wilson, 1980. Ecological theory and pest management. Ann. Rev. Entomol. 25: 287-308.

Mangel, M., 1985. Decision and Control in Uncertain Resource Systems, Academic Press, Orlando, Florida.

Mangel, M., R. E. Plant and J. R. Carey, 1984. Rapid delimiting of pest infestations: a casestudy of the Mediterranean fruit fly. J. Appl. Ecol. 21: 563-579.

Mangel, M., S. E. Stefanou and J. E. Wilen, 1985. Modeling Lygus hesperus injury to cotton yields. J. Econ. Ent.

Michalski, R. S., J. H. Davis, V. S. Bisht and J. B. Sinclair, 1983. A computer-based advisory system for diagnosing soybean diseases in Illinois. Plant Disease 67: 459-463.

Norton, G. A., 1986. Feasibility assessment of pest management. This volume.

Plant, R. E., 1986. The sterile insect technique: a theoretical perspective. This volume.

Plant, R. E. and M. Mangel, 1985. Modeling and simulation in agricultural pest management. Unpublished manuscript.

Ruesink, W. G., 1976. Status of the systems approach to pest management. Ann. Rev. Entomol. 21: 27-44.

Shortcliffe, E. H., 1976. Computer-based Medical Consultation: Mycin. American Elsevier, New York.

Starfield, A. M. and A. L. Bleloch, 1983. Expert systems: an approach to problems in ecological management that are difficult to quantify. J. Environ. Mgmt. 16: 261-268.

Starfield, A. M. and A. L. Bleloch, 1986. Building models for wildlife management. Macmillan, New York.

Stefanou, S. E., M. Mangel and J. E. Wilen, in press. Information in agricultural pest control. J. Agric. Econ.

Stone, N., 1984. Analysis of the Pest Management of Pink Bollworm, Pectinophora gossypiella (Saunders) in Southwestern Desert Cotton, Gossypium hirsutum L. Ph.D. Thesis, Univ. of Calif., Berkeley.

Taylor, L. R., 1971. Aggregation as a species characteristic. In: C. P. Patil, E. C. Pielou and W. E. Waters (eds.), Statistical Ecology, vol. 1, pp. 357-397, Pennsylvania State University Press.

Van den Driessche, P., 1986. Some analytical models for biotechnical methods of pest control. This volume.

Wagner, T. L., H. Wu, P. S. Sharpe, R. M. Schoolfield and R. W. Coulsen, 1984. Modeling insect development rates: a literature review and application of a biophysical model. Ann. Entomol. Soc. Am. 77: 208-225.

Williams, D. W., H. L. Andris, R. H. Beede, D. A. Luvisi, M.V.K. Norton and L. E. Williams, 1985. Validation of a model for the growth and development of the Thompson Seedless grapevine. II. Phenology. Am. J. Enol. Vitic. 36: 283-289.

Wilson, L. T., D. Gonzalez, T. F. Leigh, V. Maggi, C. Foristiere and P. Goodell, 1983. Within-plant distribution of spider mites (Acari: Tetranychidae) on cotton: developing an implementable monitoring program. Ens. Ent. 12: 128-134.

TRAPPING AND INFORMATION IN FRUIT FLY MANAGEMENT

Marc Mangel*
Departments of Mathematics, Agricultural Economics
and Entomology
University of California
Davis, California 95616

ABSTRACT

The role of traps as sources of information and for reduction of uncertainty is discussed. This discussion includes time scales and trap problems, interpretations of zeros, population estimation, growth rate estimation, sterile to wild ratio estimation and effects of aggregation on trap catches.

* Address correspondence to Department of Mathematics, University of California, Davis, California 95616.

NATO ASI Series, Vol. G11
Pest Control: Operations and Systems Analysis
in Fruit Fly Management. Edited by M. Mangel et al.

1. MOTIVATIONS

Over the past few years, I have worked on the development of methodologies that can be used to obtain information from trap catches. My presentation at the ARW was an example of one such methodology (it is summarized in Mangel (1985 a,b) and a little bit in section 2 of this paper). So many interesting questions, problems and issues were raised at the ARW, however, that I chose to rewrite my presentation to address some of these. In the spirit of a workshop proceeding, then, this paper represents my most current thoughts about the problems of trapping, uncertainty, and information.

The original motivation of this work came from the following kind of problems. Traps can be used for purposes of control (as, for example, is currently being done in Greece on a large scale - see below) but trapping also may provide considerable information about pest population levels. Since one of the points consistently raised at the ARW concerns reduction of uncertainty and how it can be achieved, it seems appropriate to address the informational issues in trapping. Three operational motivations are these:

1. Information Gain Per Trap. In their study of the medfly problem in California, Mangel, Plant and Carey (1984) investigated how the information about the presence of fruit flies depended upon trap density. They did it as follows. Define P_0 and P_n by

P_0 = Prior probability that fruit flies are present in the region of interest

P_n = Probability that fruit flies are present in the region of interest, given that none were caught with n traps/sq. mi. (1)

(Apologies to the scientific community for not using SI units, but the use of miles rather than kilometers is dictated by operational reality.) A measure of information Δ_n provided by trapping is then

$$\Delta_n = \frac{P_0 - P_n}{P_0} \tag{2}$$

This measure has the following properties:

1. $0 \leq \Delta_n \leq 1$ (since under any reasonable set of circumstances $P_0 \geq P_n$)
2. $\Delta_0 = 0$ corresponds to no gain in information due to trapping

3. $\Delta_n = 1$ corresponds to perfect information (i.e., the chance of files being present after negative results in trapping is zero).

Mangel, Plant, and Carey (1984) constructed a very crude analytical model for the computation of P_n. From that they computed plots of Δ_n versus n. The advantage of such plots is that they can be used to answer questions such as "how much extra information does one obtain by use of five or more traps per square mile?" Figure 1, for example, shows a hypothetical plot of Δ_n versus n. If the current trap density is n_1 traps/sq. mi., then there is a considerable amount of information

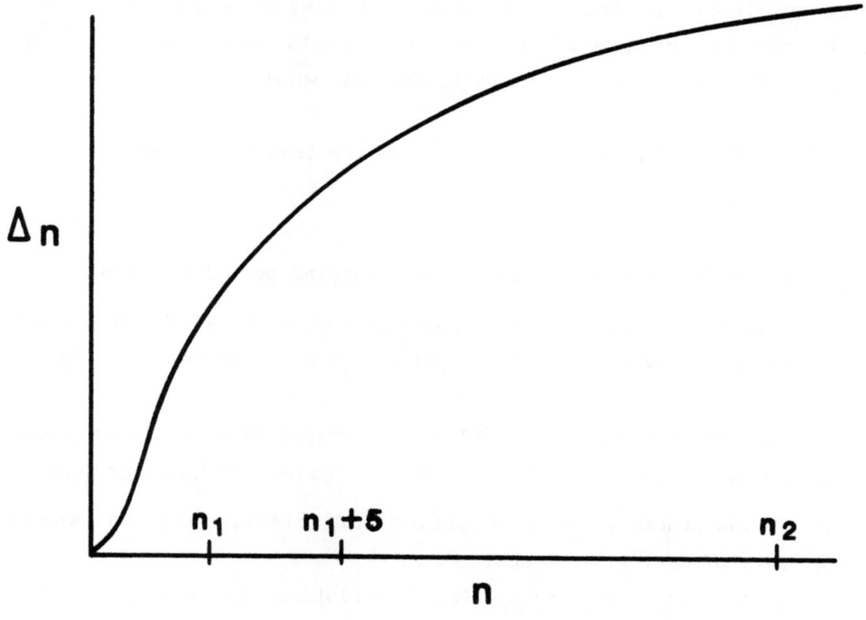

Figure 1. A Hypothetical Plot of Δ_n versus n

that can be gained by adding five or more traps per square mile (this gain must be weighed against economic costs, however). On the other hand, if the current trap density is n_2 traps per square mile, there is little point in adding five more, at least from the standpoint of a gain in information.

One motivation for this paper, then, is the ability to compute Δ_n. Calkins, Schroeder and Chambers (1984) did a similar estimation by using empirical data. Cunningham (1981) discusses the effects of spacing and dose/ trap on the estimation of P_n. One major question is how P_n depends upon n and how P_n and Δ_n are computed.

2. Early Detection of Invasive Pests. The following problem was posed by D. Chambers. Consider a number, C, of cities that are geographically separated. There are hosts for fruit flies within cities but there are no hosts between cities. The distance between cities is sufficiently great that the chance of a fruit fly migrating from one city to the next is negligible. (This kind of situation arises, for example, in the Central Valley of California where the cities are (from north to south) Sacramento, Stockton, Modesto, Fresno and Bakersfield.) Assume that fruit flies can be introduced into any one of the cities over some time period of length T. This would correspond to someone bringing infested fruit into the city. Finally assume that one has a total of N traps available and that these traps are renewable and moveable (e.g., McPhail traps). Many strategies for trapping over space and time are conceivable, but the extremes seem to be the following ones

S_1 : Use N/C traps per city for the entire trapping period of length T.

S_2 : Use N traps in one city for the trapping period of length T/C. At the end of this trapping period, move all the traps to the next city for another trapping period of length T/C, etc.

The main questions are then: (1) Which strategy gives a higher probability of detecting an invading pest? (2) If S_2 gives a higher probability of detection, is the probability of detection sufficiently high to warrant the increased costs of moving the traps?

3. The Haniotakis Experiment. Dr. G. Haniotakis is currently conducting a large scale (2,200,000 trees) experiment on olive fly control in Greece. Every other tree has a plywood trap impregnated with insecticide. These traps are used for control. At a much lower density (1 trap/1000 trees), McPhail traps are used to detect the level of olive flies. From an analytical perspective, two of the main questions associated with trapping for control in this experiment are the following ones:

1) How many traps are needed for a given level of control?

2) There are three kinds traps of traps available, with these parameters

Ratio of males:females Trapped	Cost
100 : 0	Intermediate
80 : 40	High
25 : 25	Low

Which trap should be selected?

From the point of view of trapping and information, the following questions arise

1) When does one know that the control traps have failed to do the job and it is time to spray? (For a variety of reasons - ecological, economic and operational - spraying is not desired.)

2) What is the relationship between trap catch and per-cent of infested fruit?

For example, one could envisage the following kind of plot

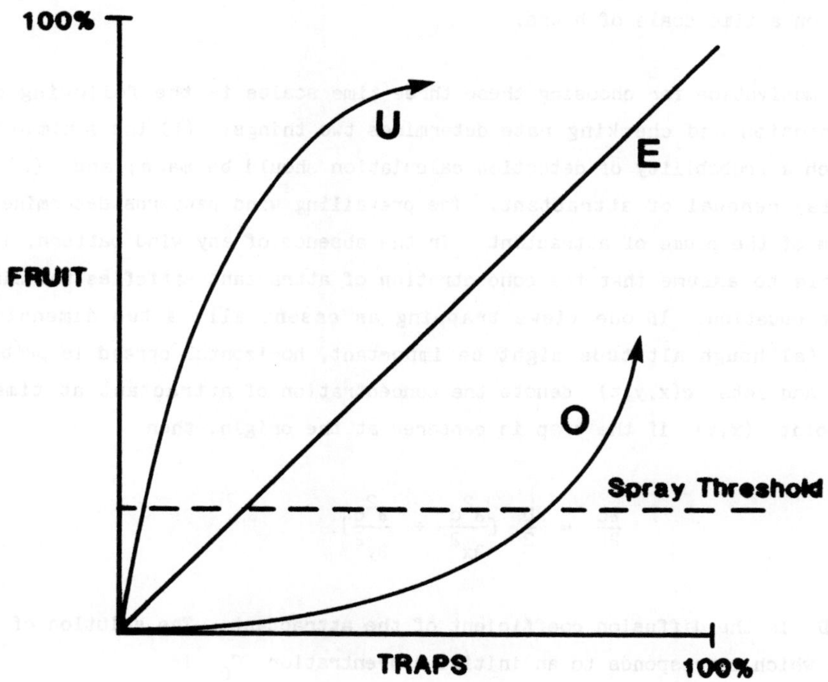

Figure 2. Trap and Fruit Infestations

If traps are equally efficient as fruit (the line marked E in Figure 2), then if x% of the traps have flies, one deduces that x% of the fruit has flies. If traps are under efficient (the line marked U), then a trap infestation of x% corresponds to a much higher fruit infestation; the reverse is true if traps are over-efficient (the line marked O). A theory of trapping should provide one with the ability to generate a plot such as Figure 2.

The rest of this paper is dedicated to a discussion of pieces of the theory of trapping and sketching out a number of important questions. In the spirit of the ARW, most of these questions are not answered in detail but are left as future areas of research.

2. TIME SCALES AND TRAP PROBLEMS

There are at least three time scales that should be considered in trapping problems. These are:

1. Inspection/Checking Rates. These are typically on the order of a week or so.
2. Prevailing Wind Patterns. One can assume that these patterns have a time scale of the order a day or so.
3. Fly Movement. One can assume that movement of the fly occurs on a time scale of hours.

The motivation for choosing these three time scales is the following one. The inspection and checking rate determines two things: (1) the minimum time over which a probability of detection calculation should be made; and (2) the (possible) renewal of attractant. The prevailing wind patterns determine the structure of the plume of attractant. In the absence of any wind pattern, it is reasonable to assume that the concentration of attractant satisfies the simple diffusion equation. If one views trapping as essentially a two dimensional process (although altitude might be important, horizontal spread is probably more so) and lets $c(x,y,t)$ denote the concentration of attractant at time t at the point (x,y) if the trap is centered at the origin, then

$$\frac{\partial c}{\partial t} = \frac{D}{2}\left[\frac{\partial^2 c}{\partial x^2} + \frac{\partial^2 c}{\partial y^2}\right], \tag{3}$$

where D is the diffusion coefficient of the attractant. The solution of this equation which corresponds to an initial concentration C_0 is

$$c(x,y,t) = \frac{c_0}{2\pi Dt} \exp\{\frac{-(x^2+y^2)}{2Dt}\} \qquad (4)$$

The main point of equation (4) is this: suppose that the fruit fly can sense an attractant concentration greater than c_{min}. At each fixed time, one obtains from (4) a circular plume corresponding to a level curve of (4).

Now, if a wind is present one does not expect the plume to be circles, but rather one expects ellipes with an orientation determined by the wind, as shown below: (Actually, ellipes may not be that reasonable either; see Figure 3 of Perry and Wall (1984).)

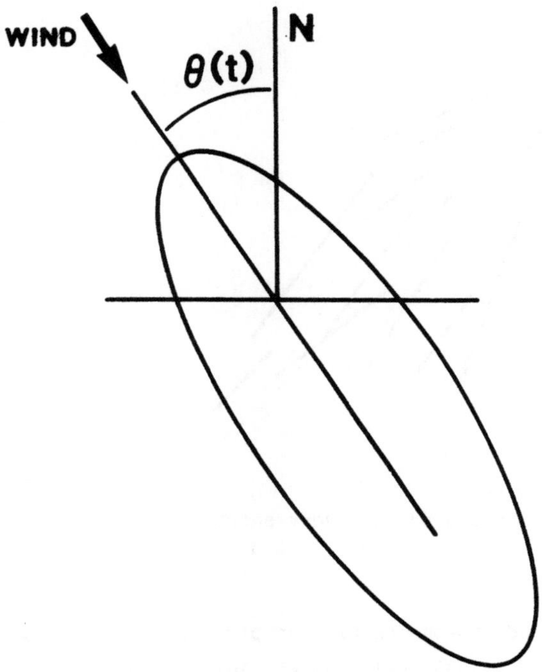

Figure 3. Plume Structure with a Wind

On the time scale of the wind, it is reasonable to use an "averaged" plume like the one shown and to characterize its orientation by the angle $\theta(t)$ where t is time. Presumably $\theta(t)$ could be predicted - at least with given probabilities - from historical wind data.

The third time scale characterizes the actual foraging behavior of the fruit flies. This is currently a hot topic of research (e.g., Kennedy et. al. (1981), Roitberg et. al. (1982), David et. al. (1982), David et. al. (1983), Perry and Wall (1984), Roitberg (1985), David (1985)). Clearly, on the time scale of an individual fly, one does not have a smooth plume such as the one shown in Figure 3. Probably one doesn't even have a smearing or "fuzzification" of the elliptical plume. Work reported in David et. al. (1982) suggests that on the fly's time scale, the plume structure probably looks like that shown in Figure 4.

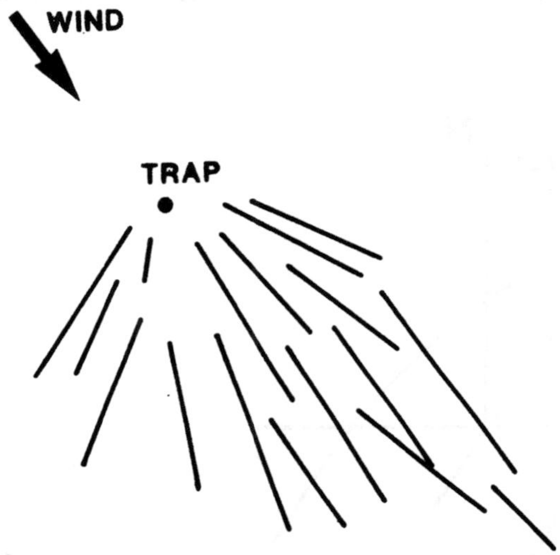

Figure 4. "Plume Structure" on a very short time scale.

Each line in Figure 4 corresponds to one particular plume of attractant. In the experiments of David et. al. (1982), using smoke plumes, the longest plume is about 5 - 7 m.

There is clearly not enough information for modeling detection on the time scale of Figure 4. Instead, the following procedure is a reasonable one to adopt:

1) Most flies execute a simple random walk (with convection due to wind currents) when outside of a plume of attractant. Some flies make large jumps, incompatible with a diffusion equation model. The percentage of flies making large jumps varies with species (e.g., about 2% for medfly or 90% for Queensland fruit fly).

2) When a fly enters a plume, there is a certain probability that it will be trapped, given its position in the plume. A useful operational technique is to divide the plume in Figure 3 into regions of low (L) median (M) and high (H) probability of trapping, as shown, for example, in Figure 5:

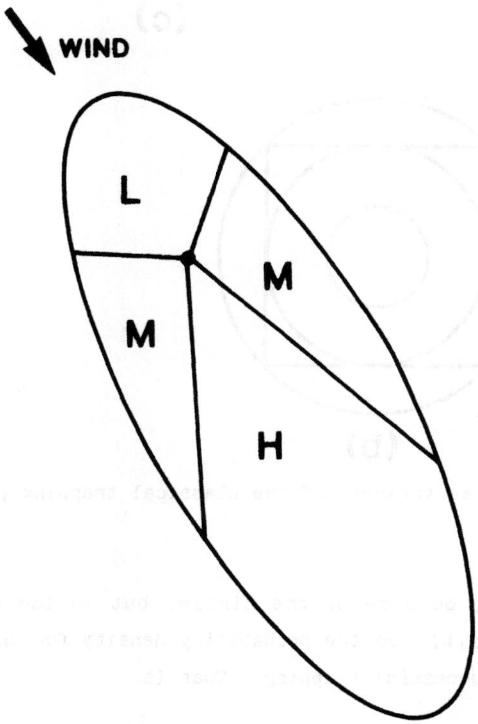

Figure 5. Trapping in the Plume.

The partition of the plume shown in Figure 5 is based on the assumption that flies fly upwind when following the scent of the attractant.

Data collected by Cunningham and Couey (1985) and other data discussed at the ARW suggest that a value for the probability of trapping in the high region is between 0.3 and 0.6 (depending upon the trap); a first guess might be that the probability in the medium region is 50% of the probability in the high region and in the low region it is 20% of the probability in the high region.

Now, the probability that a fly is trapped clearly depends on the amount of time that a fly in in the plume. This can be treated as follows. Let $p_H(t)$ denote the probability that a fly in the H region of the plume is trapped

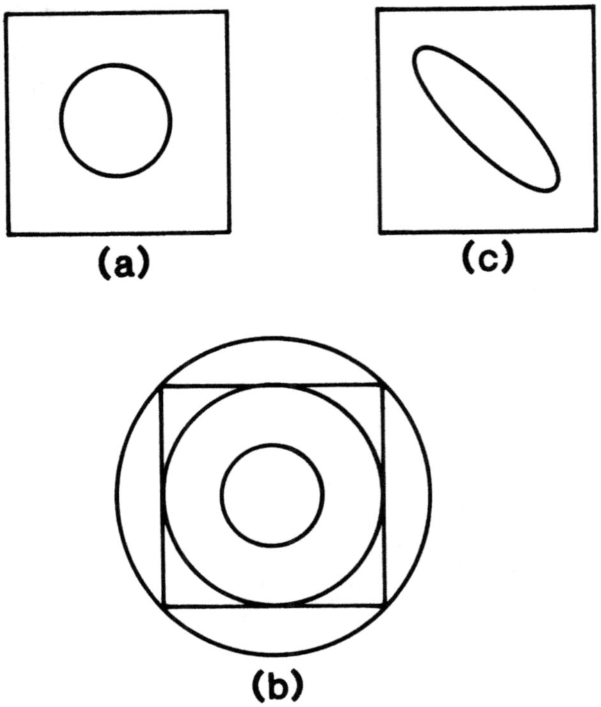

Figure 6. Three variants of the classical trapping problem.

Consider a fly starting outside of the circle, but inside of the square in Figure 6a. Let $f(x,y,t)$ be the probability density for the position of the fly at time t and unsuccessful trapping. That is

$$f(x,y,t)\,dxdy = \text{Prob}\,\{\text{fly is in the little rectangle } (x,\ x+dx) \text{ by } (y,\ y+dy) \text{ and not trapped at time } t\} \tag{9}$$

Let R denote the region outside of the circle but inside the square. If the fly executes a random walk then, in R, $f(x,y,t)$ satisfies the differential equation

$$\frac{\partial f}{\partial t} = \frac{\varepsilon}{2}\left[\frac{\partial^2 f}{\partial x^2} + \frac{\partial^2 f}{\partial y^2}\right] \tag{10}$$

where ε is the diffusion coefficient that characterizes motion of the fly. One needs initial and boundary conditions to go along with (10). A reasonable after t units of time in the plume. A reasonable model, suggested by the data of Cunningham and Couey (1985), is

$$p_H(t) = P_H(1 - e^{-\psi_H t}) \tag{5}$$

where P_H is the probability that the fly is ultimately trapped and ψ_H is a trapping rate. Note that

Prob {fly is trapped in t to $t + \Delta t$ given that it is not trapped by t}

$$= \frac{p_H(t + \Delta t) - p_H(t)}{1 - p_H(t)} \tag{6}$$

The numerator in (6) is the probability that the fly is trapped in $(t, t + \Delta t)$; the denominator is the probability that the fly is not trapped at time t. After simplification, (6) becomes

$$\frac{P_H e^{-\psi_H t}(1 - e^{-\psi_H \Delta t})}{1 - P_H(1 - e^{-\psi_H t})} \approx \frac{P_H e^{-\psi_H t}\,\psi_H \Delta t}{1 - P_H(1 - e^{-\psi_H t})} \tag{7}$$

Note that if P_H were 1 - which it is not - then (7) would simplify to $\psi_H \Delta t$. The important point, however, it that one can write

$$\text{Prob \{fly is trapped in } t, t + \Delta t\} = \bar{\psi}(t, P_H, \psi_H)\,\Delta t \tag{8}$$

where $\bar{\psi}$ is given by the right hand side of (7) and the operational variables P_H, ψ_H can be determined experimentally. The following table shows $\bar{\psi}(t, P_H, \psi_H)$ for $\psi_H = .05 \text{ hr}^{-1}$ and for three values of P_H:

$\underline{t}$ (hrs)	$\bar{\psi}(t,1,.05)$	$\bar{\psi}(t,.6,.05)$	$\bar{\psi}(t,.3,.05)$
1	.05	.0294	.0145
3	.05	.0282	.0135
5	.05	.0269	.0125
7	.05	.0257	.0116
9	.05	.0244	.0107
11	.05	.0232	.0099
13	.05	.0220	.0091
15	.05	.0207	.0084
17	.05	.0195	.0077
19	.05	.0184	.0071
21	.05	.0172	.0065
23	.05	.0161	.0059

Note that the per unit time trapping rate is strongly affected by the probability of ultimate trapping.

3. SET-UP, THE "CLASSICAL TRAPPING PROBLEM", AND ITS EXTENSIONS

Throughout the rest of the paper, it is assumed that one is dealing with a regularly spaced "lattice" of traps with center-to-center spacing given by Z_s. It is assumed that Z_s is sufficiently great that the traps do not interface with each other.

There is one mathematical problem that can be called the "Classical trapping problem". It goes as follows: If trap spacing is Z_s, then each trap has an effective area of Z_s^2 which it must "cover". Assume that there is no wind, so that the plume is a circle as in part (a) of the following figure:

initial condition is that the fly is uniformly distributed in R, so that $f(x,y,0) = 1/(Z_s^2 - \pi\rho^2)$ where ρ is the radius of the circular plume.

Boundary conditions are tougher. First consider the boundary condition on the square. A fly that crosses this boundary enters a different square, with exactly the same picture - it is as if the fly were reflected back into the original square. Thus, the appropriate condition is a reflecting boundary condition.

Now consider the inner circle. In what I call the "classical" trapping problem, one assumes that as soon as a fly enters a plume, it is trapped. Thus, one sets $f(x,y,t) = 0$ on the circle $x^2 + y^2 = \rho^2$.

The problem that has just been described is too hard to solve in closed analytical form, so an approximation to it is often considered. This approximation is shown in Figure 6b, in which one replaces the square by a circle of radius $\frac{Z_s}{2}$ (inscribed), $Z_s/\sqrt{2}$ (circumscribed) or $Z_s/\sqrt{\pi}$ (equal area). (Take your pick - this is an approximation). Then one switches to polar coordinates, so $f(x,y,t)$ is replaced by $f(r,\theta,t)$ and solves the transformed version of (10) with boundary conditions $f(\rho,\theta,t) = 0$ and $\frac{\partial f}{\partial r} = 0$ on the big circle. This problem can be solved by the method of separation of variables and the solution can be formed in the books by Crank (1975) and Carslaw and Jaeger (1959) or the papers of Muskat (1934) and Eagle (1985). This solution is good for many problems in physics and chemistry, but it is not clear how useful this "classical" solution is here. The problem is that the classical solution assumes perfect trapping inside of the plume and that is simply not true for trapping fruit flies.

In order to model imperfect trapping, it is easiest to use the formalism of stochastic differential equations (Mangel 1985 a,b,c). To begin, imagine a trap centered at the origin. Let $R(t)$ denote the distance that a fly is from the origin at time t and let $\Theta(t)$ denote the angle that the fly is at, relative to the x-axis. (That is, R and Θ are the usual polar coordinates). Consider a little increment of time dt and let $dR = R(t + dt) - R(t)$ and $d\Theta = \Theta(t + dt) - \Theta(t)$. Assume that $d\Theta$ is normally distributed with mean 0 and variance $\sigma_\Theta^2 dt$. Assume that dR is normally distributed with mean $-\alpha(R,\Theta)dt$ and variance $\sigma_R^2 dt$. The function $\alpha(R,\Theta)$ is a measure of how much the fly is attracted to the origin from the plant (R,Θ). Rough approximations of this function can be constructed from figures such as Figure 5.

Symbolically, one writes that $R(t)$ and $\Theta(t)$ satisfy the Ito equations

$$d\Theta = \sigma_0 dW_1$$
$$dR = -\alpha(R,\Theta)\,dt + \sigma_R dW_2 \tag{11}$$

where dW_1 and dW_2 are independent Brownian motion processes: dW_i is normally distributed with mean 0 and variance dt, and has continuous sample paths and independent increments.

The gist of this model is that from the point (R,Θ) the fly moves, an average, a little bit closer to the origin, but its motion is subject to random deviations as well. Let $f(r,\theta,t)$ be the probability density for $R(t)$, $\Theta(t)$, so that

$$f(r,\theta,t)dr d\theta$$
$$= \text{Prob}\,\{r \leq R(t) \leq r + dr,\ \ \theta \leq \Theta(t) \leq \theta + d\theta\}. \tag{12}$$

It can be shown (Mangel 1985 b) that $f(r,\theta,t)$ satisfies the partial differential equation

$$\frac{\partial f}{\partial t} = \frac{1}{2}\{\sigma_R^2 \frac{\partial^2 f}{\partial r^2} + \sigma_\Theta^2 \frac{\partial^2 f}{\partial \theta^2}\} + \frac{\partial}{\partial r}\{\alpha(r,\theta)f\} \tag{13}$$

(In writing (13), it is assumed that σ_R and σ_Θ are constants.) Let

$$\bar{f}(r,\theta) = \lim_{t \to \infty} f(r,\theta,t) \tag{14}$$

so that $\bar{f}(r,\theta)$ is the long time (equilibrium) limit of $f(r,\theta,t)$. The function $\bar{f}(r,\theta)$ can be found by setting the left hand side of (13) equal to 0 and solving the resulting partial differential equation.

The probability of trapping can now be identified as the probability that a fly is in a small circle, of radius δ, say, around the origin. To a first approximation, one can estimate the trapping probability by time t as $f(0,\theta,t)\pi\delta^2$ and the probability of ultimate trapping by $\bar{f}(0,\theta)\pi\delta^2$.

Thus the following program needs to be carried out:

1. Devise a procedure for estimating σ_R^2 and σ_Θ^2.
2. Construct $\alpha(R,\Theta)$.

3. Compute the functions $f(r,\theta,t)$ and $\bar{f}(r,\theta)$.
4. Extend steps 2 and 3 to multiple traps.

This program will not be done here but will be carried out in a subsequent paper.

To close this section, it is worth noting that trapping could be characterized in other ways as well. For example, one could consider a function

$$u(r,\theta) = \text{Prob}\,\{\text{fly reaches the origin before the boundary of the square} \mid \text{it starts at } (r,\theta)\}$$

and use this function (or a time dependent version of it) to characterize trapping.

4. INTERPRETATION OF ZERO TRAP CATCH

In an eradication program or an early detection program, one has a situation in which the lattice of traps produces no catch. The question is then: what can one say about the absence of fruit flies?

The simplest model, and one which is currently used as a rule of thumb, is the following. Assume that p is the probability that an individual trap in a unit area will catch at least one fly over one generation given that flies are present. (The exact way in which p is calculated or estimated is not specified here. That is, it is presumed that p is known.) Suppose that this lattice of traps contains T traps and that one is fairly confident that there are traps everywhere in the habitat of the fruit fly. Then

$$\text{Prob}\,\{\text{none of the traps catch a fly given that flies are present}\} = (1 - p)^T \tag{15}$$

To obtain equation (15) one assumes that traps catch independently; this need not be the case - see Mangel, Plant, and Carey (1984) for alternative modeling ideas. From (15) one has

$$\text{Prob}\,\{\text{at least one trap would catch a fly given that flies are present}\} = 1 - (1 - p)^T \tag{16}$$

Using (16), a common rule of thumb result is the following one. First, fix p (the ever elusive trapping probability). Second, choose a confidence level α close to 1 (e.g., $\alpha = .99$). Third, determine the number of traps $T(\alpha)$ by setting

$$\alpha = 1 - (1 - p)^T$$

so that (17)

$$T(\alpha) = \frac{\ln (1 - \alpha)}{\ln (1 - p)}$$

The interpretation of (16) and (17) is now the following: If p is the probability the one trap per unit area will catch at least one fly and one has $T(\alpha)$ traps per unit area, then the probability is α that at least one fly is trapped. For $\alpha = .99$, here are the values of $T(\alpha)$ for different p's:

p	T(α)	p	T(α)
.9	2	.05	90
.8	3	.01	458
.7	4		
.6	5		
.5	7		
.4	9		
.3	13		
.2	21		
.1	44		

Now one wants to argue as follows. Suppose that $p = .6$ and that 5 traps are placed per unit area. If no flies are caught over the time corresponding to one generation, then one is 99% confident that flies are not present. If no flies are caught over the time corresponding to three generations, then one is confident at a level of $1 - (.01)^3$ (or in general $1 - (1 - \alpha)^3$) that flies are not present.

What is wrong with this type of argument? There could be a number of difficulties; some of the most important are the following:

1) Traps may not trap independently.

2) The per unit area, per generation trapping probability p certainly depends upon population size, aggregation characteristics of the population, and movement.

Simply put, one might assert that we really don't know which p to use in (16) or (17); this will make a big difference in the resulting calculations and operational interpretations. In order to truly interpret zero trap catches, one must have a good idea about the value of p.

In spite of a lack of a firm knowledge of the value of p, however, one might proceed as follows. First, define a "cell" to have an area such that there is one trap per cell. Assume that the population of fruit flies in the ith cell, N_i, follows a negative binomial distribution with parameters M and k. That is

$$\text{Prob } \{N_i = x\} = \frac{\Gamma(k + x)}{\Gamma(k)x!} \left(\frac{k}{k + M}\right)^k \left(\frac{M}{k + M}\right)^x \tag{18}$$

The negative binomial (NB) distribution is chosen because it exhibits aggregation - most cells will have $N_i = 0$ if k is small and a few cells will have exceptionally large values of N_i. It is assumed that M is unknown, but that k is known.

In order to include trapping, let ρ be the probability that a fly is not trapped. Then $1 - \rho$ is the probability that a fly is trapped. It can be shown that the number of flies trapped in the ith cell, Y_i, is negative binomial with mean $m = (1 - \rho)M$ and aggregation parameter k.

The probability that the ith cell has no catch is thus (see (18))

$$\Pr \{Y_i = 0\} = \left\{\frac{k}{k + (1 - \rho)M}\right\}^k ; \tag{19}$$

and the likelihood L_T that none of T traps has a catch is then

$$L_T = \left\{\frac{k}{k + (1 - \rho)M}\right\}^{kT} \tag{20}$$

In Mangel (1985 d), it is shown how L_T can be used as the likelihood in a Bayesian analysis to estimate the posterior distribution on M. This analysis requires knowledge of the aggregation parameter k, which can be found as follows. For the NB distribution, the variance V of the population is related to the mean M by

$$V = M + \frac{M^2}{k} \qquad (21)$$

so that

$$\begin{aligned} \log\ V &= \log(M + \frac{M^2}{k}) \\ &= \log(\frac{M^2}{k}\ (1 + \frac{k}{M})) \qquad (22) \\ &= \log(\frac{M^2}{k}) + \log(1 + \frac{k}{M}) \end{aligned}$$

For large values of M, then, (22) becomes

$$\log\ V \sim 2 \log M - \log k \qquad (23)$$

and thus k can be determined as the intercept in a mean - variance plot. Baker (1985) provides the necessary data for such a computation. The results of this computation, and the Bayesian computation of the distribution on M will be reported in a subsequent paper.

5. POPULATION LEVEL ESTIMATION

Now consider the situation in which the lattice of traps is set out at time 0 and at time s, the trap catch is n flies, where n is small (e.g., n = 5). What can be said about the population level? Clearly, the smallest value of the wild population, N, that could generate a trap catch of n is $N_{min} = n$. What about the upper limit on N ? Very large values of N are conceivable, but one would like to have an assessment of the likelihood of differing values of N.

In this section, it will be shown how the methods of Mangel and Beder (1985), for fishing problems, can be adapted for the fruit fly problem. This also provides an opportunity to introduce Smith's Corollary, due to Paul Smith of the Southwest Fisheries Center, La Jolla, Ca.

Smith's Corollary: Any advance in entomological sampling theory is an advance in plankton sampling theory, and vice-versa.

That is, the analytical problems associated with fisheries and entomology are virtually identical - except that in fisheries are often does not know if the site being sampled is a habitat for the species of interest.

The analysis of the distribution of N is a Bayesian one, which proceeds as follows. Consider the trapping interval 0 to s in which T traps operate independently. Assume that

$$\text{Prob \{any one of these traps catches a fly in s hours of operation\}} = \beta(1 - e^{-\varepsilon Ts}) \tag{24}$$

Here β is the value of the ultimate probability of trapping ($\beta \sim 0.3$ to 0.6), ε is the trapping rate, and T is the number of traps.

Based on (24), if N flies are present in the wild population, then

$$\text{Prob \{n of N flies are caught in s hours of trapping\}} = \binom{N}{n} [\beta(1 - e^{-\varepsilon Ts})]^n \; [1 - \beta(1 - e^{-\varepsilon Ts})]^{N-n} \tag{25}$$

Equation (25) is based on the assumptions that i) flies are trapped once or not at all and ii) different flies are trapped independently of each other (for alternate views, see Mangel, Plant and Carey (1984)).

Using (25), the (classical) maximum likelihood estimate for N, is (Mangel and Beder, 1985)

$$\hat{N} = \frac{n}{\beta(1 - e^{-\varepsilon Ts})} \tag{26}$$

which is not a very surprising result. Equation (26) provides a value for $\hat{N}$, but says nothing about its likelihood. For this, a Bayesian analysis is used. Observe that (25) can also be interpreted as the likelihood $L(N,n;\beta,\varepsilon,T,s) = L(N,n)$ of a value of N given that n flies were trapped in trapping time s with other parameters β, ε and T.

If $g(N)$ is the prior distribution of N, then a Bayesian analysis can be used to construct the posterior distribution $g(N|n)$ (with the rest of parameters suppressed). Mangel and Beder (1985) show that a good choice of $g(N)$ is the uniform prior

$$g(N) = \begin{cases} 1 & N < 0, 1, 2, .. \\ 0 & \text{otherwise} \end{cases} \Big\} \tag{27}$$

The prior given in (27) is improper (it does not have a finite integral) and behaves quite similarly to the non-informative prior (that is, one in which data change the location, but not the shape of the posterior distribution). Using (27), the posterior distribution is

$$g(N|n) = \frac{L(N,n)}{\sum_{N=n}^{\infty} L(N,n)} \qquad N = n,\ n+1,\ldots \tag{28}$$

After some computations (Mangel and Beder (1985), pg. 153) one can show that

$$g(N|n) = \binom{N}{n} [\beta(1 - e^{-\varepsilon Ts})]^{n+1} \quad [1 - \beta(1 - e^{-\varepsilon Ts})]^{N-n} \tag{29}$$

for $N = n, n + 1,\ldots$ and $g(N|n) = 0$ otherwise.

Bayesian confidence intervals for the wild population are easily constructed from (29). Finally, the number of flies remaining in the wild, W has a distribution which is the same as the distribution of $N - n$. Thus

$$\begin{aligned} \Pr\{W = w\} &= \Pr\{N = n + w\} \\ &= \binom{w + r - 1}{w} [\beta(1 - e^{-\varepsilon Ts})]^{r} \quad [1 - \beta(1 - e^{-\varepsilon Ts})]^{w} \end{aligned} \tag{30}$$

where $r = n + 1$. Readers may recognize (30) as yet another form of the negative binomial distribution; this time characterizing the remaining wild population.

Although (29) allows one to easily compute the distribution of N and thus answer the question that began this section, it is based on two assumptions which are worth discussing in detail. These are

1. The value of ε is known. Note that ε characterizes the rate of trapping in some spatially averaged way. For accurate computations, the value of ε is needed, unless T or s is so large that $1 - e^{-\varepsilon Ts}$ is effectively 1. But it is likely that ε will be known only inpercisely and thus one must simultaneously estimate N and ε. Such an estimation can not be done with only one data point but can be done with a sequence of trapping periods and trap catches.

2. The binomial assumption (25) is valid. This assumption essentially corresponds to the model of independent trapping with removal upon trapping. Aggregation characteristics of the population are not taken into account. One

can extend the binomial distribution (25) to include analog of aggregation in the same way that the NB distribution is an extension of the Poisson that includes aggregation. This extended NB distribution is described in Mangel (1985 a); but work remains to be done on the use of it in population estimation similar to that reported in this section.

6. TWO OPEN PROBLEMS: r AND S:W ESTIMATION

Part of the objective of the ARW was to identify future areas of research which may have a considerable impact on the management of fruit flies. In that spirit, this section contains a discussion of two open problems.

r-estimation. A hotly debated issue at the ARW was the value of the population growth rate. That is, an effective management program often requires a reasonably accurate estimate of

$$r(t) \equiv W(t + 1)/W(t) \tag{31}$$

where $W(t)$ is the wild population in the field at time t. Until one knows clearly the relationships between field and laboratory growth rates (and possibly even afterwards), traps will be a major source of information about the growth rate.

In analogy with the problem discussed in the last section, there is now a triplet of parameters $(W(t), r(t), \varepsilon)$ to estimate. It is conceivable that the methods described in the previous section can be extended to include the estimation of this triplet, but the work still needs to be done.

S:W estimation. At the heart of decision making in a sterile insect program is the ratio of sterile to wild flies (the overflooding ratio). Once again, a major way for determining this ratio on a periodic (e.g., weekly) basis is by monitoring trap catches, assuming that one can separate sterile from wild flies easily (e.g., that sterile flies are somehow marked).

Using trap catches to estimate the ratio

$$R(t) = S(t)/W(t) \tag{32}$$

(where $S(t)$ and $W(t)$ are the sterile and wild populations at time t respectively) presents a number of interesting analytical problems. One approach would be to estimate $S(t)$ and $W(t)$ separately and then compute $R(t)$. This is probably not a good approach since it requires two separate estimates when what one wants is really one estimates and, even if the estimates of S(t) and W(t) are unbiased, then the estimate of R(t) need not be unbiased.

A second approach would use the ratio of sterile to wild flies found in traps. This presents some thorny questions for the case in which no wild flies are found in a trap or in which no flies are found in a trap (i.e, what does a ratio of 0/0 mean?). This should be an area of productive research in the future.

7. CONCLUSION

Effective trapping and understanding the meaning of trap catches lies at the heart of successful management of fruit flies. Although traps can be used to control fruit flies, they also can be used to provide information about location, population levels and growth rates.

The problems discussed in this paper highlight the need for understand two things associated with trapping. The first is the effectiveness of a single trap. Experimental and theoretical work are in progress, but more needs to be done until a point is reached where one can accurately predict the probability that a fly in a given distance away from a trap will be caught. The second area for future research is the study of the effects of population aggregation on trap catches and how differing levels of aggregation modify the predictions of single trap effectiveness.

ACKNOWLEDGEMENTS

This work was partially supported by the Agricultural Experiment Station of the University of California and by the California Department of Food and Agriculture through contract 6822.

I thank M. Aluja and R. Cunningham for discussions at the ARW, and A. Ludlow and C. David for discussions at Silwood Park, which caused me to rethink the problem, and thus rewrite the paper.

REFERENCES

Baker, P. 1985. Report on fruit fly trapping experiments, Tapachula, Mexico., unpublished report issued by MOSCAMED Program.

Calkins, C.O., W.J. Schroeder, and D.L. Chambers. 1985. The probability of detecting the Caribbean fruit fly, Anastrepha suspensa (Loew), (Diptera: Tephritidae) with various densities of McPhail traps. J. Econ. Ent., forthcoming.

Carslaw, H.S. and J.C. Jaeger. 1959. Conduction of Heat in Solids, Oxford University Press, Oxford, UK.

Crank, J. 1975. The Mathematics of Diffusion, Oxford University Press, Oxford, UK.

Cunningham, R.T. and H.M. Couey. 1985. The mediteranean fruit fly: distance-response curves to trimedlure to measure trapping efficiency, Environ. Ent., to appear.

David, C.T. 1985. Visual control of the partition of flight force between life and thrust in free-flying Drosophila. Nature 313:48-50.

David, C.T., J.S. Kennedy, A.R. Ludlow, J.N. Perry, and C. Wall. 1982. A reappraisal of insect flight towards a distant point source of wind-borned odor. J. Chem. Ecol. 8:1207-1215.

David, C.T., J.S. Kennedy, A.R. Ludlow. 1983. Finding of a sex phermone source by gypsy moths released in the field. Nature 303:804-806.

Eagle, J.N. 1985. Estimating the probability of a diffusing target encountering a stationary sensor. Naval Postgraduate School Report NPS55-85-013, Department of Operations Analysis, Naval Postgraduate School, Monterey, CA.

Kennedy, J.S., A.R. Ludlow, and C.J. Sanders. 1981. Guidance flying of male moths by wind-borne sex phermone. Physiol. Ent. 6:395-412.

Mangel, M. 1985a. Search models in fisheries and agriculture. to appear in Proceedings of the R. Yorque Conference on Resource Management (M. Mangel, ed.), Springer Verlag, New York.

Mangel, M. 1985b. Search Theory. Unpublished manuscript, available from the author

Mangel M. 1985c. Decision and Control in Uncertain Resource Systems, Academic Press, New York.

Mangel, M. 1985d. Sampling surveys for high aggregated populations, Biometrics, submitted.

Mangel, M. and J.H. Beder. 1985. Search and stock depletion: theory and applications. Can. J. Fish. Aq. Sci. 42:150-163.

Mangel M., R.E. Plant, and J.R. Carey. 1984. Rapid delimiting of pest infestations: a case study of the mediteranean fruit fly. J. Appl. Ecol. 21:563-579.

Perry, J.N. and C. Wall. 1984. A mathematical model for the flight of pea moth to phermone traps through a crop. Phil. Trans. Roy. Soc. Lond., B 306:19-48.

Roitberg, B. 1985. Search dynamics in fruit-parasitic insects. J. Insect. Physiol., to appear

Roitberg, B.D., J.C. van Lenteren, J.J. M. van Alphen, F. Galis, and R.J. Prokopy. 1982. Foraging behavior of Rhagoletis pomonella, a parasite of hawthorn (Crataegus viridis) in nature. J. Anim. Ecol. 51:307-325.

PESTS AND POPULATION MODELS: FLUCTUATIONS, EQUILIBRIUM AND PERSISTENCE

A. Meats
School of Biological Sciences, University of Sydney, N.S.W. 2006, Australia

ABSTRACT

A survey is made of the various possible causes of fluctuations, equilibrium and persistence in populations. Particular attention is given to the dynamical behaviour of systems which have apparently no stabilizing mechanisms but which nevertheless persist. Problems posed by the population dynamics of thrips, a cranefly, and tephritid fruit flies are used as illustrations.

INTRODUCTION

A pest system involves at least one type of relationship between a consumer (the pest) and a resource (usually a crop). Relationships involving other species either from the same trophic level or other trophic levels may complicate the situation and thus affect the relationship of primary interest between the pest and its resource. The dynamical properties of a pest system can therefore be due to a number of causes, the relative importance of which may vary from time to time and from system to system. It is the purpose of this article to survey the various causal mechanisms that have been proposed with a view to illuminating unresolved problems which require investigation by experiments. It will be shown that the possibility of alternative causes can be overlooked. In pest ecology, a failure to identify important mechanisms has economic as well

NATO ASI Series, Vol. G11
Pest Control: Operations and Systems Analysis
in Fruit Fly Management. Edited by M. Mangel et al.

as scientific consequences.

CLASSIFICATION OF CONSUMER-RESOURCE INTERACTIONS

The Monro (1967) classification of the possible relationship between consumers and their resources can encompass all existing hypotheses or population dynamics. Caughley and Lawton (1981) use it extensively to discuss relationships both within and between trophic levels. The brief outline given below is given as a basis for further discussion.

The classification is dichotomous as shown in Fig. 1. The first branching (1,2) distinguishes whether the consumer affects the rate of supply of the resource. In the first case (1) it does not in the second (2) it does. The second branching distinguishes whether intraspecific influences (such as competition, density dependent dispersal etc.) occur among consumers (1a, 2a) or whether it does not (1b, 2b).

A classification with only four endpoints seems too restrictive. The latter should be regarded, however, as idealized points towards which a situation tends at any one time. Caughley and Lawton (1981) point out that some consumer-resource systems move from one category to another according to circumstance. Fig. 2 shows the changes that will be discussed at various stages in the present paper.

Non-interactive relationships

Relationships in category 1(a) are best exemplified by (i) populations of micro-organisms, insects, etc., grown in culture with a constant supply of food and space, (ii) consumers of leaf litter, fruit, etc., (iii) plants and sessile animals which compete for space, oxygen, light, soil moisture, etc. The consumer population is constrained by competition in such a way that it can be modelled by a

Fig. 1 CONSUMER-RESOURCE RELATIONSHIPS

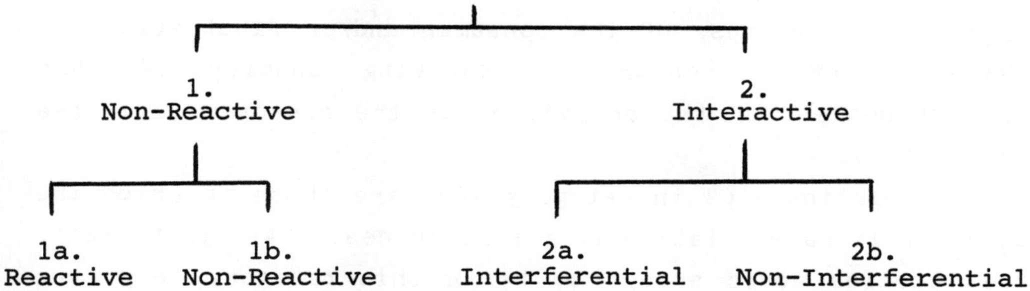

Fig. 2 CHANGES IN CONSUMER-RESOURCE RELATIONSHIPS

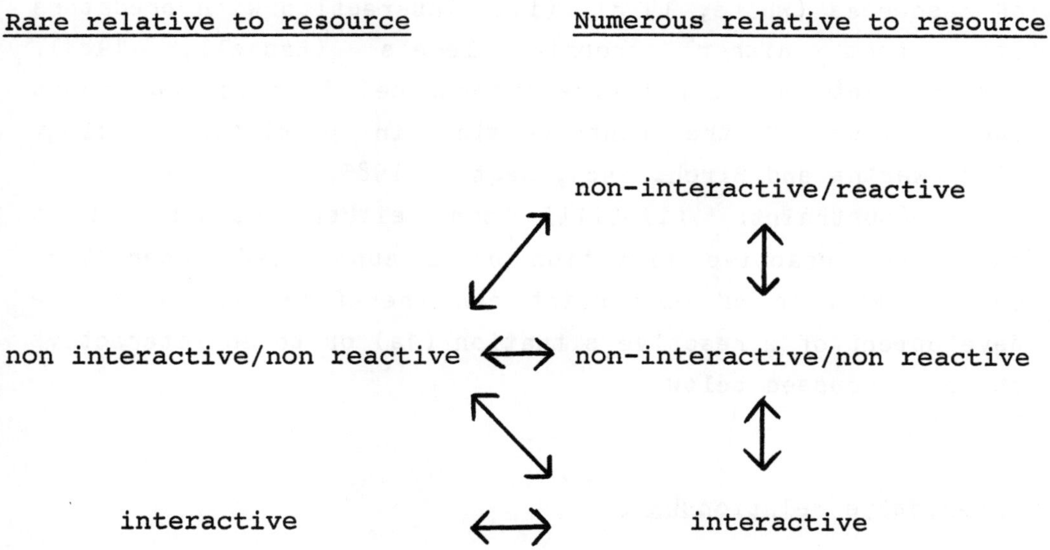

density-dependent equation such as the logistic. The latter may have to be modified to account for the existence of time lags in the response of the consumer and/or fluctuations in the supply of the resource or 'carrying capacity' (K) that are independent of the activities of the consumer (e.g. see May, 1981a).

Relationships in category 1(b) are those in which the consumer is rare relative to its resources. At first sight, most species would appear to be in this category (e.g. see the species/abundance curves given by Williams, 1964). We must, however, be cautious about such a conclusion since (because of their rarity) the ecology of most rare species is not studied and we do not know what their real resources are. However it is possible to conceive that a consumer may never become numerous because of the supervention of factors (independent of its resource) which reduce its ability to maintain a positive rate of increase in numbers for long enough.

Such causes of rarity include (i) frequent adverse weather (Birch, 1957; Milne, 1984); (ii) relative shortage of resources (White, 1978); (iii) interaction with predators etc. from higher trophic levels (Hassell, 1981). Disagreement on the relative importance of these causes has fueled most of the controversies in population ecology (Andrewartha and Birch, 1984; Lawton, 1985).

Constraints (i)-(iii) can either result in a suppressed reactive situation or a suppressed interactive one. Release of constraint can therefore result in the development of a reactive situation (1a) or to an interactive one as discussed below.

Interactive relationships

Here, consumption of a living resource affects the ability of that resource to replace itself; this in turn affects the availability of the resource to the consumer.

Models of this situation indicate that such systems, in the absence of other influences, would tend towards either a stable equilibrium or (more rarely) to a stable 'limit-cycle'; 'chaotic' dynamical behaviour is also possible in theory (Hassell, 1981, 1985; May, 1981b).

Caughley and Krebs (1983) make the best distinction available between the two types of interactive relationship. In the first kind (2a) the consumer populations have some self-regulating (intrinsic) features such as territoriality, breeding or feeding hierarchies, which act in a density dependent manner, hence the term 'interferential'. With the second kind (2b) the consumers are governed by a relative shortage and do not contemporaneously pre-empt one anothers' supply (i.e. they are non-interferential). The rate of increase of such consumers are thus governed only by the supply of the resource and not by consumer numbers; Caughley and Krebs (1983) give examples of models where the rate of increase of grazers is governed by the sparseness of vegetation rather than the population density of the grazers.

EQUILIBRIUM, FLUCTUATION AND EXTINCTION

If a system has a potential equilibrium state it would be useful to know what it is if one is going to manage that system.

Interactive systems will generally tend to equilibrium or cycles as mentioned earlier. An equilibrium state may involve both the consumer and resource holding each other at low densities with the resource being at a much lower density than it would be in the absence of the consumer (see later).

Non-interactive consumers may come into equilibrium with the rate of supply of the resource; if this rate fluctuates, the density of the consumers may track these fluctuations (May, 1981a). Consumers like this (non-interactive/reactive) tend to push up to the carrying capacity of the resource and if the consumer is a pest then levels of damage may be unacceptable.

Non-reactive systems have potential stability if the reason for non-reaction is that the consumer has an interactive relationship with a natural enemy. Hassell (1981) gives examples of natural enemies (biological control agents) keeping pests to around 1% of the levels attained in the absence of biological control.

If either climatic factors or relative shortages are proposed as causes of non-reaction then it is very much harder to explain why the densities of consumers persist around certain levels because such mechanisms are incapable of causing an equilibrium. Models driven entirely by changes in climatic factors or relative shortages show that these influences may be very important in determining whether populations increase or decrease but they also show that there is no guarantee that a population increase will stop short of reaction or that a decrease will stop short of extinction.

Stability and relative shortage

When a consumer is governed by a relative shortage then the level of that shortage determines the rate of increase of the consumer regardless of the latter's numbers. Different densities of resource (e.g. food items available per day's foraging; grams of protein obtained per day's eating) permit different rates of increase. A certain density of resource permits zero increase; the higher the resource density is above this critical level the higher the rate of increase of the consumer; conversely the lower the resource density is below the critical level the higher the rate of decrease of the consumer. Interactive systems involving relative shortage usually have a tendency for equilibrium (Caughley and Krebs, 1983) but there is no inherent tendency in non-interactive systems.

A good example of the latter is the famous case of *Thrips imaginis* which has generated much controversy (see summary by Andrewartha and Birch, 1984). The latter hold that Thrips imaginis fluctuates chiefly according to fluctuations in the relative shortage of flowers. Abundance of flowers varies with season with the highest abundance in spring. The spring peak of flowers in some years is higher than others, depending on the weather, with the result that thrips numbers can be predicted by an equation derived from a multiple regression analysis of the relationship between thrips numbers, time of year and weather factors. Unfortunately, the resulting model links weather and thrips directly (which is quite valid for predictive purposes) but the omission of the proximal causal link (relative shortage of flowers) has led most readers to overlook the overriding importance of relative shortage.

The following is an heuristic model for a hypothetical thrips population which is governed solely by a relative shortage of flowers. It is not a strategic model for Thrips imaginis; it does not attempt to mimic exactly the numbers of thrips or flowers in any recorded situation but merely serves to illustrate the driving power of relative shortage.

The model is as follows .....

$$B_t = B_{t-1} \quad \lambda_{t-1,t} \; P_t$$

Where B_t, B_{t-1} are the numbers of thrips in flowers at times t and t-1 respectively; $\lambda_{t-1,t}$ is the rate of increase between t-1 and t; thrips produced in one interval disperse to find new flowers in the next; P_t is the probability of a thrip finding a flower and is found by $P_t = F_t C$ where F_t is the number of flowers and C is the chance of a thrip finding a flower when $F_t = 1$.

An example of the results of the model is given in Fig. 3. The value of C is set at 0.01 and λ is set at 10. Fluctuation in the density of flowers (F_t) drives the model. The outcome illustrates the effect of an increase of flowers in spring. There is no reactive (density dependent) mechanism in the model for the thrips and the number of thrips has no influence on flower density. The value of λ is arbitrary and held constant so that the effect of F_t alone is made clear. 'Generations', however, are plotted on a physiological time scale in Fig. 3 so that they are shorter in the warm weather, as might be expected.

If the number of thrips per flower was directly proportional to the density of 'searching' thrips (i.e. if thrips fell like rain with no real ability to locate flowers) then the plots in Fig. 3 could equally apply to 'thrips per flower'. If, however, thrips had some searching ability then we would expect the summer peaks to be sharper and perhaps we could expect the winter lows would not be so low (also see later).

Note that the 'population' would only remain constant if F_t had a certain constant value. In this case, the critical value is $F_t = 10$. This is determined by the value of P_t required to balance λ, and is found by $F_{t(critical)} = 1/\lambda C$. For true stability, in the current example, F_t must fluctuate on either side of 10 in such a way that upward trends are balanced by downward ones. This does not happen in Fig. 3 where there is tendency for downward trends to be greater than upward ones; if the model kept using the same sequences of F_t values, the summer peaks and winter lows would progressively decline. A set of F_t values could have been chosen to continue a sequence of identially high summer peaks and identically low spring troughs but we should not expect F_t values to follow such a fortuitous chain of events. The trajectory of flower availability will vary from year to year and we can expect the trajectories of thrips numbers to vary also. In the case of Thrips imaginis there was a 100 fold difference between the highest and lowest spring numbers during the 14-year study (Davidson and Andrewartha, 1948). A 10-fold range was observed in summer peaks.

There is nothing in the hypothetical model to guarantee freedom from extinction or to prevent an astronomic rise in numbers by relative shortage. The model only

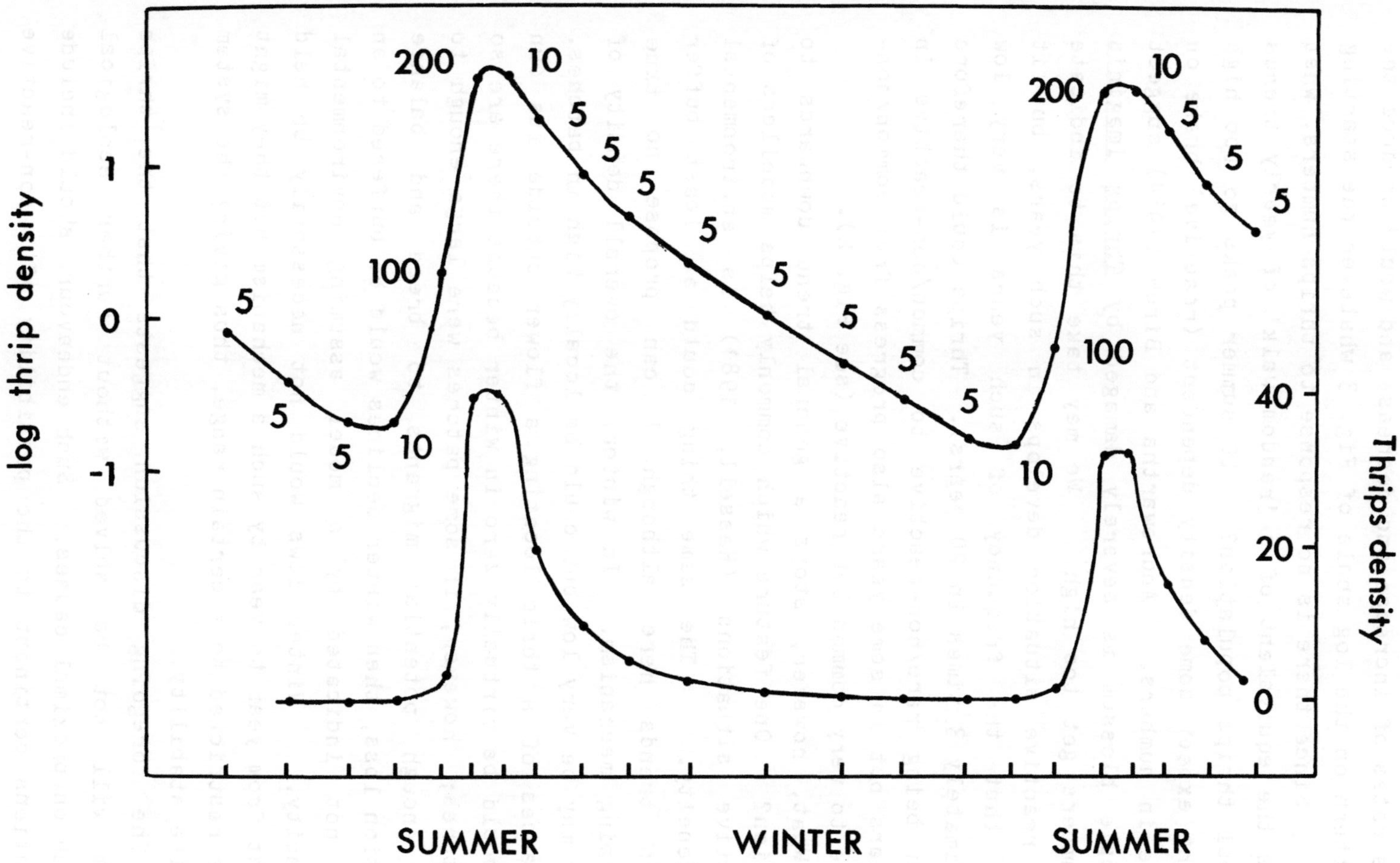

<u>Fig. 3</u>. A hypothetical thrips population affected only by a relative shortage of flowers. Density of flowers at any time is given by figures along the logarithmic plot.

dictates rates of increase and decrease and would produce the same pattern on the log scale of Fig. 3 whatever the starting numbers. Since there is no response to thrips numbers, what prevents the equivalent of a 'random walk' of yearly trends in a real thrips population? If summer peaks go too high one might expect some density dependent (reactive) brake on the rise in numbers. Andrewartha and Birch (1984) suggest that apple blossom is severely damaged by Thrips imaginis when numbers get too high. We may take this to indicate that a reactive situation develops in such years, but it appears that the frequency of such years is very low (approximately 3 times in 20 years). Thrips could therefore go from being rare/non-reactive to common/non-reactive in most years but in some years also progress from common/non-reactive to very common but reactive (see Fig. 2).

What, however, stops a general trend downwards to extinction? One feature which commonly helps modellers of interactive situations (Hassell, 1981) is environmental heterogeneity. The same thing could at least buffer downward trends here although I can propose no true stabilizing mechanism. In winter, the overall density of flowers may be very low but could be locally high in patches. The chances of a thrip locating a flower outside its own patch would be virtually zero in winter because there are so few patches; however, if some patches were large enough to retain enough potential migrants to breed and balance emigration loss, then winter declines would be buffered to an extent not indicated by a model assuming environmental homogeneity. Winter lows would not necessarily be held constant from year to year by such a mechanism but they might well be restricted to a certain range, thus giving the system effective stability.

The foregoing discussion suggests that the thrips problem will not be solved without further ecological research on proximal causes. Such endeavours should include observations pertinent to the possibility of a non-reactive system persisting through mechanisms (such as reponse to patches of resource, dormant stages, iteroparity) which result in resistance to extinction in unfavourable periods coupled with a tendency to 'saturate' the resource in favourable periods. If (as in the hypothetical model above) there is a tendency for increase in the consumer to proceed faster than increase in resource once the resource goes above a critical density, then there will be a tendency to catch up on losses even before saturation (reaction resulting in a density dependent brake) sets in. If this were the case then density dependence would not be needed as a mechanism for stability except in the occasional very favourable year.

If full saturation of the resource is required to balance losses incurred in unfavourable periods then a reactive situation would be found regularly in a persistent population

Another possibility to be looked for is that the thrips are potentially more reactive than previously thought and they are continually chasing a fluctuating 'K' or carrying capacity (e.g. see May 1981a). If May's model is appropriate for thrips we should find that competition occurs frequently and is particularly severe at breeding sites during downward phases in population numbers.

Stability and climate

Climatic variation can be a challenge to population stability. Favourable conditions may allow a non-reactive consumer to increase to levels where density dependent growth (reaction) occurs. Reaction is therefore the ultimate stabilizing mechanism at higher densities (Milne, 1962). But what saves non-reactive populations from extinction in prolonged periods of adverse weather?

This question is never posed in the case of a consumer which is controlled by an <u>interactive</u> predator. Models for interaction (e.g. Hassell, 1981) generally omit terms for climatic influence (but see Hassell, 1985). Without weather terms they imply that weather is never unfavourable and the only thing which prevents increase of the prey to reactive levels is interaction with natural enemies. Occasionally, weather is acknowledged to be of influence at the edges of the range of a consumer. Here we may infer that numbers are augmented by diffusion from the central area where weather is no problem.

An analogous argument can be put for a species without significant natural enemies and which is not restricted by climate in the centre of its range. It could be stabilised by reaction in favourable areas and maintain lesser but relatively stable numbers in marginal areas by diffusion.

But what if there is only rare evidence of reaction in the most favourable part of the range? Such an example is provided by the nematoceran cranefly, Tipula paludosa (Milne et al., 1965). This species is a univoltine pest of cereals and pasture which spends its immature life in the top layer of the soil. The strategic model for its population dynamics (Meats, 1974a,b) is one driven by climatic factors and offers no guarantee of persistence. The model can explain geographic range (Meats, 1975) by identifying places where persistence is definitely not indicated. But how does the species resist extinction within its endemic range?

The answer probably involves environmental heterogeneity (Meats, 1974b). Population crashes are caused by dry weather when the insect is in the egg or early larval stages which are very prone to desiccation in dry soil. However, there will always be some ground in marshy areas and at the sides of streams where the soil remains favourable. Birch (1957) put forward a similar argument concerning the grasshopper, Austroicetes cruciata. In both cases, increase to competitive (reactive) levels is usually prevented by the advent of adverse weather and it is proposed that crashes are buffered by persistence in 'climatic refuges'.

If crash conditions do not affect isolated 'climatic refuge' areas why do not populations in these pockets rise to reactive levels in favourable times? One possible answer is that they do. However passive (non-reactive) diffusion from such areas may buffer this trend and prevent reaction while serving to recolonize the areas in which the population crashed. Another possibility is that an area which permits survival in dry periods may be unfavourable in wet ones (Birch, 1957; Meats, 1970).

Weather, competition and stability in tephritid fruit flies

Tephritid fruit flies are apparently not amenable to biological control at economically acceptable levels even where many parasites have been tried and relatively high rates of parasitism have been maintained (Haramoto and Bess, 1970). Rate of parasitism or predation is obviously a naive measure of biological control. The death rate imposed by useful natural enemies must be sufficient to reduce the rate of increase of the pest to low levels when the pest is at low density. The rate of parasitism of the Queensland fruit fly, Dacus tryoni, is negligible (Bateman, 1968).

In destroying only the fleshy parts of fruits, tephritid fruit flies are non-interactive with their hosts. Is there any stabilizing mechanism?

At first sight they appear to be non-reactive because fruit is rarely seen to be overloaded to the point of food competition (Gibbs, 1967). However, the reason for this appears to be either that females can be deterred by the 'marking' activities of previously ovipositing females or that the larvae that hatch first in a fruit can make conditions lethal for those which hatch later (Prokopy, 1972; Fitt, 1983).

If these reactive mechanisms are important they should result in a situation at high population levels where most fruits are infested yet none are overloaded with larvae. If a large proportion of fruit is uninfested then the above mechanisms have a significance not immediately related to stabilization of numbers.

This leads us to the problem of explaining the persistence of fruit flies at low densities towards the periphery of their ranges. The chief determinant of range in polyphagous species is weather (Bodenheimer, 1951; Bateman, 1968; Meats, 1981; O'Loughlin et al. 1984).

The most documented example is that of the Queensland fruit fly, Dacus tryoni (see Meats, 1981; O'Loughlin et al., 1984). It has a range in eastern Australia covering the full latitudinal extent of the continent (11°S to 38°S). Its most southerly consistent outpost would appear to be Melbourne; populations within 500 km of Melbourne are extremely low compared to those in regions 500-2000 km to the north (see later). Is it possible for a species to maintain itself at low densities, in a non-reactive state, subject to the vagaries of weather influences, without the assistance of long distance diffusion from the north?

In the southern part of its range D tryoni has a 'breeding gap' in winter of up to 5-6 months which causes the adult population to decline without new recruits (it is the only stage capable of surviving in these regions). Thus summer increases have to balance winter declines. The more south a population is the more it is hampered in achieving this balance. This is because summers are cooler towards the south with the result that rates of increase per generation and numbers of generations per year get less the further to the south we go; furthermore, the 'breeding gap' in winter is longer in the south with the result that the decline in that season requires more of a summer increase to balance it (Meats, 1981).

At what point in the southerly distribution does the balance become negative and indicate that persistent populations are no longer possible without immigration? This is an important type of question in the population dynamics of any species but it is doubly important in the case of a pest. This is because the answer affects policy on quarantine and eradication methods. The market conditions for some crops require zero infestation but for many markets there is a finite threshold. In the latter case it would be foolish to expend money on eradication procedures (on the assumption that the discovery of a pest in an area would lead to serious outbreaks) if it could be shown that pest colonization was doomed by bioclimatic constraints.

The following exercise is an example of how these questions can be answered. More data are required for a confident conclusion; what data we do have is interpreted tendentiously, based on my assumption of what further studies might show.

The rate of increase during the summer, λ_s, must equal or exceed the reciprocal of the rate of decline in winter, λ_w. Using Fletcher's (1979) figures we can assume a winter decline in all but the coldest places of around 8% per week. Thus for a 12 week breeding gap the value of λ_s must exceed 2.72; for a 24 week breeding gap λ_s must exceed 7.4 and so on.

The extent of the 'breeding gap' and hence the required λ_s in any place can be calculated by the method given in O'Loughlin et al. (1984). The potential rate of summer increase, λ_s in any place can be estimated from the figures given by Bateman (1968), Fletcher (1974) and Meats (1981). Assuming favourable rainfall, Meats (1981) estimated the likely increase λ_8 for various sites in an 8 week period staring in December. Using the figures of Fletcher (1974) we conservatively assume potential λ_s is three times λ_8 in all districts with favourable rainfall as defined by Meats (1981). For areas of medium and marginal rainfall we use the figures of Bateman (1968) to divide this estimate by 3 and 10 respectively.

The results are shown in Fig. 4 where 'required' λ_s, that is $1/\lambda_w$, is plotted against 'potential' λ_s. It appears that the sites on the mainland within 500 km of Melbourne where monthly trapping has been done all have values of potential λ_s near or just below the line of equality, indicating that potential for summer increase would just compensate or slightly overcompensate for winter

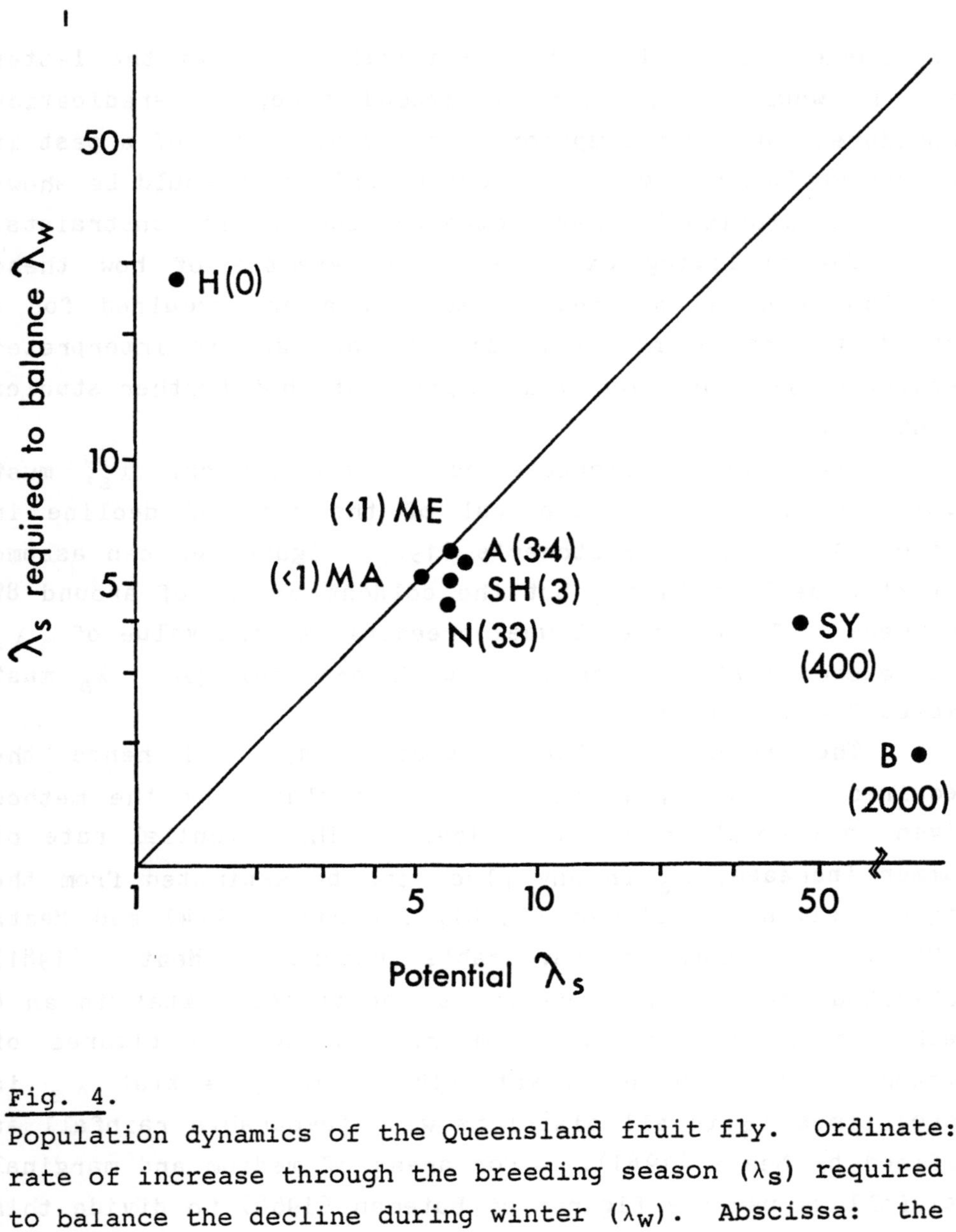

Fig. 4.
Population dynamics of the Queensland fruit fly. Ordinate: rate of increase through the breeding season (λ_s) required to balance the decline during winter (λ_w). Abscissa: the potential λ_s permitted by the environment (calculations as in text).
The points pertain to the sites indicated: ME, Melbourne; MA, Mallacoota; A, Albury; SH, Sheparton; N, Narrandera; SY, Sydney; B, Brisbane. The figures in brackets pertain to the maximum value of flies/trap/month obtained at these sites.
References: O'Loughlin, 1983; O'Loughlin et al, 1983, 1984; Jones and Skepper, 1965; Fletcher, 1974; Drew and Hooper, 1983.

declines. The point for Hobart, to the south, with an effective breeding gap of 10 months (Meats, 1981) is well above the line of equality and indicates that this site is highly unfavourable for persistence. The point for Sydney, on the other hand, is well below the line of equality with potential λ_s being around ten times that required. The point for Brisbane is even further below the line of equality because potential λ_s is vastly more than required.

Districts with favourable rainfall from around and to the north of Sydney are thus expected to reach reactive levels for at least part of the season. Drew and Hooper (1983) consider that D. tryoni numbers in Brisbane are governed chiefly by seasonal fruit availability and thus we can conclude that the species is reactive for most of the year with peak numbers tracking 'K'.

However, in southern N.S.W. and across the state of Victoria, we still have the problem of explaining why numbers are never high (see Fig. 4). A balance of increase and decrease could maintain a high population just as much as a low one yet Fig. 4 suggests that maximum peaks of populations get less the nearer we get to the line of equality. The trend is too strong to be accounted for by variation in trap efficiency.

If the primary constraint on summer increase is lack of rainfall (points A, SH and N on Fig. 4) then persistence might be explained by 'climatic refuges' in patches of commercial or domestic irrigation (Meats, 1981). However where the constraint is chiefly one of temperature in the south and south-east of Victoria (typified by points ME and MA on Fig. 4) it is highly unlikely that there are localized areas of climatic refuge in which temperatures are atypically high for the region.

The most ready explanation is that the populations with points on or near the line should be above the line and that their numbers are only maintained by immigration from the north. The potential for immigration into this vast area is currently under investigation by the Victorian

Department of Agriculture (McFarlane, unpublished).

Should the immigration hypothesis prove unlikely the problem of low marginal populations which cover vast areas will continue to be a challenge to both theoretical and practical ecologists.

REFERENCES

Andrewartha, H.G. and Birch, L.C., 1984. The Ecological Web. University of Chicago Press, Chicago, xiv + 506 pp.

Bateman, M.A., 1968. Determinants of abundance in a population of the Queensland fruit fly. Symp. R. Entomol. Soc. Lond., 4:119-31.

Birch, L.C., 1957. The role of weather in determining the distribution and abundance of animals. Cold Spring Harb. Symp. Biol. 22:203-218.

Bodenheimer, F.S., 1951. Citrus Entomology in the Middle East. W. Junk, The Hague, 663 pp.

Caughley, G. and Krebs, C.J., 1983. Are big mammals simply little mammals writ large? Oecologia (Berl.) 59:7-17.

Caughley, G. and Lawton, J.H., 1981. Plant-herbivore systems. In: R.M. May (ed.), Theoretical Ecology, 2nd ed., Blackwell, Oxford, pp.132-166.

Davidson, J. and Andrewartha, H.G., 1948. The influence of rainfall, evaporation and atmospheric temperature on fluctuation in the size of a natural population of *Thrips imaginis* (Thysanoptera). J. Anim. Ecol., 17:200-22.

Drew, R.A.I. and Hooper, G.H.S., 1983. Population studies of fruit flies (Diptera: Tephritidae) in south-east Queensland. Oecologia (Berl.) 56:153-159.

Fitt, G.P., 1983. Factors limiting the host range of tephritid fruit flies. PhD thesis, University of Sydney, ix + 291 pp.

Fletcher, B.S., 1974. The ecology of a natural population of the Queensland fruit fly, *Dacus tryoni*. VI. Seasonal changes in fruit fly numbers in the areas surrounding the orchard. Aust. J. Zool., 22:353-363.

Fletcher, B.S., 1979. The overwintering survival of the adults of the Queensland fruit fly, Dacus tryoni, under natural conditions. Aust. J. Zool., 27:403-411.

Haramoto, F.H. and Bess, H.A., 1970. Recent studies on the abundance of the Oriental and Mediterranean fruit flies and the status of their parasites. Proc. Hawaiian Ent. Soc., 20:551-566.

Hassell, M.P., 1981. Arthropod predator-prey systems. In: R.M. May (ed), Theoretical Ecology (2nd ed). Blackwell, Oxford, pp 105-131.

Hassell, M.P., 1985. Insect natural enemies as regulating factors. J. Anim. Ecol., 54:323-334.

Gibbs, G.W., 1967. The comparative ecology of two closely related sympatric species of Dacus (Diptera) in Queensland. Aust. J. Zool. 15:1123-39.

Jones, E.L. and Skepper, A.H., 1965. Suppression of Queensland fruit fly, Dacus Strumeta) tryoni in Narrandera, New South Wales. Agric. Gaz. N.S.W., 76:501-503.

Lawton, J.H., 1985. An independent view. Nature, 315:81.

May, R.M., 1981a. Models for single populations. In: R.M. May (ed) Theoretical Ecology (2nd ed.). Blackwell, Oxford, pp 5-29.

May, R.M., 1981b. Models for two interacting populations. In: R.M. May (ed) Theoretical Ecology (2nd ed.). Blackwell, Oxford, pp 78-104.

Meats, A., 1970. Susceptibility of the leatherjackets Tipula oleracea and T. paludosa to soil flooding. Ann. appl. Biol., 65:25-30.

Meats, A., 1974a. A population model for two species of Tipula (Diptera, Nematocera) derived from data on their physiological relations with their environment. Oecologia (Berl.), 16:119-38.

Meats, A., 1974b. Simulation of population trends of Tipula paludosa using a model fed with climatological data.

Oecologia (Berl.), 16:139-47.

Meats, A., 1975. The developmental dynamics of Tipula paludosa and the relation of climate to its growth pattern, flight season and geographical distribution. Oecologia (Berl.), 19:117-28.

Meats, A., 1981. The bioclimatic potential of the Queensland fruit fly, Dacus tryoni, in Australia. Proc. Ecol. Soc. Aust., 11:156-61.

Milne, A., 1962. On a theory of natural control of insect population. J. Theoret. Biol. 3:19-50.

Milne, A., 1984. Fluctuation and natural control of animal population, as exemplified in the garden chafer Phyllopertha horticola (L.). Proc. R. Soc. Edin., 82B:145-199.

Milne, A., Laughlin, R. and Coggins, R.E., 1965. The 1955 and 1959 population crashes of the leatherjacket, Tipula paludosa Meigen in Northumberland. J. Anim. Ecol., 34:529-44.

Monro, J., 1967. The exploitation and conservation of resources by populations of insects. J. Anim. Ecol., 36:531-47.

O'Loughlin, G.T., 1983. Queensland fruit fly in Victoria - a synopsis of observations on its biology and control. Research Project Series, No. 177, Department of Agriculture, Victoria, Melbourne, 71 pp.

O'Loughlin, G.T., East, R.W., Kenna, G.B. and Harding, E., 1983. Comparison of traps for the Queensland fruit fly, Dacus tryoni (Diptera: Tephritidae). Gen. appl. Ent., 15:3-6.

O'Loughlin, G.T., East, R.A. and Meats, A., 1984. Survival, Development Rates and Generation times of the Queensland fruit fly, Dacus tryoni, in a marginally favourable climate: experiments in Victoria. Aust. J. Zool., 32:353-61.

Prokopy, R.J., 1972. Evidence for a marking pheromone deterring repeated oviposition in apple maggot flies. Env. Ent., 1:326-32.

White, T.C.R., 1978. The importance of a relative shortage of food in animal ecology. Oecologia (Berl.) 33:71-86.

Williams, C.B., 1964. Patterns in the balance of nature and related problems in Quantitative Ecology. Academic Press, New York.

THE STERILE INSECT TECHNIQUE: A THEORETICAL PERSPECTIVE

Richard E. Plant
Department of Mathematics
University of California
Davis, CA 95616

ABSTRACT

We present a theoretical study of the management of a campaign to eradicate an invasive population of Mediterranean fruit flies, Ceratitis capitata (Wied.), by application of the sterile insect technique, or SIT. The model for the process consists of three interacting populations: wild females, released sterile males, and females in the rearing facility. We begin by examining the properties of each of the models individually, and then we conduct simulation studies of the three populations together.

INTRODUCTION

In any program to eradicate or control an insect pest population by application of the sterile insect technique (SIT), three populations of insects interact. These are the wild population, the released sterile population, and the population in the mass rearing facility. In this paper we present a modeling study of a program to eradicate an invasive population of Mediterranean fruit fly, Ceratitis capitata (Wied.), or medfly, in which each of these populations is simulated. The first purpose of this study is to determine any general properties of medfly eradication programs that might not be transparent without such a study. The second purpose is to estimate some of the parameter values of such a program, for example, time to eradication, as they depend on the size and age structure of the wild population.

Theoretical analysis of the SIT has played a crucial role since the inception of this technique (Knipling, 1955). Ito (1977) credits modeling studies with suggesting modifications in the successful campaign to eradicate the melon fly Dacus cucurbitae Coq. from the island of Kume.

NATO ASI Series, Vol. G11
Pest Control: Operations and Systems Analysis
in Fruit Fly Management. Edited by M. Mangel et al.

Other early theoretical studies of the SIT include those of Berryman (1967), Lawson (1967), and Monro (1973). With the exception of the latter, all of the early studies used models not incorporating age structure, and focused on the effect of multiple female mating on the effectiveness of the program. Ito (1977) found that for parameter values typical of the melon fly the effect of multiple mating is small.

Carey (1982, 1984) has done extensive research on the demographics of the medfly, and our model is based strongly on his work. Our models for each of the three populations are age structured, Leslie matrix type models, with modifications introduced to account for the capacities of the rearing facility, the harvest of pupae from this facility, and the effect of sterile males on the reproductive ability of the wild population.

One of the major concerns in the development of any model for the dynamics of insect populations is the accuracy of the model as a representation of the actual population. In our particular system the confidence that we can have in the model as an accurate representation depends on the subpopulation. The model for the population in the rearing facility is probably a fairly good representation of this population. The models for the two field populations, however, are more uncertain. We have tried to base these models on the best available data. Our philosophy has been to consider the "worst case" scenario; that is, we give the field populations demographic parameters that we feel would correspond to those in conditions most conducive to medfly development and thus least conducive to eradication.

In Section 2 of the paper we develop the basic models. In Section 3 we examine some of the properties of the model for the wild population in isolation. The examination focuses on how these properties may provide a guide to the proper management of an eradication campaign using the SIT. In Section 4 we present simulation studies of an eradication campaign under various conditions. In Section 5 we give a brief discussion of our results.

DEVELOPMENT OF THE MODEL

The model simulates the course of an eradication campaign through the evolution of three subpopulations: those of the wild females, the released sterile males, and the females in the rearing facility. We assume a one to one sex ratio. Each of the three subpopulations is described by a vector of age classes. We use classical Leslie matrix methods (Lewis, 1942, Leslie, 1945) to model the dynamics of the populations. The model is, however, nonlinear; the nonlinearity occurs because of the effect of constraints on the size of the rearing facility population, because of the harvest of a portion of this population, and because of the interference of the sterile males with the reproduction of the wild population.

Each age class in each of the three population vectors has a duration of two days; thus, age class j consists of all those individuals in the population between the ages of $2j$ and $2(j+1)$ days. We use Julian rather than physiological time units for convenience of interpretation of the results. Temperature is assumed constant at 25°C. Let $x_j(t)$ denote the number of wild females in age class j at time t, $y_j(t)$ the number of released sterile males, and $z_j(t)$ the number of females in the rearing facility. The variable t is discrete and has units of two days.

We begin our development with the rearing facility. The demographics of insect mass rearing have been studied by Carey and Vargas (1985), and many of our ideas are drawn from this work. In the absence of harvest, we assume that the dynamics of the female population are described by a Leslie matrix equation, modified to take factory capacities into account. The age classes of the stages are as follows. Age class 0 is the egg stage, 1 through 4 are the larvae, 5 though 9 are the pupae, and 10 through 39 are the adults.

The factory capacities are incorporated into the model as follows. We assume that at the start of each two day period new members of each life stage (egg, larva, pupa, and adult) are placed in a holding container. After the initial deposit, no further individuals are placed in that container. The individuals in a container incur ordinary mortality within the container. To simulate this we fix values E_{max}, L_{max}, P_{max}, and A_{max}, and require that $z_0(t) \leq E_{max}$, $z_1(t) \leq L_{max}$, $z_5(t) \leq P_{max}$, and $z_{10}(t) \leq A_{max}$ for all t. . The means of establishing the values of E_{max}, L_{max}, P_{max}, and A_{max} are described below.

Pupae are harvested from age class 9, the last preadult age class. We define $u(t)$ to be the fraction of the population of this age class harvested at time t, and $H(t)$ to be the total harvest at time t, so that $H(t) = u(t)z_9(t)$. The complete factory model is then

$$z_0(t) = \min\{\sum_{XP5}^{\delta} F_{zj}z_j(t-1), E_{max}\},$$

$$z_1(t) = \min\{P_{z0}z_0(t-1), L_{max}\},$$

$$z_{j+1}(t) = P_{zj}z_j(t-1), \quad j=1,\ldots,3, \tag{2.1}$$

$$z_5(t) = \min\{P_{z4}z_4(t-1), P_{max}\},$$

$$z_{j+1}(t) = P_{zj}z_j(t-1), \quad j=5,\ldots,9,$$

$$z_{10}(t) = \min\{[1-u(t)]P_{z9}z_9(t-1), A_{max}\},$$

$$z_{j+1}(t) = P_{zj}z_j(t-1), \quad j=10,\ldots,\delta,$$

where δ is the discard age.

Vargas et al. (1984) have published a life table for a factory population of medflies in Hawaii, and we use this to determine the fecundities F_{zj} and survival probabilities P_{zj} of the factory population. Figure 2.1 shows the m_j data of Vargas et al., converted to two day units, and a curve fitted to these data by nonlinear regression. The curve used to fit the data is

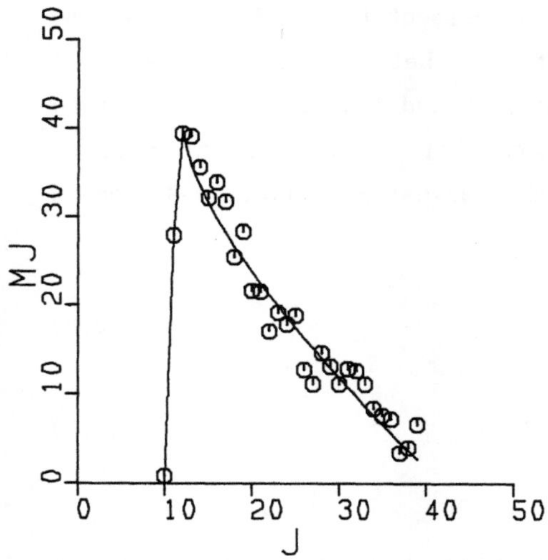

Figure 2.1

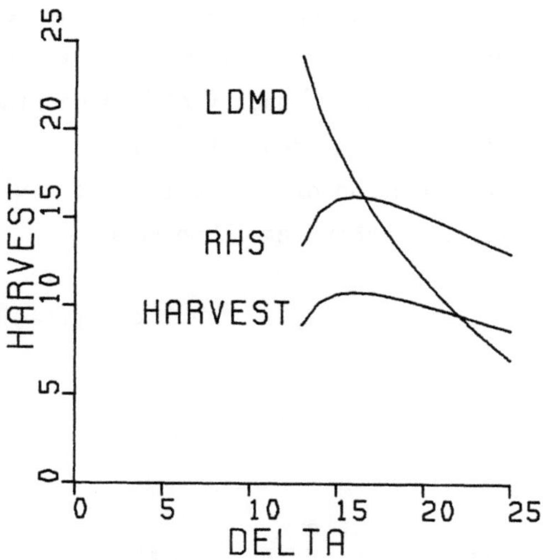

Figure 2.2

Figure 2.1. Circles: life table m_j values from Vargas et al. (1984). Curve: fitted function of equation (2.2).

Figure 2.2. Curves for determining discard age. LDMD: left hand side of equation (2.6); RHS: right had side of this equation; HARVEST: harvest function of equation (2.5).

$$m_z j = \begin{cases} 0: & j=0,\ldots,10 \\ 27.9(j-110)^{0.5}: & j=11,12, \\ 39.46-3.66(j-12)^{0.7}: & j=13,\ldots,\delta \end{cases} \qquad (2.2)$$

As is usual in converting life table date to a Leslie matrix model (e.g., Emlen, 1984), we take $F_{zj} = P_{zj} m_{z(j+1)}$. The values of the P_{zj}, taken from the life table of Vargas et al. (1984), are $P_{zj} = 0.96$, $j=0,\ldots,5$; $P_{zj} = 0.99$, $j=6,\ldots,\delta$.

We may determine the discard age by expanding upon an idea of Monro and Osborn (1967). Namely, we compute the optimal discard age to be the discard age such that the number of pupae harvested per adult female is maximized. Since we wish to find a maximum it is useful to be able to compute the derivative of the harvest with respect to the discard age, so we temporarily switch to continuous time. Let A denote the total female adult capacity of the rearing facility and let E denote the rate of female egg production at full capacity. Let I denote the rate of introduction of new adults. If ν is the harvest age (i.e., ν=9) and δ is the discard age then we have

$$A = I(\delta-\nu), \tag{2.3}$$

and

$$E = \frac{A}{\delta-\nu} \int_{\nu}^{\delta} m_a \ell_a da \tag{2.4}$$

where $\ell_a = \prod_{j=0}^{a} P_{xj}$. Letting the harvest rate at a given discard age be denoted $H(\delta)$, we have $I = \ell_\nu E - H(\delta)$. Therefore, substituting (2.3) and (2.4) yields

$$H(\delta) = \frac{A}{\delta-\nu} (\ell_\nu \int_{\nu}^{\delta} m_a \ell_a da - 1) \tag{2.5}$$

Differentiating (2.5) with respect to δ and setting the derivative equal to zero yields an implicit equation for δ in the form

$$\ell_\delta m_\delta = \frac{1}{\delta-\nu} (\int_{\nu}^{\delta} m_a \ell_a da - \ell_\nu^{-1}) \tag{2.6}$$

Figure 2.2 shows plots of the left side and the right side of equation (2.6), together with the harvest, for a given δ. Based on this plot, we set δ at 17 periods, or 34 days.

We can now set values for the various capacities of the rearing facility. We take as a basic figure that the maximum harvest rate of the facility is H_{max} = sixty million male pupae per two day time period. This figure roughly corresponds to the capacity of 400 million pupae per week proposed for the APHIS facility in Hawaii. Using this figure for H in equation (2.5) with $\delta = 17$ we obtain a total adult capacity of A = 3.25 million. Therefore the value of A_{max}, the maximum number of adults that can be accepted from larvae in a two day period, is

A_{max} = 408,000.

We can now determine the capacities for the other life stages by establishing for each life stage just enough capacity so that when the facility is at maximum harvest rate there is no wasted space. This means that the values of E_{max}, L_{max}, and P_{max} should correspond to the levels maintained at those respective age classes in a population that is held stable by harvesting from the last pupal age class.

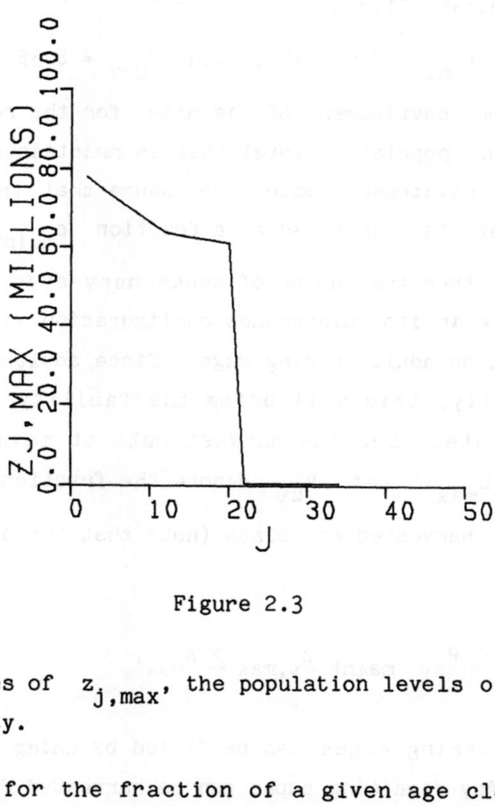

Figure 2.3

Figure 2.3. Values of $z_{j,max}$, the population levels of each age class at maximum capacity.

The formula for the fraction of a given age class that must be harvested to maintain a population at a fixed level is well known, and is given for example by Beddington and Taylor (1973) and by Carey and Vargas (1985). In our notation it is

$$u_{ss} = 1 - \left(\sum_{j=\nu+1}^{\delta} \left[\prod_{k=1}^{j-1} P_{zk} \right] F_{zj} \right)^{-1} \tag{2.7}$$

where u_{ss} is the fraction of age class ten that must be harvested to maintain the rearing facility population at a steady state. Substituting in the values from the life table data yields a value of 0.9932 for u_{ss}. Using this value and the life table data for the rearing facility we may compute the values of $z_{j,max}$, the maximum steady state values of each life stage in the rearing facility. A plot of these values is shown in Figure 2.3. The values of E_{max}, L_{max}, and P_{max} then correspond to the values of $z_{0,max}$, $z_{1,max}$, and $z_{5,max}$, respectively. These are $E_{max} = 7.8 \times 10^7$, $L_{max} = 7.5 \times 10^7$, and $P_{max} = 6.35 \times 10^7$.

To complete our development of the model for the rearing facility we must establish the population level that is maintained when the factory is in a idling, or maintenance mode. We assume that in the maintenance mode each life stage is maintained at a fraction c_{maint}. If c_{maint} is sufficiently large then the number of pupae harvested and discarded to keep the facility at its maintenance configuration will be large enough to completely fill an adult rearing cage. Since adult rearing cages are filled sequentially, this will bring the facility to capacity at the fastest possible rate. Let the harvest rate at maintenance level be $H_{maint} = c_{maint} H_{max}$. Let $P_{z\nu}$ denote the fractional survivorship of age class ν, the harvested age class (note that for our facility $\nu=9$) Then if

$$P_{z\nu} c_{maint} z_{\nu,max} \geq A_{max}, \tag{2.8}$$

then the adult rearing cages can be filled by using the pupae that had been discarded. Any remaining pupae can be harvested for sterilization. Substituting the relationship $A_{max} = (1 - u_{ss})z_{\nu,max}$, we see that inequality (2.8) is satisfied if

$$c_{maint} \geq (1 - u_{ss})/P_{z\nu}. \tag{2.9}$$

Therefore, if the rearing facility is operated at a maintenance configuration that satisfies inequality (2.9) then the plant may be brought to capacity by simply diverting pupae that would otherwise have been discarded into the adult rearing cages, and harvesting any surplus pupae for sterilization and release. From that data given earlier we have $u_{ss} = 0.9932$ and $P_{x\nu} = 0.99$, so that we require $c_{maint} \geq 0.007$. The actual value of c_{maint} that is being proposed for the APHIS facility is 0.05, which is well within this range.

This completes the development of the model for the rearing facility. Let us now consider the wild population. We assume that in the absence of sterile males the population displays density independent growth. We may write

$$x_0(t+1) = \sum_{j=0}^{J} F_{xj}x_j(t),$$
$$x_{j+1}(t+1) = P_{xj}x_j(t), \quad j=0,\ldots,J-1, \tag{2.10}$$

where F_{xj} is the fecundity of the jth age class, measured in female eggs per female per two days, P_{xj} is the proportion of age class j surviving to age $j+1$, and J is the index of the oldest age class. The model permits wild females to live at most 80 days, so J equals 39.

No life tables have been published for field populations of medflies, although some cage data are available (Monro, 1973). As mentioned in the Introduction, our philosophy is to formulate a "worst case" model in which the wild population is assumed to be in the most favorable possible environment. This will at least provide some sort of bound on the dynamics of the population. Our F_{xj} are based on the data of Vargas at al. (1984), modified as suggested by the observations of Wong and Nakahara (1978). They found that the preovipositional period of the wild population is extended for about eight days beyond that of the factory population, and the portion of mated females in the wild population after the normal adult life expectancy is about one half. We therefore take the birthrate curve of the wild population to be identical to that of Figure 2.1 except that it is shifted four units to the right and multiplied by a factor of 0.5. The equation of the m_{xj} curve for the wild population is

$$m_{xj} = \begin{cases} 0: & j=0,\ldots,14 \\ 13.95(j-14)^{0.5}, & j=15,16. \\ 19.73 - 1.83(j-16)^{0.7}, & j=17,\ldots,39 \end{cases} \tag{2.11}$$

The equation for F_{xj} is $F_{xj} = P_{xj}m_{x(j+1)}$.

The values of P_{xj}, the portion of survivors form age class j, are based on the host specific studies of Carey (1984). The age classes of the various life stage are the same as those of the rearing facility. We assume that within stage survivorship is independent of age. Egg, larval, and pupal survivorships are estimated by averaging Carey's data for lime, lemon, peach, nectarine, and pear, the most common hosts in California. This results in average egg and larval survivorship of 47 percent and average pupal survivorship of 91 percent. The corresponding values of P_{xj} are $P_{xj} = 0.83$, $j=0,\ldots,4$, and $P_{xj} = 0.96$, $j=5,\ldots,9$. Finally, following Cunningham et al. (1980), we assume an adult survivorship of $P_{xj} = 0.96$, $j=10,\ldots,39$.

We now incorporate the effect of the presence of sterile males on the dynamics of the wild population. Letting $y_j(t)$ denote the number of sterile males in age class j at time t, let $Y(t)$ denote the total population of sterile males, that is, $Y(t) = \Sigma_j\, y_j(t)$, the sum being taken over all the age classes present in the sterile population. Let $X(t)$ similarly denote the total adult wild female population (which is assumed equal to the total fertile adult male population). We assume that the presence of sterile males reduces the fertility, but has no effect on the fecundity, of the wild population. We therefore let $P_{x0} = 0.83S(R(t))$ where $R(t) = Y(t)/X(t)$ and

$$S(R) = \frac{1}{1 + \alpha R}\,. \tag{2.12}$$

The parameter α is a factor taking into account the reduced mating effectiveness of the sterile males. Based on the results of Denny et al. (1985), we set α at 0.5. The model represented by equation (2.5) is a very simple one containing a number of assumptions, including random, age independent mating. There do not, however, seem to be enough appropriate data to justify a more complex model.

The model for the released sterile males is very simple. Recall that $H(t)$ is the number of pupae harvested from the rearing facility at time t. We assume that pupae are harvested at age twenty days (i.e. ten time periods), that they enclose in their canisters, and that they are released two days after harvest as adults. We therefore have

$$y_{11}(t) = \sigma H(t-1),$$

$$y_{j+1}(t) = P_{yj} y_j(t-1), \quad j=11,\ldots,22, \tag{2.13}$$

where σ is the proportion of the males that survive sterilization, transshipment and release. Due to the assumed lower survivorship of the sterile flies, only eleven age classes are assumed to be present. The value of P_{yj} is assumed to be 0.9 for each j, and the value of σ is assumed to be 0.8 (R. Cunningham, personal communication).

This completes the development of the model. In the next section we discuss the use of the model to design optimal release rate strategies for population eradication.

ANALYSIS OF THE MODEL FOR THE WILD POPULATION

In this section we analyze the properties of the model for the wild population, particularly as they relate to questions of concern in an eradication program. We begin this section by discussing the properties of the Leslie matrix model of the wild population in the absence of sterile flies. This discussion is based on standard techniques that are given, for example, by Pollard (1973).

The dominant eigenvalue of the Leslie matrix has the value $\lambda_x = 1.217$, corresponding to a growth rate at stable age distribution of 10.3 percent per day. Figure 3.1 shows a plot of the components of the right eigenvector corresponding to this eigenvalue; this gives the stable age distribution relative to the egg stage. At stable age distribution, the population is 29.6 percent eggs, 49.7 percent larvae, 14.4 percent pupae, and 6.3 percent adults. Figure 3.2 shows a plot of the components of the left eigenvector corresponding to this eigenvalue. These components give the reproductive value, relative to the egg stage, of each age class. As usual, the highest reproductive value is in the young adult age classes.

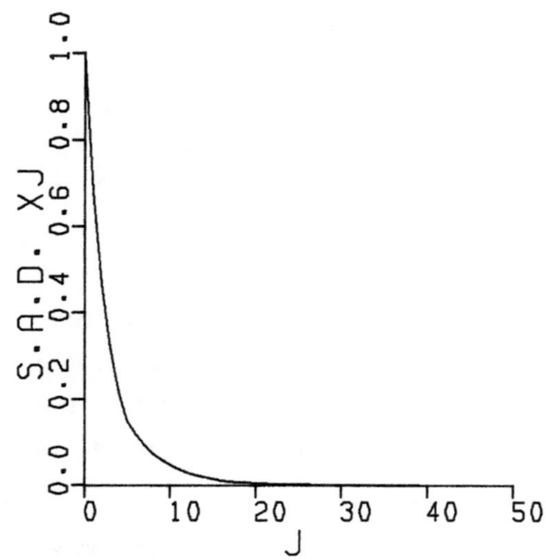

Figure 3.1

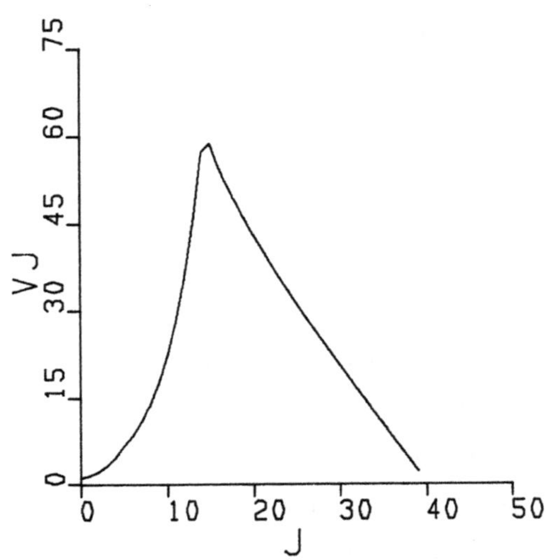

Figure 3.2

Figure 3.1. Components of the right eigenvector of the Leslie matrix for the wild population, giving the stable age distribution.

Figure 3.2. Components of the left eigenvector of the Leslie matrix for the wild population, giving the reproductive values.

We may use the formulas given by Caswell (1978) to determine the values of $d\lambda_x/dF_{xj}$ and $d\lambda_x/dP_{xj}$. These formulas are as follows. Let u and v be the right and left eigenvectors corresponding to λ_x, normalized so that $u^T v = 1$. Then

$$\frac{d\lambda_x}{dP_{xj}} = u_{j+1} v_j \tag{3.1}$$

$$\frac{d\lambda_x}{dF_{xj}} = u_1 v_j.$$

The values of $d\lambda_x/dF_{xj}$ are not of particular concern since we cannot modify the population's fecundity in an age dependent way.

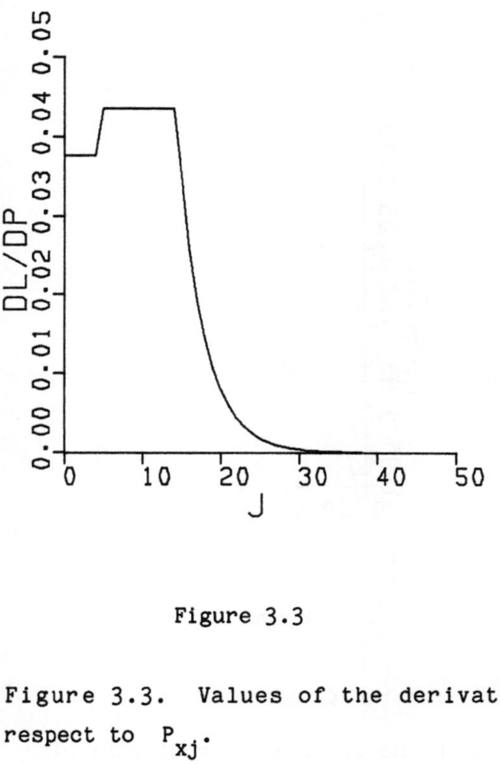

Figure 3.3

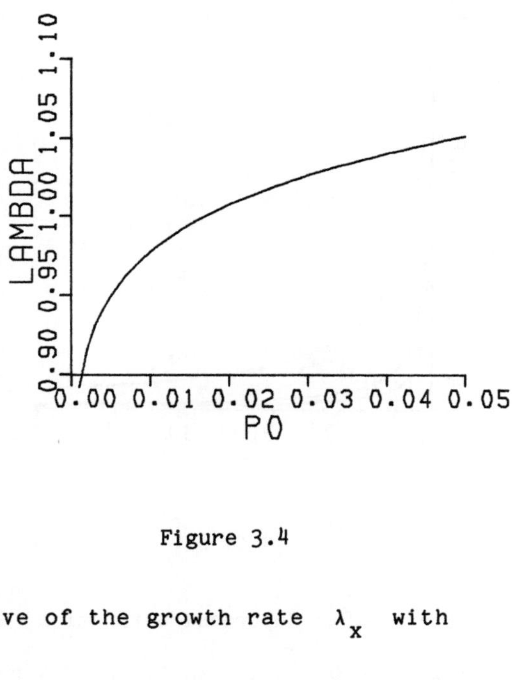

Figure 3.4

Figure 3.3. Values of the derivative of the growth rate λ_x with respect to P_{xj}.

Figure 3.4. Plot of λ_x as a function of P_0.

Figure 3.3 shows a plot of $d\lambda_x/dP_{xj}$. Although the reproductive value of the egg and larval stages is very low, reducing the value of P_{x0} has a relatively great effect on the growth rate of the population. This is because such a large fraction of the population is in these stages. In our model the effect of the sterile males is to reduce P_{x0}.

Figure 3.4 shows a plot of λ_x as a function of P_{x0}. It is evident that P_{x0} must be reduced significantly (to approximately 0.01) before the growth rate λ_x is reduced to unity. The reason for this is the high fecundity of the population; the value of R_0, the average number of offspring produced by a female over her lifespan, is 97.0. Therefore, roughly speaking, if only one percent of the offspring survive an eradication effort, the population will maintain itself (I am grateful to J.R. Carey for this observation).

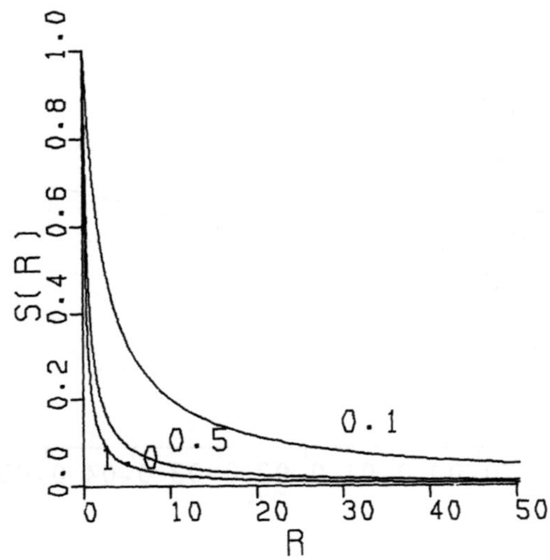

Figure 3.5

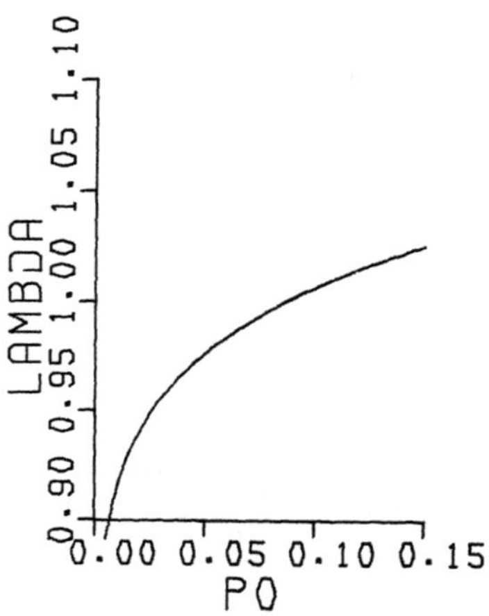

Figure 3.6

Figure 3.5. Suppression function S(R) for three values of α.

Figure 3.6. Same as Figure 3.4 but with the effects of fruit stripping and ground spray included.

Figure 3.5 shows plots of the suppression functions S(R) defined in equation (2.12) as a function of the sterile/fertile ratio $R = Y/X$ for several values of the effectiveness parameter α. Since the value of P_{x0} in the model is 0.83, the fraction of eggs hatching into larvae (the ordinate in Figure 3.4) is 0.83S(R). We may determine the value of R necessary to reduce the growth rate λ_x to unity (i.e., to stop the population from increasing) to be that value of R such that $S(R) = 0.01/0.83 = 0.012$. Taking the value of α to be 0.5, this value is $R = 160$. Therefore, in the absence of any ancillary eradication effort a fertile/sterile ratio of 160:1 is necessary even to stop the growth of the wild population. This figure is rather high but probably of the right order of magnitude, bearing in mind that it represents a population growing under ideal conditions.

In many sterile insect release eradication campaigns ancillary efforts are included in the campaign. In this study we focus on ground spray with insecticide and fruit stripping. The effect of ground spray is to kill pupae; the effect of fruit stripping is to kill eggs and larvae and to remove oviposition sites. Since during the 1980-1982 Northern California infestation the fraction of infested fruit was quite small even in the heavily infested areas (J.R. Carey, personal communication), we neglect the latter effect of fruit stripping and focus on its effect on egg and larval mortality.

Both ground spray and fruit stripping operations are carried out in different locations every time period. Each operation induces an additional mortality into the target population. Therefore we model the effect of ground spray by multiplying the survival rates P_{xj} of the pupal age classes by a factor $(1-k_p)$, where k_p is the fraction of the pupal population killed by ground spray in each two day time period. Similarly, we model the effect of fruit stripping by multiplying the survival rates P_{xj} of the egg and larval age classes of the by a factor $(1-k_e)$, where k_e is the fraction of the egg and larval age classes killed per two day time period by fruit stripping.

Figure 3.6 shows a plot of λ_x vs. P_{x0}, similar to Figure 3.4, for the case $k_p = 0.15$ and $k_e = 0.15$. One way to interpret this level of mortality is that it corresponds to taking about two weeks (1/0.15 = 6.667 time periods) for crews to completely cover the infested region. For this case value of P_{x0} needed to achieve a stable wild population is $P_{x0} = 0.07$. Therefore for $\alpha = 0.5$ the fertile/sterile ratio needed to stabilize the population is about 25:1.

A rule of thumb often used in medfly eradication campaigns is to strive for a sterile/fertile ratio of 100:1 (e.g., Hooper, 1982). Our results indicate that this figure is probably appropriate for all but the most extreme circumstances. However, the main point of this section is that the proper overflooding ratio depends strongly on the growth rate of the population and will be different for an invasive population than for one that is already established and near its carrying capacity. In the next section we examine simulation results for an eradication campaign using SIT against an invasive population.

A MODEL OF AN ERADICATION CAMPAIGN

Having developed models for the three populations (wild, sterile, and factory), we now turn to the question of using these models in a theoretical study of an eradication campaign. We formulate the models as follows. At time $t = 0$ the wild population is assumed to be in a stable age distribution with a total adult female subpopulation of $X(0)$. The rearing facility is at a stable maintenance configuration in which each age class is maintained at a fraction $c_{maint} = 0.05$ of its capacity at full production. The model simulates the process of bringing the factory up to full capacity and releasing sterile insects, and the impact of this process on the wild population. The simulation continues until eradication is achieved, as determined by the total wild female population falling below unity, or until a maximum permitted number of two day time periods has elapsed.

We assume that the wild population is in the stable age distribution because it is likely that an invasive population will have been present for some time (say, two or three generations) before being detected. As will be shown below, the evolution of the eradication campaign is not very sensitive to the age structure of the wild population. We generally specify the initial wild adult female population since it is the same as the wild adult male population, and this is the portion of the population that would be observed in practice.

Our definition of eradication is somewhat crude. There is an extensive theory of extinction of populations that has been used by Costello and Taylor (1975) to study the sterile insect technique. We do not feel that the incorporation of this theory would add anything significant to our particular problem. This is because at the time of eradication the wild population is decreasing rapidly; if the rate of decrease at eradication were gradual then a more precise study of the eradication process might be required.

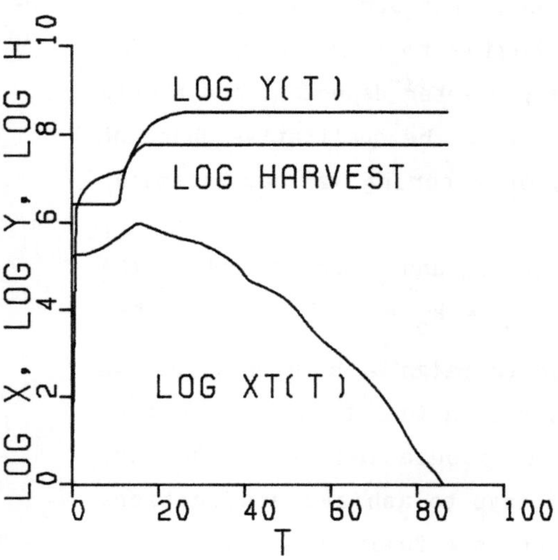

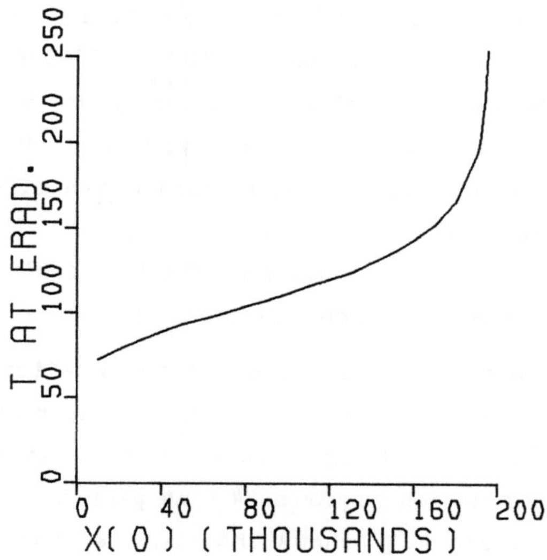

Figure 4.1

Figure 4.2

Figure 4.1. Simulated eradication campaign. Y(t) = total sterile male population, HARVEST = harvest rate, XT(t) = total wild female population (all life stages).

Figure 4.2. Time to eradicate vs. initial wild adult female population.

Figure 4.1 shows the results of the simulation with an initial wild adult population of 10,000. Note that common logarithms of all of the variables are plotted. the figure shows the number of male pupae harvested per two day time period, the total sterile male population in the field, and total wild female population. As evident from Figure 4.1, there is an initial phase in the mass rearing facility in which the harvest rate is established by the number of preadult insects "in the pipeline." This initial phase is followed by a rapid transition to full capacity, which is then maintained for the duration of the program. The most apparent effects of the released sterile males on the wild population are observed after a lag of about ten time periods, the time required for an egg to reach adulthood. Eradication of the wild population is achieved in 87 time periods, or 174 days. Figure 4.2 shows the number of time periods T required to achieve eradication as a function of X(0), the wild adult population at time 0. For moderate values of X(0) the time to eradication is not strongly dependent on X(0); then at about X(0) = 175,000 it begins to rise sharply. The specific values in these curves are highly parameter dependent and should not be taken too seriously; what is important is the qualitative shape of the curves, and the comparative effect of altering various parameter values.

We can model the effect of fruit stripping and ground drench in the manner discussed in Section 3. We set $k_p = k_e = 0.15$, as in that section. The course of the eradication campaign with these values in this case is not shown since it is virtually identical to that of Figure 4.1. With an initial wild adult female population of 10,000, the simulation requires 74 time periods or 148 days to achieve eradication. Figure 4.3 shows the time to eradication as a function of the initial wild adult female population. In comparison with Figure 4.2 it is evident that although the time to eradication is not changed much for small wild populations, the effect of fruit stripping and ground drench is to greatly increase the eradication threshold.

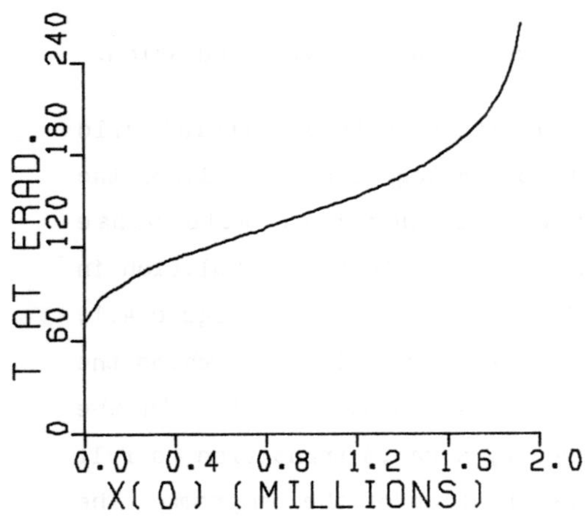

Figure 4.3

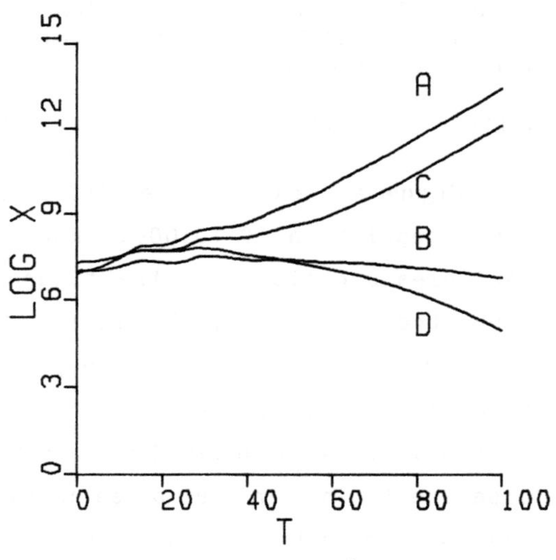

Figure 4.4

Figure 4.3. Same as Figure 4.2, but with fruit stripping and ground drench included.

Figure 4.4. Effect of various fruit stripping programs. See text for details.

Further simulations indicate that for fruit stripping and ground drench to be effective, they must be continued for virtually the entire duration of the program. If, for example, these operations are carried on for only the first month, the effect on the T at which eradication vs. X(0) curve is to roughly double that value of X(0) for a given T. Figure 4.3 shows that if fruit stripping and ground drench are continued throughout the eradication campaign then the effect for a given T is to increase the value of X(0) about tenfold.

If it is possible to do so, however, some fruit stripping and ground drench can have a positive effect on marginal situations, even if only done for a brief period. Figure 4.4 shows the effects of fruit stripping in greater detail. This figure shows the value of the total wild population as a function of t for the first hundred periods for four different combinations of parameter values. Curve (a) shows the total population vs. t when the initial adult female population is X(0) = 500,000, a value above the eradication threshold. The population rises rapidly. Curve (b) shows the same initial population, but with fruit stripping and ground drench for the first 15 time periods. This is just enough to hold the population until the rearing facility can be brought to capacity. Curve (c) is the same as curve (b) except that the initial wild population is doubled. The initial fruit stripping and ground drench are no longer sufficient to hold the population. Curve (d) shows the same situation except that fruit stripping and ground drench are maintained for the entire campaign.

To test the effect of the age structure of the wild population on the eradication process we ran the following simulation. Let $\bar{x}_{js}$ represent the fraction of the female population in age class j at the stable age distribution, so that if $X_T(0)$ is the total (all life stages) female population at time 0 and if the population is at a stable age distribution then the population in age class j is $x_j(0) = X_T\bar{x}_{js}$. To vary the age structure we multiply $\bar{x}_{js}$ by a factor γ^{j-1} and then renormalize so that the total (all life stages) population is again fixed at X_T. If $\gamma < 1$ then the population will be younger than that at a stable age distribution. This might be expected to be the case in an invasive, rapidly expanding population (Keyfitz, 1977). We ran a simulation with the same initial total wild population as in Figure 4.1, but with γ set at 0.8, so that the adult population is about ten percent of that at stable age distribution. The time to eradication is 83 time periods. Thus we may say that the age structure of the population has little effect on the outcome of the eradication campaign.

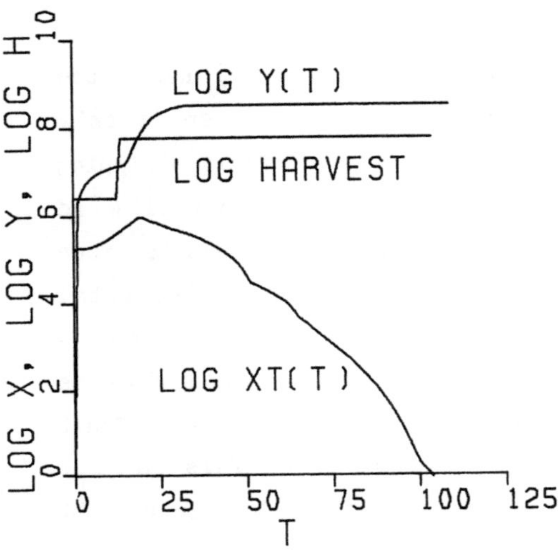

Figure 4.5

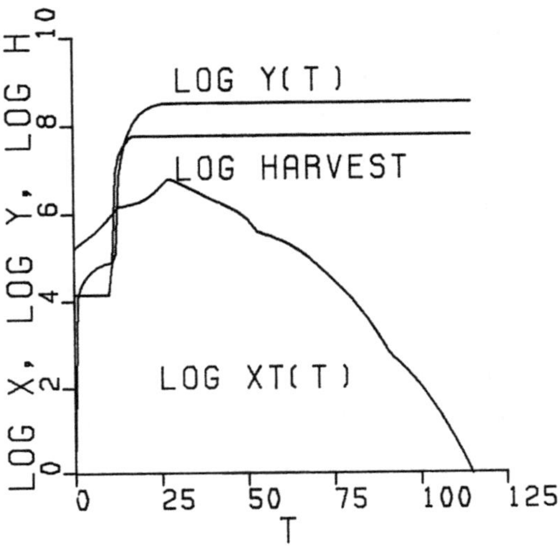

Figure 4.6

Figure 4.5. Simulation of an eradication campaign in which the adult cages in the rearing facility are full in the maintenance configuration.

Figure 4.6. Simulation of an eradication campaign in which the rearing facility is maintained at a value $c_{maint} = 0.007$.

Next let us consider the effect of the initial factory configuration. Since the marginal cost of maintaining adults in the rearing facility is quite small (N. Tanaka, personal communication), one relatively inexpensive way to increase the responsiveness of the facility is to always maintain the adult cages at full capacity, while maintaining the larval and pupal stages at fractional capacity. This increases the egg production rate of the facility, and thus speeds up the process of bringing the facility to full capacity. On the other side of the coin, one could economize by running the facility at a much lower value of c_{maint}, the maintenance configuration. Provided that inequality (2.9) is satisfied the adult cages can still be filled immediately as the plant is being brought to full capacity. Figure 4.5 shows the results of a simulation in which the adult cages are maintained at full capacity. The only change from Figure 4.2 is that the transition from maintenance to full capacity occurs more sharply, and the course of the eradication campaign is almost unchanged.

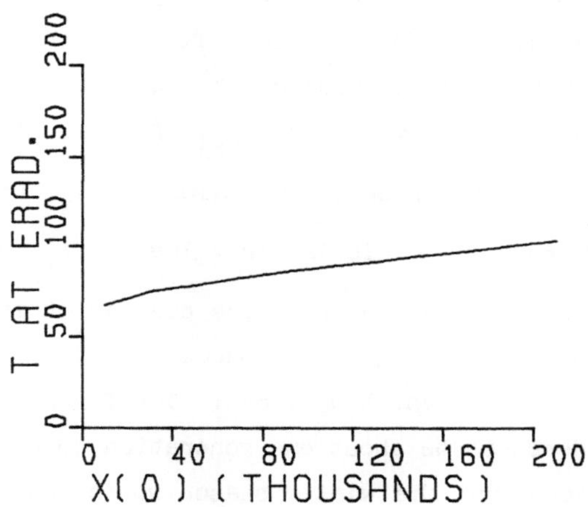

Figure 4.7

Figure 4.7. The effect of a temperate climate on the time to eradicate.

Figure 4.6 shows the outcome of an eradication campaign in which all stages of the rearing facility are maintained at a capacity c_{maint} = 0.007, just enough to fill the adult cages with the surplus. It is evident that the lack of surplus flies to release early in the campaign has a strong effect on the amount of time needed to achieve eradication. Comparison of Figures 4.1 and 4.5 shows the value of a rapid and strong response to a detected infestation. Note that this response is particularly evident in the second generation of the wild population, which declines in Figure 4.1 but increases in Figure 4.5. This effect is due to the released sterile males in the first generation.

It is possible in principle to maintain the rearing facility at a level below that of inequality (2.9). In this case there is no surplus to release in the first generation, and, indeed, there is a decision problem involving the portion of flies to harvest in the first generation, and the portion to hold back for breeding. This problem has been studied by Plant (1985). It may be applicable to mass rearing of other, less fecund species, but it probably does not have any relevance to the mass rearing of medflies.

One objection to all of these simulations as they relate to eradication campaigns in California is that they assume a constant temperature throughout the campaign. In a temperate climate the population may be expected to decline during the winter due to lack of food and oviposition sites. To test the effect of this decline we ran the following simulation. At a value of t equal to 75, representing an elapsed time of 150 days, the values of P_{xj} for all j were multiplied by a factor of 0.75. This reduced the value of λ_x, the finite rate of increase, from 1.22 to 0.91. This latter value of λ_x would correspond to the population's declining to about one quarter of its previous value after about one generation. Figure 4.7 shows the effect of this decline on the T at eradication vs. X(0) curve. The figure extends to a time to eradicate considered to be about one generation beyond $t = 75$, since at a reduced temperature the winter season would consist of only about one generation. It is evident by comparison with Figure 4.2 that the temperate climate does have an effect on the eradication threshold, although it is difficult to quantify the strength of this effect.

DISCUSSION

In this section we list a number of conclusions that may be drawn from the study. We focus on qualitative conclusions that are more or less independent of the particular parameter values of the model. Of course, any conclusion is to some extent model dependent. We do not necessarily list the conclusions in the same order as they arise in the preceding sections.

1. Life stage for most efficient eradication. Based on Figure 3.2, the reproductive value as a function of age, one might be tempted to say that an eradication campaign can have its greatest impact by killing young adults. This figure is somewhat misleading, however, since the reproductive value actually represents the proportional contribution of

an individual to the future gene pool. A more useful indication, perhaps, of the effectiveness of eradication measures is Figure 3.3, which shows the derivative of the growth rate λ_x with respect to conditional survivorship P_{xj}. According to this figure, the greatest reduction in growth rate of the population can be achieved by killing in immature stages. Even this figure is misleading, however, because it says nothing about the relative cost and effectiveness of the available methods of reducing the survivorship of the various life stages.

2. Discard age of the rearing facility. Based on Figure 2.2, near the optimal discard age the harvest rate per female is not strongly dependent on the precise value of the discard age; in other words, the harvest rate curve is relatively flat near its peak. Therefore, if for reasons such as imprecise knowledge of the parameter values or logistical necessity the discard age is not precisely the optimal, as long as it is close the rearing facility will produce harvestable pupae at nearly the maximal rate.

3. Rearing facility configuration. Based on Figure 4.3, the precise configuration of the rearing facility during maintenance periods is not crucial. Regardless of the adult configuration, there will be a transition period lasting about three weeks during which the number of harvested pupae is fixed by the larval configuration. The facility should be maintained at a level sufficient to stock the adult cages with surplus pupae. A relatively high maintenance level does permit some initial growth suppression of the wild population during the transition of the rearing facility to full capacity. This growth suppression may have a significant effect on the growth rate of the next generation of the wild population.

4. Response to a detected infestation. In the case of an invasive infestation, a rapid response is crucial. The population may be growing at a rate of as much as ten percent per day at the time of detection. By the time the rearing facility can be brought to full capacity, the wild population would in this case have increased almost tenfold. Rapid response is particularly important if auxiliary control measures such as fruit stripping and ground drench are not planned.

Another important point in this regard is that if the SIT fails, then it fails dramatically. This point can best be seen in Figure 4.4. Note that logarithms of the wild population are plotted in this figure.

Although the two cases in which the SIT fails represent borderline cases in which successful control is almost achieved, nevertheless the population, once it escapes control, grows very rapidly.

5. The overflooding ratio. A point that is often overlooked is that the overflooding ratio necessary for successful eradication depends on the rate of population growth. The classical ratio of 100:1 will not, according to the calculations presented here, provide successful control if the population is growing at a sufficiently rapid rate. Similarly, the eradication threshold is critically related to the population growth rate. Auxiliary control measures such as fruit stripping and ground spray, which reduce the growth rate of the population, greatly increase the eradication threshold. In general, the effectiveness of an eradication campaign using the SIT against an invasive insect pest will depend on both the current size of the population and on the growth rate of the population.

Acknowledgements

I am very grateful to J. Carey, R. Cunningham, R. Dowell, M. Mangel, N. Tanaka, and R. Vargas for many useful discussions. This research was supported by the California Department of Food and Agriculture Exotic Pest Research Program.

REFERENCES

Beddington, J.R. and Taylor, D.B., 1973. Optimum age specific harvesting of a population. Biometrics. 29: 801-809.

Berryman, A.A., 1967. Mathematical description of the sterile male principle. The Canadian Entomologist 99: 858-865.

Carey, J.R., 1982. Demography and population dynamics of the Mediterranean fruit fly. Ecological Modelling 16: 125-150.

Carey, J.R., 1984. Host-specific demographic studies of the Mediterranean fruit fly Ceratitis capitata. Ecological Entomology 9: 261-270.

Carey, J.R. and R.I. Vargas, 1985. Demographic analysis of insect mass rearing: a case study of three tephritids. Journal of Economic Entomology 78: 523-527.

Caswell, H., 1978. A general formula for the sensitivity of population growth rate to changes in life history parameters. Theoretical Population Biology 14: 215-230.

Costello, W. and H. Taylor, 1975. Mathematical models of the sterile male technique of insect control. Lecture Notes in Biomathematics, vol. 5, A. Charnes and W.R. Lynn, ed., Springer-Verlag, New York, pp. 318-359.

Cunningham, R., W. Routhier, E. Harris, G. Cunningham, L. Johnson, W. Edwards, R. Rosander, and W.G. Vettel, 1980. Eradication of medfly by sterile-male release. Citrograph 65: 63-69.

Denny, C.L., T. Wong, N. Tanaka, and R.V. Dowell, 1985. Size as a factor in the mating propensity of laboratory-reared, unirradiated Mediterranean fruit flies. Preprint.

Emlen, J.M., 1984. Population Biology. Macmillan Publishing Co., New York.

Hooper, G.H.S., 1982. The sterile insect release method for suppression or eradication of fruit fly populations. Economic Fruit Flies of the South Pacific Region, R.A.I. Drew, G.H.S. Hooper, and M.A. Bateman, ed., Queensland Dept. of Primary Industries, Brisbane, Australia, pp. 99-114.

Ito, Y., 1977. A model of sterile insect release for eradication of the melon fly, Dacus cucurbitae Coquillet. Applied Entomology and Zoology 12: 303-312.

Keyfitz, N., 1977. Introduction to the Mathematics of Populations. Addison-Wesley, New York, 490 pp.

Knipling, E.F., 1955. Possibilities of insect control or eradication through the use of sexually sterile males. Journal of Economic Entomology 48: 459-462.

Lawson, F.R., 1967. Theory of control of insect population by sexually sterile males. Annals of the Entomological Society of America 60: 713-722.

Leslie, P.H., 1945. On the use of matrices in certain population mathematics. Biometrika 33: 183-212.

Lewis, E.G., 1942. On the generation and growth of a population. Sankhya 6: 93-96.

Monro, J., 1973. Some applications of computer modelling in population supression by sterile males. Computer Models and Application of the Sterile-Male Technique. International Atomic Energy Agency. Vienna, pp. 81-94.

Monro, J. and A.W. Osborn, 1967. The use of sterile males to control populations of Queensland fruit fly, Dacus tryoni (Frogg) (Diptera: tephrididae) I. Methods of mass-rearing, transporting, irradiating, and releasing sterile flies. Australian Journal of Zoology 15: 461-473.

Plant, R.E., 1985. An algorithm for the time optimal discrete nonlinear control problem. Preprint.

Pollard, J.H., 1973. Mathematical models for the growth of human populations. Cambridge University Press, 186 pp.

Vargas, R.I., D. Miyashi, and T. Nishida, 1984. Life history and demographic parameters of three laboratory-reared tephritids. Annals of the Entomological Society of America 77: 651-656.

Wong, T.T.Y. and L.M. Nakahara, 1978. Sexual development and mating response of laboratory-reared and native Mediterranean fruit flies. Annals of the Entomological Society of America 71: 592-596.

ANALYSIS OF SPATIAL DISTRIBUTION IN FRUIT FLY ERADICATION

Masakazu Shiga
Fruit Flies Laboratory, Okinawa Prefectural Agricultural Experiment Station, Sakiyama, Naha, Okinawa 903, Japan

ABSTRACT

Spatial pattern of distribution of the fruit flies caught in traps under the implementation of the male annihilation method (MA) and sterile insect technique (SIT) was analysed. As for MA applied to *Dacus dorsalis*, the index of dispersion of individuals, I_δ , and the density-contagiousness coefficient, β , in the frequency distribution of flies caught in traps were useful to predict a successful eradication. In SIT for *D. cucurbitae*, sterile flies released could be conveniently dispersed to local patches where wild flies were abundant. However, the correlation coefficient between the number of sterile flies and that of the wild flies caught in traps in the area where SIT was being implemented was often low. This observation suggests that the effect of SIT is not homogeneous among local patches. These facts suggest that the analysis of the spatial distribution of fruit flies in the target area of the eradication project is important to devise a strategy for the implementation of SIT and also for the evaluation and analyses of the eradication processes.

INTRODUCTION

To evaluate the effects of eradication techniques on insect pests, changes in the abundance of the target pests must be studied in time and in space. In this context, analyses were conducted on some of the spatial problems encountered in the course of the implementation of the male annihilation method (MA) and sterile insect technique (SIT) for the eradication of the oriental fruit fly, *Dacus dorsalis*, and the melon fly, *D. cucurbitae*, respectively, in Okinawa, Japan. Three problems will be dealt with here:

(1) Changes in spatial distribution of the target insect under MA implementation.

(2) Dispersal and resultant spatial distribution of released

NATO ASI Series, Vol. G11
Pest Control: Operations and Systems Analysis in Fruit Fly Management. Edited by M. Mangel et al.

sterile flies in a field experiment.

(3) Spatial relationships in the abundance of wild and released flies in the early phase of SIT implementation.

SPATIAL DISTRIBUTION OF *D. DORSALIS* UNDER MA IMPLEMENTATION

1. Analysis with I_δ-index

Both on the Miyako Islands and the Yaeyama Islands, Okinawa, the eradication project of *D. dorsalis* with the application of MA was initiated in April, 1982. The value of the index of dispersion of the individuals, I_δ(Morisita, 1959a), was calculated to detect the changes in the frequency distribution of the male flies caught in traps baited with poisoned methyleugenol.

In this case, I_δ is defined as follows:

$$I\delta = q \frac{\sum_{i=1}^{q} n_i \, (n_i - 1)}{N \, (N - 1)} , \qquad (1)$$

where n_i is the number of flies caught in the i-th trap, N is the total number of flies caught ($= \sum_{i=1}^{q} n_i$), and q is the number of traps observed. The value of I_δ varies in case of underdispersed, random (Poisson), and overdispersed distribution as $I_\delta<1$, $I_\delta=1$ $I_\delta>1$, respectively.

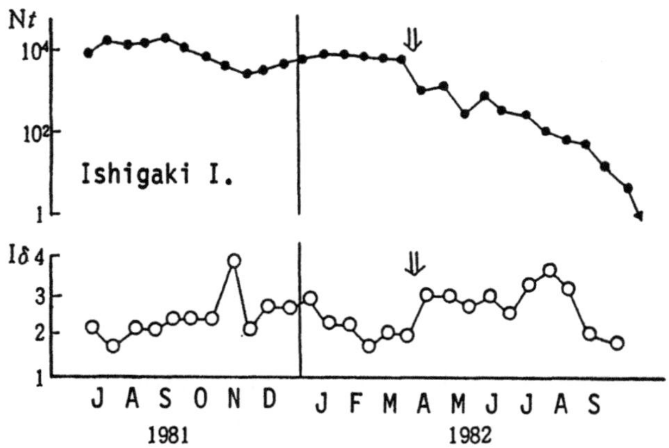

Fig. 1. Temporal changes in the number of *D. dorsalis* caught in traps (top) and index of dispersion of individuals, I_δ (bottom) on Ishigaki Island.
Nt : No. of flies caught / 1000 traps / day
⇓ : Initiation of MA implementation

In the Yaeyama Islands, I_δ value seemed to be slightly increased or not to be changed fairly immediately after the innitiation of the eradication program on Ishigaki Island (Fig. 1, bottom) and Yonaguni Island where the number of flies caught in traps was continuously decreasing to zero (Fig. 1, top).

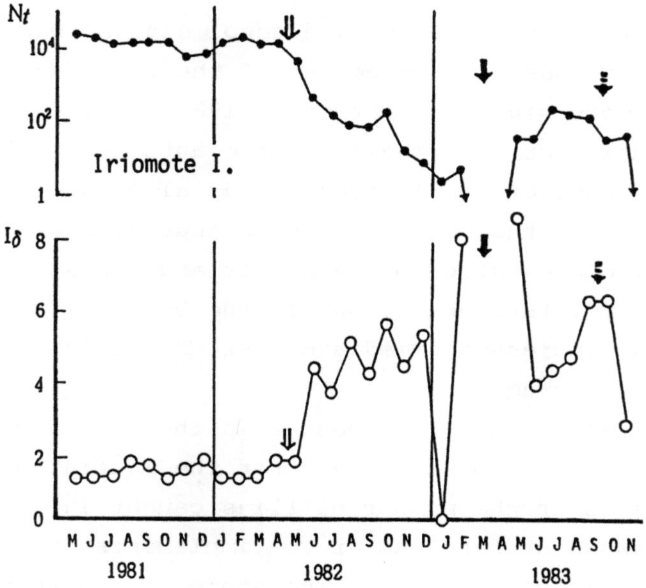

Fig. 2. Temporal changes in the number of *D. dorsalis* caught in traps (top) and index of dispersion of individuals, I_δ (bottom) on Iriomote Island.

N_t : No. of flies caught / 1000 traps / day

⇓ : Initiation of MA implementation (excl. inland and southern coastal areas)

⬇ : Initiation of MA implementation throughout the island

⬇ : Initiation of additional application of poisoned methyl-eugenol to the high density areas of the fly.

On Iriomote Island, where the environmental structure is markedly heterogeneous as the island is covered with dense primary forest over wide areas, and where some difficulties in the eradication had occurred, although the density of *D. dorsalis* decreased continuously immediately after the initiation of the MA implementation, the number of flies caught in the traps increased again in 1983 (Fig. 2, top). In this island, I_δ value increased significantly immediately after the initiation of MA implementation (Fig. 2, bottom). This fact suggests that there were some local patches where MA was less effective than in the other areas. This conclusion does not necessarily imply that the limited effect of MA may be caused by the relaxation of control efforts on the inland and in the southern coastal areas of the island in 1982, because the I_δ value was still high after the initiation of control throughout the island after April, 1983 (Fig. 2, bottom). Actually, on the island, there were some areas where the amount of poisoned methyleugenol applied was deficient was detected. Increase in the amount of the poisoned bait applied to such spots since the autumn of 1983 resulted in the successful eradication of *D. dorsalis* by the autumn of 1984.

A similar trend of I_δ fluctuation was also found in the Miyako Islands, i.e., the value did not change even after the initiation of MA implementation on Tarama Island where the eradication was successful throughout, but in the Ueno Village of Miyako Island the value increased immediately after the initiation of the eradication program.

It must be noted that the changes in the I_δ value, when occurring, were observed immediately after the initiation of MA implementation, even if the number of flies caught in the traps continuously decreased at first as if the implementation of MA had been highly successful (Fig. 2). Therefore, the possible occurrence of difficulties in the eradication in such areas can be predicted in the early phase of the eradication project with MA by observing the trend of I_δ as well as N_t.

2. Analysis with $\overset{*}{m}$ - m method

Iwao's method of distribution analysis (Iwao, 1968) using the regression of mean crowding ($\overset{*}{m}$) on mean density (m) is also applicable. Lloyd's (1967) mean crowding, $\overset{*}{m}$, is defined as,

$$\overset{*}{m} = \frac{\sum_{i=1}^{q} n_i (n_i - 1)}{N} = \frac{\sum n_i^2}{N} - 1 \qquad (2)$$

where symbols are the same as those in equation (1)

Iwao (1968) observed that linear relationships can be obtained when $\overset{*}{m}$ is plotted on m , i.e.,

$$\overset{*}{m} = \alpha + \beta m \qquad (3)$$

In this equation, α and β are constants and β is defined as the 'density-contagiousness coefficient' by which contagiousness of distribution can be measured (Iwao, 1968).

On Ishigaki Island where MA was highly effective throughout the island, β increased only slightly during the period of the implementation of MA (Fig. 3). However, on Iriomote Island, where MA was only partly effective at first, β increased significantly under the implementation of MA (Fig. 4). In the latter case, the fact that the values in the middle and right graphs of Fig. 4 are nearly equal suggests that the difficulties were not solved after MA implementation was conducted throughout the island since April, 1983 (See also Fig. 2).

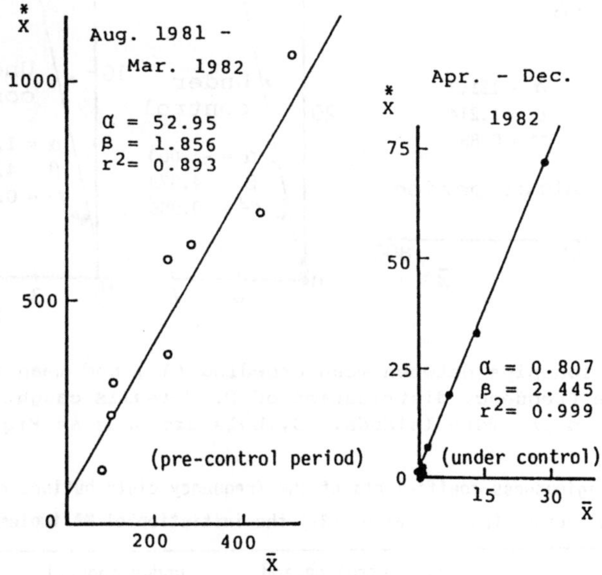

Fig. 3. Relationships between mean crowding ($\overset{*}{x}$) and mean ($\bar{x}$) of the frequency distribution of *D. dorsalis* caught in traps. $\overset{*}{x}$ and $\bar{x}$ are the sample estimates of $\overset{*}{m}$ and m, respectively.

The values of β before the initiation of MA implementation (pre-control period), β_0 , and those under MA implementation, β_1, in several areas of the Miyako Islands and the Yaeyama Islands are shown in Table 1. The ratio β_1/β_0 are much higher than unity in the areas where the eradication of *D. dorsalis* was successful only after additional applications of poisoned methyleugenol were made in certain areas or local patches where the flies still remained (Iriomote Island and Ueno Village). On the other hand, the ratio was nearly unity in the areas where MA was successfully implemented (Ishigaki Island, Yonaguni Island and Tarama Island).

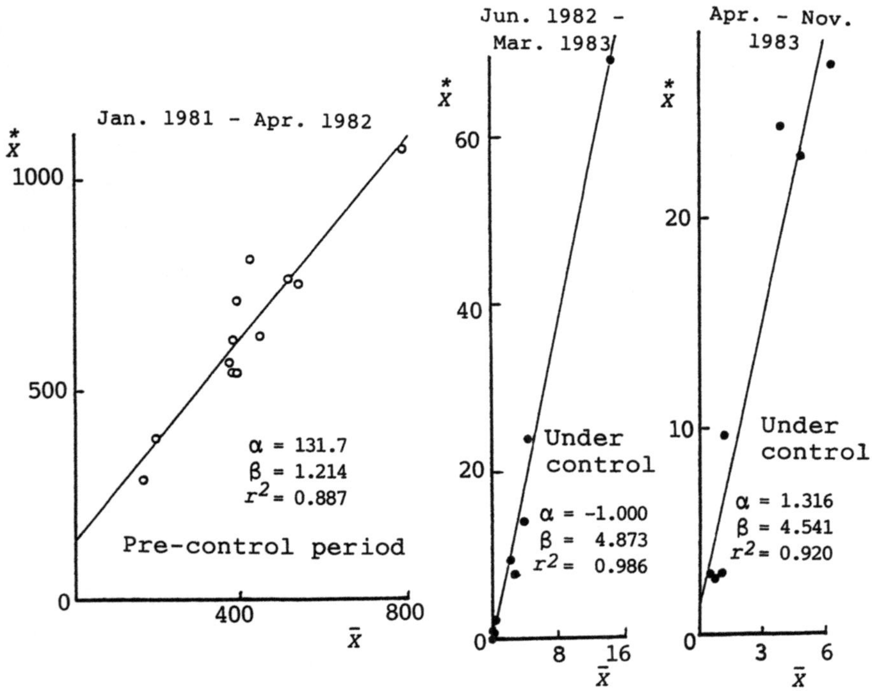

Fig. 4. Relationships between mean crowding ($\overset{*}{X}$) and mean ($\bar{X}$) of the frequency distribution of *D. dorsalis* caught in traps on Iriomote Islands. Symbols are same as Fig. 3.

Table 1. Density-contagiousness coefficients of the frequency distributions of *D. dorsalis* caught in traps before (β_0) and after (β_1) the initiation of MA implementation.

Locality	No. of traps	pre-control period		under control		β_1/β_0
		β_0	r	β_1	r	
Ishigaki I.	32	1.856	0.945	2.445	0.999	1.317
Iriomote I.	30 - 40	1.214	0.942	4.802	0.986	3.956
Yonaguni I.	15	3.548	0.957	3.267	0.971	0.921
Ueno V., Miyako I.	20	3.021	0.962	8.096	0.844	2.680
Tarama I.	15	2.120	0.899	2.116	0.991	0.998

β_0,β_1: Density-contagiousness coefficient; r: Correlation coefficient

If a linear regression of $\overset{*}{m}$ on m of the pre-control period could have been obtained prior to the initiation of MA implementation and if successive data after the initiation of MA implementation were plotted on the graph, there would be another mean of predicting the success of eradication attempts.

SPATIAL DISTRIBUTION OF RELEASED STERILE FLIES

Spatial problems in SIT seem to be far more complicated than those in MA, since the effect of sterile insect release is achieved through a large number of interrelated factors between the wild population and released sterile insects. Whether released sterile flies can successfully infiltrate the wild population is of primary importance to the effectiveness of SIT.

Iwahashi (1972) released fertile *D. dorsalis* which were artificially reared and marked to investigate their movement. He found that the number of marked males caught in each trap was positively correlated with that of wild flies. Hamada (1980) also reported similar results in a mark-and-recapture experiment of fertile mass-reared males of *D. cucurbitae*.

Whether sterile flies released in the field can reach the local patches where wild flies are abundant remains to be determined. To solve this problem, sterilized melon flies were experimentally released in the center of Miyagi Island (ca. 3 km in diameter) and the number of released and wild flies both caught in the traps were counted each week. After two weeks or more, the values of the correlation coefficient, and indices of overlapping (C_δ) and coexistence (R_δ) (Morisita, 1959b) calculated for the frequency distributions of the released and wild flies caught in the traps were high (Fig. 5). This was not the case when the number of flies caught in the traps set in the neighbouring islets which are more than 3 km apart from the release point were included in the calculation. These results suggest that sterile flies which were released in the center of the islet infiltrated the wild fly population in the islet and coexisted with the wild flies in the suitable habitats or patches where wild flies were abundant and that mass-reared, released sterile flies had similar dispersal and orientation abilities to those of the wild flies. This consideration should be useful for a theoretical or fundamental approach, i.e., the analysis of SIT process. Practically these data may also be used as a measure to evaluate the ability of sterile males, to determine the appropriate distance among the releasing points within the target area, etc.

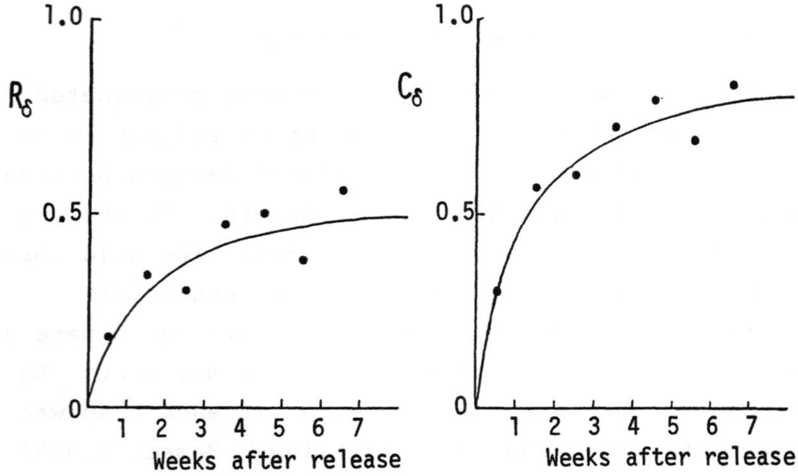

Fig. 5. Changes in values of C_δ and R_δ-index between the number of wild flies and that of released (sterile) flies caught in traps (Iwaizumi and Shiga, in preparation).

SPATIAL RELATIONSHIPS IN THE ABUNDANCE OF STERILE AND WILD FLIES UNDER SIT

The previous section dealt with the spatial distribution of sterile and wild flies in a field experiment in which sterile flies were released at a single point at one time. Spatial distribution of both sterile and wild flies should be highly complicated under SIT implimentation in which sterile flies are released at many points in the target area, repeatedly with certain intervals, because the interactions between sterile and wild fly populations would became extremely complex.

In the early phase of SIT implementation in the Miyako Islands where ca. 30 million sterilized flies (males and females) have been released every week since August, 1984, the correlation coefficients between the number of sterile flies (*S*) and that of the normal (wild) ones (*N*) caught in traps exceeded 0.5 in the autumn, 1984, decreased in the winter and in the early spring, 1985, and increased again thereafter. This pattern may result from the spatial overlapping of sterile and wild flies, as mentioned above, and also from the effect of SIT on the wild population which could be achieved differently among the local patches in the entire population of the fly. Correlation coefficients between *S* and *N* in several areas of the Miyako Islands and also in the total area are shown in Table 2. Correlation coefficients are positive in all the cases and some of them are statistically significant. However, the value of the coefficient in the total area is lower than the respective values in each area. These results can be ascribed to the fact that heterogeneous data were combined in this case. However, these data are important from the biological point of view, as they indicate the heterogeneity in the effect of SIT among various areas in the Miyako Islands. *S/N* ratio in each area is quite different in some cases (Table 2).

Table 2. Correlation coefficient between the number of normal flies and that of sterile flies caught in the traps on the Miyako Island (April 11 - 25, 1985)

Locality	No. of traps	Total no. of sterile flies (S)	Total no. of normal flies (N)	S / N	Correlation coefficient (r)
Miyako Island					
Hirara C.	23	742	17	43.6	0.642**
Gusukube T.	15	369	6	61.5	0.269
Shimoji T.	8	85	31	2.7	0.584
Ueno V.	6	72	11	6.5	0.351
Irabu Island (Irabu T.)	11	216	66	3.3	0.665*
Total	63	1485	131	11.3	0.166

$* \ 0.01 < P \leq 0.05$, $** \ P \leq 0.01$

Under a uniform environment in which the effect of SIT can be achieved evenly throughout the target area, the correlation between *S* and *N* may correspond to that shown in Fig. 6. In the initial phase of SIT implementation, i.e., shortly after the initiation of the sterile fly release, sterile flies are not abundant because the cumulative effect of periodical release on *S* is not yet fully realized and wild flies are abundant especially in the suitable habitats (Fig. 6, A). Thereafter, the number of sterile flies increased and the effect of SIT is partly realized (Fig. 6, B). In this phase, if sterile males have similar abilities of dispersal and orientation and viability to those of the wild ones, a positive correlation would be expected. Moreover, the number of wild flies in each patch may decreased and the slope of regression may reach zero (Fig. 6, C). In this scheme, it is assumed that sterile flies can disperse easily to the patches where wild flies are abundant.

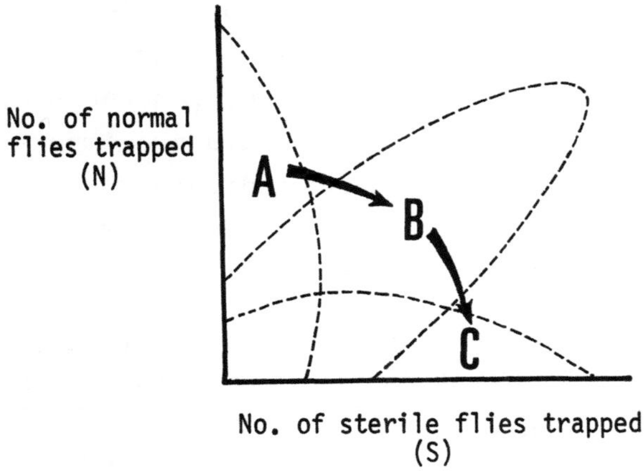

Fig. 6. A schematic illustration of correlation between the number of normal (wild) flies (*N*) and that of sterile flies (*S*) caught in traps.

In the case of SIT implementation in the Miyako Islands, the points on the correlation diagrams are highly scattered and the correlation coefficient are often low (Fig. 7), just as the various patterns shown in Fig. 6 are superimposed. Therefore, SIT may act quite heterogeneously among different patches in the area where SIT is being conducted.

Considering such heterogeneity in environment and its effect on SIT, a basic strategy for the implementation of SIT in wide areas can be devised. Implementation of SIT may be more effectively achieved in considering the following three steps.

Step 1: Sterile insects are released rather uniformly throughout the target area.

Step 2: Local patches which are still highly populated by the wild flies ('hot spot') should be detected.

Step 3. Additional eradication efforts should be directed to such hot spots.

It must be noted that hot spots of wild flies may be partly determined prior to the release of sterile insects (e.g., certain orchards for several kinds of fruit flies). However, others may not be localized and detected until most of the local populations in the target area are successfully suppressed or eliminated by SIT. As the requirements of step 2 are essential to achieve the total eradication of the pest, it is important to localize and detect hot spots adequately by analysing the spatial distribution of the pest.

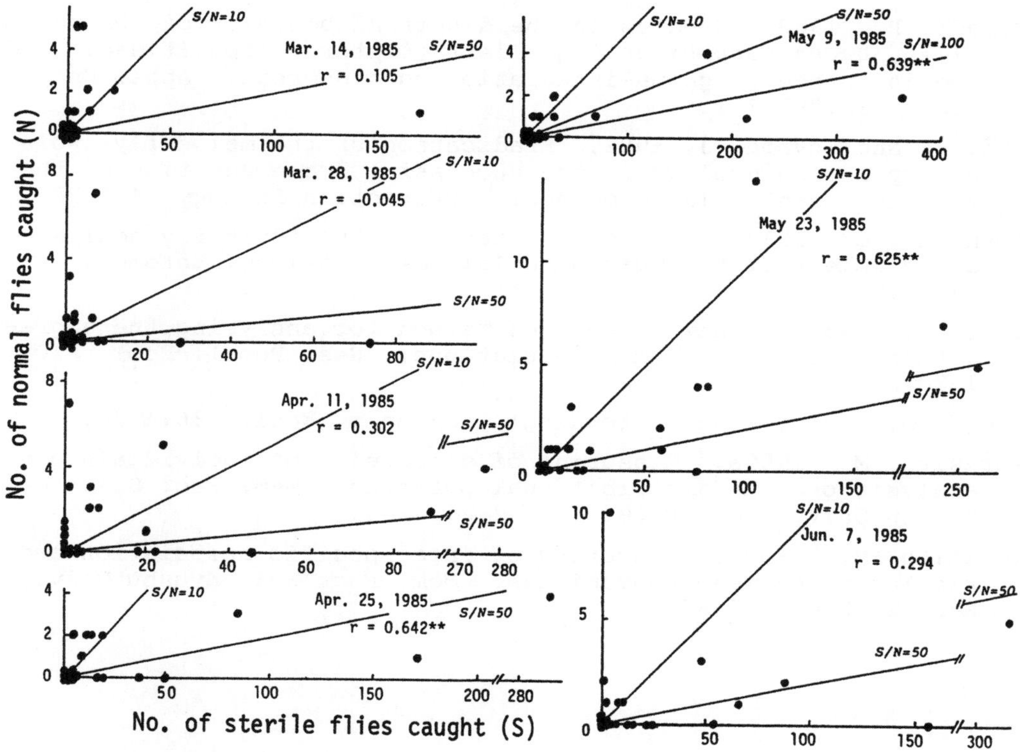

Fig. 7. Correlations between the number of sterile flies and that of normal flies caught in traps in Hirara City, Miyako Island. *r* : Correlation coefficient, ** $P < 0.01$.

CONCLUDING REMARKS

Eradication efforts with MA, and even with SIT, cannot be realized uniformly throughout a target area, unless the area is exceptionally small and/or uniform. In this context, spatial or spatio-temporal aspects are of fundamental importance to implement, analyse and evaluate the eradication processes. This conclusion is reasonable since biological populations in the field occur under heterogeneous environments and populations are affected by such environments in time and in space. As Itô and Koyama (1982) stated, SIT is essentially an exercise in population dynamics. It is also the case in other eradication methods of insect pests, e.g., MA.

REFERENCES

Hamada, R. 1980. Studies on the dispersal behavior of the melon flies, *Dacus cucurbitae* Coquillett (Diptera: Tephritidae) and the influence of gamma-irradiation on dispersal. Appl. Ent. Zool. 15:363-371.

Itô, Y. and Koyama, J. 1982. Eradication of the melon fly: Role of population ecology in the successful implementation of the sterile insect release methods. Protection Ecology 4:1-28.

Iwahashi, O. 1972. Movement of the oriental fruit fly adults among islets of the Ogasawara Islands. Environ. Entomol. 1: 176-179.

Iwao, S. 1968. A new regression method for anslysing the aggregation pattern of animal populations. Res. Popul. Ecol. 10: 1-20.

Lloyd, M. 1967. 'Mean crowding'. J. Anim. Ecol. 36:1-30.

Morisita, M. 1959a. Measuring of dispersion of individuals and analysis of the distributional patterns. Mem. Fac. Sci. Kyushu Univ., Ser. E (Biol.) 2:215-235.

Morisita, M. 1959b. Measuring of interspecific association and similarity between communities. Mem. Fac. Sci. Kyushu Univ., Ser. E (Biol) 3:65-80.

CONTROL STRATEGIES DESIGNED TO REDUCE THE CHANCE OF RESISTANCE WITH SPECIAL REFERENCE TO TEPHRITID FRUIT FLIES

R.J. Wood,
Department of Zoology,
Manchester University,
Manchester M13 9PL, U.K.

ABSTRACT

Laboratory selection experiments on the medfly Ceratitis capitata have produced resistance to several insecticides. Yet resistance has never become a practical problem in medflies nor in other tephritids. This state of affairs is analysed. Considerable benefit probably arises from natural movements of flies between treated and untreated trees or areas. The integration of insecticide usage with other methods of control is also considered to be beneficial. But probably the most important factor is the way insecticides are combined with baits or lures which maximises the dose received and (in the case of lures) restricts exposure mainly to males. This assessment is supported by a review of published results of computer simulations.

The possibility of resistance developing to SIT is also considered. This would depend on behavioural or physiological barriers evolving between released males and wild females. Strategies to avoid or overcome resistance are considered.

NATO ASI Series, Vol. G11
Pest Control: Operations and Systems Analysis
in Fruit Fly Management. Edited by M. Mangel et al.

INTRODUCTION

Insecticide resistance has never become a problem in tephritid fruit flies despite widespread applications of a variety of different chemicals. This desirable state of affairs deserves close examination for it is clearly important that it be assured of continuing. Moreover its analysis may provide guidelines for the avoidance of resistance in other pests.

The potential for resistance to non-insecticidal control measures, notably the sterile insect technique (SIT) is also worthy of consideration. SIT is now being used increasingly against fruit fly species, especially the medfly Ceratitis capitata. Resistance is expected to occur through genetic variants favouring an increase in assortative mating between wild females and wild males at the expense of factory bred males. Up to now there is no evidence of resistance but the danger exists. As in the case of insecticides, it seems worth considering possible ways in which resistance might arise and how to forstall it.

INSECTICIDE RESISTANCE

Insecticides are a diverse collection of chemicals classified into eight or more major groups (Wood 1981a). Until the late 1950's the insecticides most widely used for controlling fruit flies were the chlorinated hydrocarbons, including DDT, HCH and various cyclodienes. After this time, increasing use was made of organophosphate compounds, among which were malathion, brought into use for fruit fly control in Hawaii in 1956 and in Israel in 1957 (Keiser 1968, Roessler, personal communication), dimethoate, first used against the medfly in Italy in 1958 (de Pietri-Tonelli et al 1960) and fenthion, applied against the same species in South Africa in 1959 (Myburgh 1961). Meanwhile, use of organochlorines declined although applications of DDT for medfly control were made in

Morocco in 1963 (Haltebourg 1966), of dieldrin against the same species in Egypt in 1965 (El Sawy et al 1965) and of HCH, chlordane and endrin as late as 1974 (Azab et al 1974, Awadallah 1974).

Despite extensive applications of these and related insecticides, and also lesser use of the carbamate insecticide carbaryl (Haltebourg 1966, Keiser 1968) very few indications of resistance in fruit flies has been reported in fruit flies (Keiser et al 1971, Georghiou 1981) and resistance has never become a practical problem. Yet in some of the same fruit growing areas, on the same crops attacked by non-resistant fruit flies, resistance has arisen in other pests. Thus, on citrus, resistance to various insecticides has been found in members of the Hemiptera (Aonidiella aurantii: Coccidae) in South Africa, the Thysanoptera (Scirtothrips citri: Thripidae) in California and the Acarina (Panonychus citri: Tetranychidae) in California, Florida, South Africa and Japan.

It is interesting, therefore, to consider what may have protected fruit flies from becoming resistant. If it is a matter of control strategy, this should be understood and continued.

1 Resistance potential

Several investigators have used laboratory selection programmes to look for resistance potential in the medfly. The experiments have been reviewed by Wood and Harris (1986). Orphanides et al (1980) submitted different colonies of medflies to pressure with dimethoate or DDT for 13 years. After 5 years (54 generations) of almost continuous pressure at the adult stage, the colonies were outcrossed to wild stock and reselected on an intermittent basis, also at the adult stage, for a further 84 generations. Intermittent selection was thought to reproduce conditions found in the field.

Resistance to DDT increased only slightly in the first period of selection but reached 7 x (ratio of LD 50) (median lethal dose) between selected and unselected strains in the second period after outcrossing. The increase in LD50 was associated with an increase in heterogeneity of response to the insecticide (indicated by a shallower slope on the ld-p lines) which may be seen as evidence that the limit of selection had not yet been reached in this strain. In the case of dimethoate, however, the ratio of LD50s never exceeded 2 x and the ld-p lines gave no indication of any potential for greater resistance.

An attempt to increase resistance to a second organophosphate insecticide, malathion, was reported by Koren et al (1984). As in the earlier study, adults of both sexes were exposed to the insecticide. Eighteen generations of selection resulted in a 2-3 x increase in LD50 in females but no change in males. It might be concluded from the small change in LD50 in females that even in this sex there was no real evidence of resistance. However there were significant changes observed in the esterases of the females similar to those associated with greater resistance in other species: (1) higher carboxylesterse activity, (2) reduced sensitivity of acetylchlorin-esterase to malaoxon. Male flies showed neither of these changes.

A study on the potential for resistance to dieldrin in medfly larvae was reported by Wood and Busch-Petersen (1982) and Busch-Petersen and Wood (1983). Selection was applied to third instar larvae which were exposed for successive generations to ethanolic dispersions of the insecticide in water, mortality being recorded at emergence. The resistance ratio at EC50 (effective median lethal concentration) increased to 45 x in one strain after 21 generations of selection and to 24 x in a second strain after 26 generations of selection. Increase in EC50 was accompanied by an increase in heterogeneity of response, and resistance was unstable when selection was relaxed. A more homogeneous strain (DiR) which was relatively stable in resistance, was produced by using single family sib selection (Bennet 1960), breeding from unexposed sibs within a particularly resistant F_2 progeny among a series of progenies derived from single pair matings.

The same technique was used to produce a homogeneous susceptible strain (DiS) (breeding from unexposed sibs within a particularly susceptible F_2 progeny). The DiR and DiS strain differed in EC50 by 49-67 x. Crosses between them showed that resistance was due to a single major gene of intermediate dominance, the expression of which was subject to modification by the background genotype (Busch-Petersen and Wood 1983, Fig. 5).

Tests with other insecticides revealed a wide spectrum of cross resistance in DiR larvae, extending not only to other cyclodienes and HCH but also to malathion and to the pyrethroid insecticide, permethrin (Busch-Petersen and Wood 1983). A similarly wide spectrum of cross resistance has been reported in *Drosophila melanogaster* larvae due to the gene $\underline{R}_I$ (Ogaki and Tsukamoto 1953, Kikkawa 1961, 1964). The mechanism for such an unusually wide spectrum of resistance is unknown but a restriction upon the uptake of insecticide is one possibility, as with the *pen* gene of the housefly *Musca domestica* (Sawicki 1973). Another possibility based on research with houseflies is a generalized "receptor gene" controlling several pathways of detoxication (Plapp 1984).

With potential for resistance both in larvae and adults, the absence of resistance in the field is a matter for comment. Orphanides et al (1980) quote a statement attributed to Professor G.P. Georghiou (expressed in an unpublished document dated 1974) that the non-appearance of resistance in this species is because of discontinuous insecticide pressure, as in the case of non-spraying or deficient spraying of trees which are poor fruit bearers. Non-spraying of some trees might indeed be an important factor in delaying resistance because of providing a means for a sample of the population to escape and thus maintain a reservoir of susceptible genes. The value of deficient spraying is more difficult to assess because, under some circumstances, it could increase the likelihood of resistance. This is because at intermediate dosages of insecticide, the heterozygous carriers of resistance genes might survive to reproduce. These conclusions gain support from computer simulations (Curtis et al 1978, Taylor and Georghiou 1979, Wood and Mani 1981) to be discussed below.

2 Rate-determining factors in the evolution of resistance

Whether the aim of fruit fly control is total eradication, or whether there is a management plan, or whether control is based entirely on the discretionary choice of the individual farmer, the obvious question about resistance remains the same, namely, which are the rate-determining factors in the evolution of resistance? Georghiou and Taylor (1976) classified factors known or suspected of influencing the kinetics of resistance into three categories - genetic, biological and operational. Genetic and biological factors are intrinsic to the species and the population, and determine the ongoing risk of resistance. The operational factors are subject to human control and may be given greater or lesser emphasis according to what the risk is judged to be. The situation in fruit flies is now considered.

2.1. Genetic factors

Laboratory selection experiments with the medfly, reported above, indicate genetic potential for resistance to several quite different insecticides both in larvae and adults. But in no experiment was resistance attained without difficulty. In the case of DDT selection by Orphanides et al (1980) it took many generations for a significant degree of resistance to appear, even after outcrossing to a freshly caught wild strain. After malathion selection carried out by Koren et al (1984), resistance was observed only in females and even then at a low level, described by the authors as "tolerance". In the case of larval exposure to dieldrin, the resistance produced by mass selection was unstable (Wood and Busch-Petersen 1982, Busch-Petersen and Wood 1983). Even in the strain DiR produced by single pair sib-selection, resistance began to decline after less than one year (Saaid and Wood unpublished). Each of these experiments pointed to a problem of coadaption, a problem related either to the degree of penetrance or expressivity of resistance or to the fitness of the resistant phenotype in the absence of the insecticide.

Problems of coadaptation are nothing new in resistance studies and have been observed in field populations as well as in laboratory strains. Different levels of adaptiveness may characterise resistance to different insecticides even in the same area. Thus, whereas diazinon resistance in the Australian sheep blow fly Lucilia cuprina is particularly well coadapted (McKenzie et al 1982), dieldrin resistant populations show a tendency to revert to susceptibility when the insecticide is removed.

Poor coadaptation may result from insufficient genetic variability. In the case of an invasive species such as the medfly, it seems likely that the gene pools of invading populations will be more restricted than the populations from which they are drawn and may therefore have a lower capacity for resistance. It is worth recalling the massive aerial insecticide programme against the medfly in California in 1981-82 under conditions in which the species passed through 6-8 generations per year. In the South San Francisco Bay region, repeated aerial applications of malathion were made at a rate of 877 ml of spray (composed of two parts of 91% malathion and eight parts of Stayley's Protein Bait) per hectare. Many regions of the infested area received over 20 applications at this rate during 1981-82 (Ehler et al 1984). No resistance appeared and the species was apparently completely eliminated.

An even more remarkable case of non-appearance of resistance in the medfly is found in the Israeli control programme. Control is achieved by malathion bait spraying combined with the use of traps incorporating trimedlure and DDVP (dichlorvos). Intensive malathion spraying has been undertaken for 28 years (1957-1985), 4-7 sprays per year being applied at a rate of 250 ml malathion (91%) and 750 ml protein hydrolysate per hectare. Rössler (personal communication) believes that the rate of selection is $>$99% although there are certain wild plants and abandoned orchards to which flies may escape. Immigration from outside the country is probably minimal. The flies go through 4-6 generations per year and there has thus been intense selection with malathion for 112-168 generations with no resistance.

Most examples of resistance in insects, including larval resistance in the medfly, may be interpreted in terms of an allelic change at a single locus. This is shown by the results of crosses between susceptible (SS) and resistant (RR) strains where the F_1 (RS) is usually intermediate in resistance ad the backcrosses reveal evidence of segregation (Fig. 1, Fig. 5). The model situation is shown in Fig. 1 where the three genotypes are absolutely distinguishable. In a polymorphic population it is clear that the three genotypes would be killed differentially by dosages A and B. This means that the "effective dominance" of the R gene (Wood and Cook 1978, Curtis et al. 1978) would vary according to the dosage

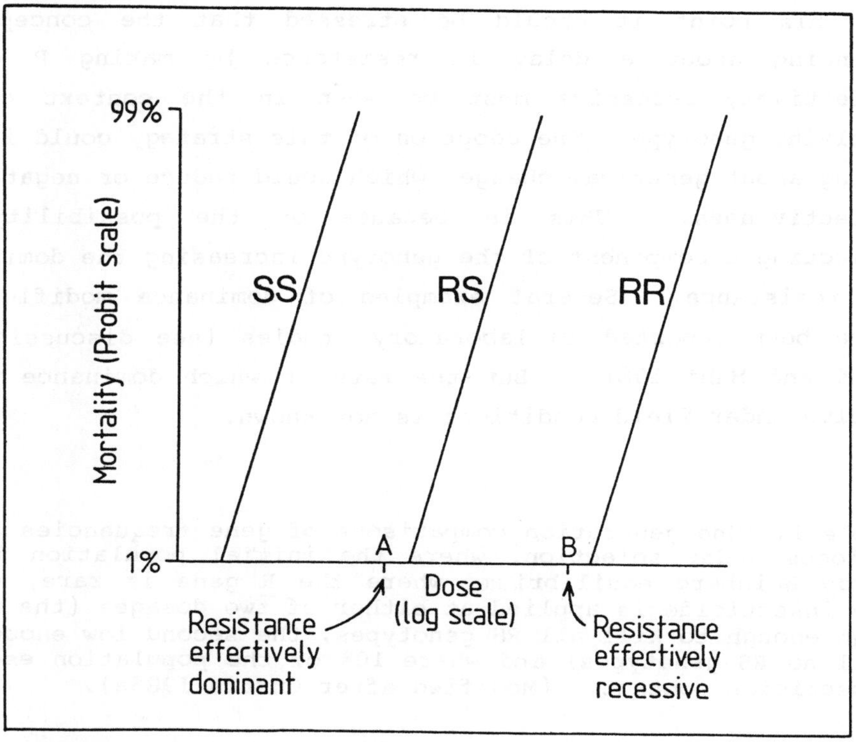

Fig. 1 Diagram to illustrate the kind of relationship generally found between dose and mortality in homozygotes for resistance and for susceptibility, and in heterozygotes.

of insecticide received. If the dosage were high (e.g. at B in Fig. 1), only homozygous resistant genotypes (RR) would survive. Resistance would then be effectively recessive (RS and SS indistinguishable) If the dose were low (e.g. at A in Fig. 1), the heterozygotes RS would remain alive, in which case the resistance gene would assume full dominance (RS and RR indistinguishable). This relationship between dosage and dominance appears to be highly significant to the evolution of resistance (Curtis et al 1978, Taylor and Georghiou 1979, Wood and Mani 1981). This is because a rare recessive gene is selected less rapidly than a rare dominant gene (Falconer 1981). The potential for manipulating R gene dominance, by regulating insecticide dosage, is an operational matter to be considered later.

At this point it should be stressed that the concept of bringing about a delay in resistance by making R genes effectively recessive must be seen in the context of an evolving genotype. The adoption of this strategy could itself bring about genetical changes which would reduce or negate its effectiveness. This is because of the possibility of selecting a component of the genotype increasing the dominance of resistance. Several examples of dominance modification have been reported in laboratory studies (see discussion by Wood and Mani 1981). But the rate at which dominance would evolve under field conditions is not known.

Table 1. One generation comparisons of gene frequencies at an R locus under selection, where the initial population is in Hardy Weinberg equilibrium, where the R gene is rare, where the insecticide is applied at either of two dosages (the first high enough to kill all RS genotypes, the second low enough to kill no RS genotypes) and where 10% of the population escapes insecticide contact. (Modified after Curtis 1985a).

	Number of insects			Freq. of R
	SS	RS	RR	(%)
Initial population	998,001	1,998	1	0.1
90% encounter insecticide	~~898,201~~	(1,798)	1	
10% escape	99,800	200	0	
Survivors if:				
a) RS are killed	99,800	200	1	0.101
b) RS not killed	99,800	1,998	1	0.982

2.2 Biological factors

Any avenue of escape from exposure to an insecticide is expected to contribute towards reducing selection pressure towards resistance and thus delay its appearance. This can be demonstrated by a simple calculation of gene frequencies over one generation (Table 1) or over several generations in a computer simulation (Figs 2,3). The delaying effect on resistance brought about by an R gene being made effectively recessive is seen to be conditional upon there being some degree of escape (a closed population is assumed i.e. no immigration).

Escape may occur by temporary migration out of a treated area. Or it may result from flies having access to trees left unsprayed because of low productivity, as suggested by Georghiou (see above). Or it may come about because of the availability of alternative wild hosts. The latter is likely to be particularly important in a highly polyphagous species such as C. capitata. The extent of escape has not been quantified in fruit fly species.

Equally unknown in fruit flies is the significance of migration of unselected flies into treated areas, another potentially important influence on the evolution of resistance which has attracted attention (Comins 1977, Taylor and Georghiou 1979). Table 2 summarises some representative results of the computer simulations of Taylor and Georghiou (1979) applied to mosquitoes, according to which resistance can be delayed under certain conditions of migration and effective dominance with no degree of escape assumed.

Table 2. Adult population size and frequency of resistant (R) allele (in parentheses) for a simulated mosquito population of 1000 initial size. Values shown are the average for the last 10 weeks of the year. The parameter h (dominance) = proportion of RS heterozygotes surviving the applied dose. (Data from Taylor and Georghiou 1979).

Dominance (h)	Number of migrants: 10	20	100
		No insecticde	
	13,900 (0.00)	8,720 (0.00)	15,660 (0.00)
		Larvicide applied	
0.5	14,000 (1.00)	14,100 (0.99)	15,000 (0.95)
0.1	14,000 (1.00)	14,200 (0.99)	670 (0.00)
0	14,300 (1.00)	14,300 (0.99)	670 (0.00)

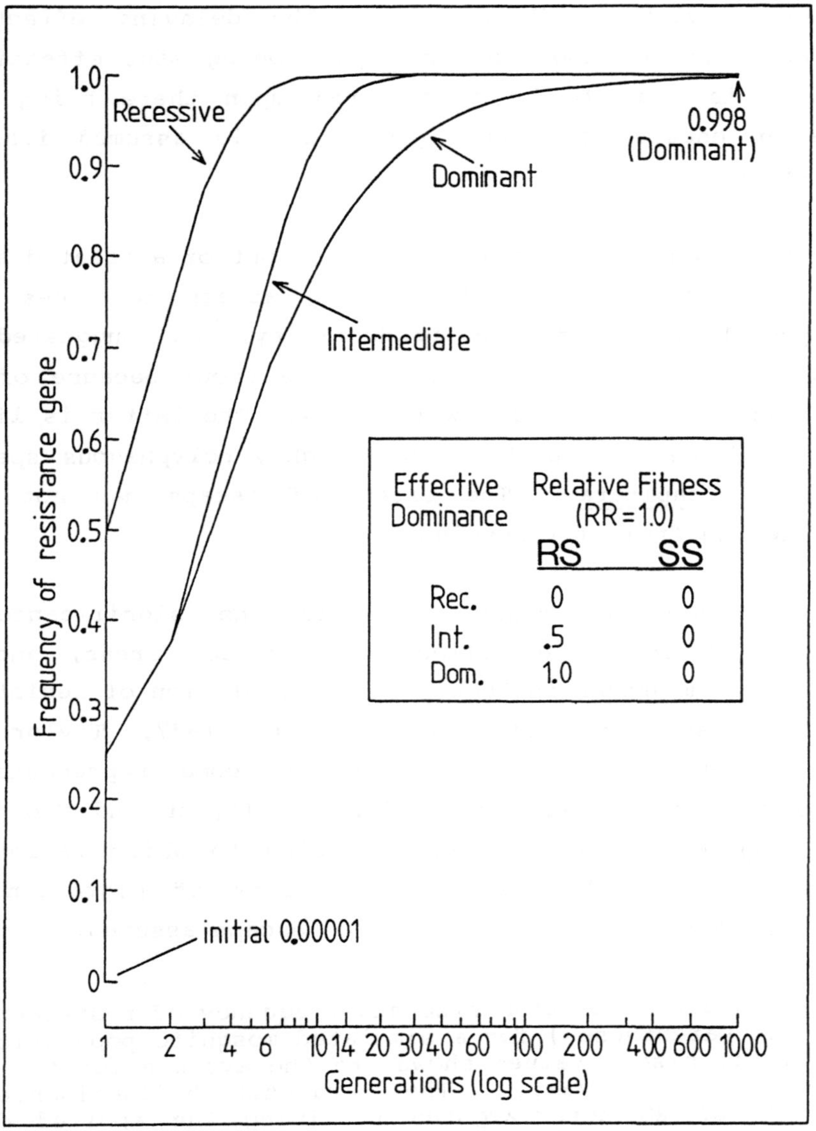

Fig. 2 Computer simulation of the number of generations required for a resistance gene to reach fixation or near fixation in an infinite population starting with an initial frequency of 10^{-5}, for the three cases where the gene is effectively dominant, recessive or has intermediate dominance (where half the hetero-zygotes survive exposure). In this example mating takes place before selection and no insect escapes exposure. (Modified after Wood and Mani 1981).

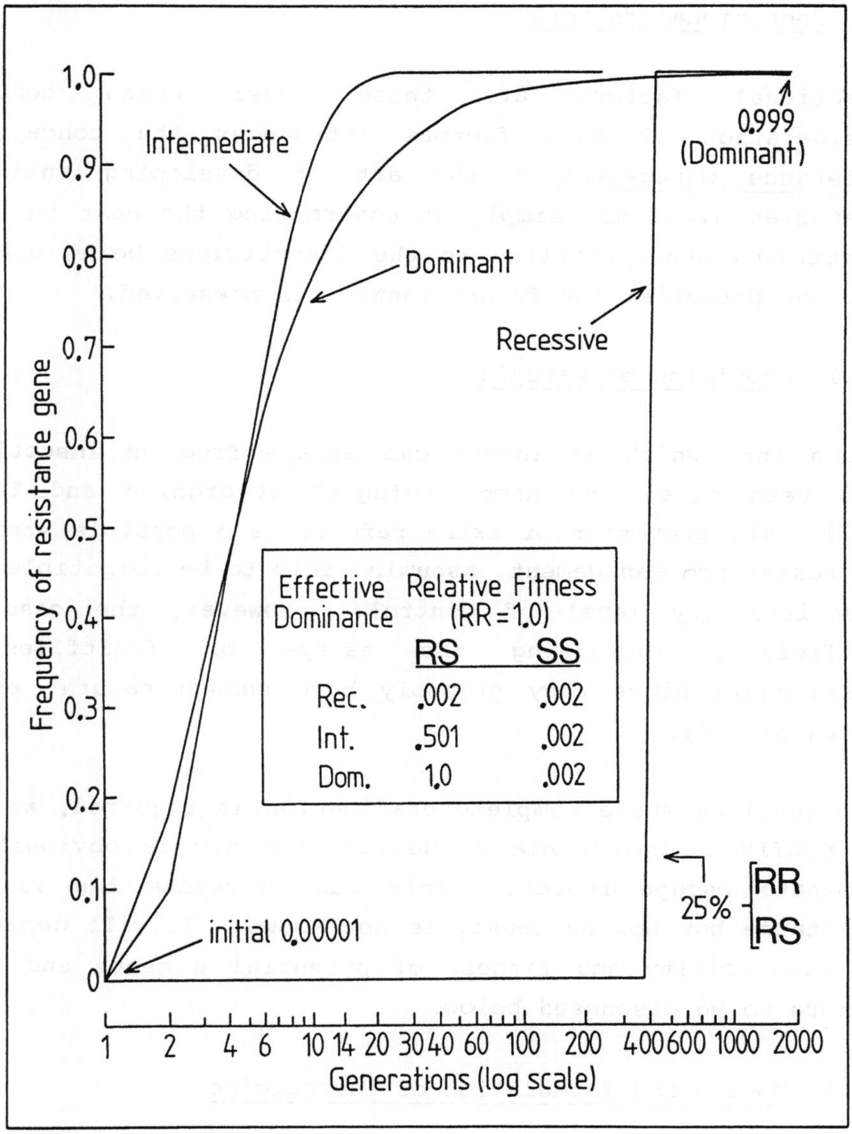

Fig. 3 Computer simulation of the number of generations required for a resistance gene to reach fixation or near fixation in an infinite population starting with an initial frequency of 10^{-5}, for the three cases where the gene is effectively dominant, recessive or has intermediate dominance (where half the heterozygotes survive exposure). In this example mating takes place before selection and 0.2% of the insects (all genotypes) escape exposure. (Modified after Wood and Mani 1981).

2.3 Operational factors

Operational factors are those under human control. Consideration of such factors introduces the concept of resistance management - the art of developing anti-pest strategies aimed not simply at controlling the pest but also maintaining susceptibility to the insecticides being used so that the potential for future control is preserved.

2.3.1 Provision of refugia

Places into which an insect can escape from an insecticide have been given the name "refugia" (Georghiou and Taylor 1977). The provision of extra refugia is a possible strategy for resistance management, assuming this to be compatible with a satisfactory level of control. However, the case for artificially increasing the escape of fruitflies is questionable since they probably have enough natural escape routes already.

In situations where complete eradication is required, as with the medfly in California or Mexico, the aim is obviously to eliminate escape routes. This must increase the risk of resistance but how seriously is not known. It will depend on the availability and fitness of potential R genes and other factors to be discussed below.

2.3.2 Dosage and formulation of insecticide

The dosage of insecticide received by the individual fly is clearly an important matter in the regulation of the effective dominance of potential R genes. If the dosage is insufficient to kill heterozygous carriers of the R gene (when the gene is still rare), ie the gene is effectively dominant, there is scope for a rapid build up of resistance. If the dosage is high, the R gene tends to become effectively recessive and resistance is delayed. The dosage will obviously be related to the total amount of insecticide sprayed into the environment. However, the proportion of this which enters the fly will depend on its formulation and mode of application.

Insecticides for fruit fly control are commonly formulated with an attractant. If this is a feeding stimulant, the formulation is referred to as a bait spray. Sprays incorporating protein hydrolysate as a bait are widely used against the medfly and are also effective against other fruit fly species including the Mexican fruit fly and melon fly (Knipling 1979). It is not necessary for the bait spray to be ingested in order to be effective. This is the case with the olive fly _Dacus oliae_ (Haniotakis, personal communication). The insecticide which is mixed with bait is normally one of the organophosphate compounds, often malathion. Bait sprays have been thought to be quite selective for controlling fruit flies but there is now some doubt about this (Ehler et al 1984). A more certain advantage claimed for bait sprays is that they increase the amount of active compound taken up by the fly so that less need be sprayed (Knipling 1979). Clearly the ideal is to keep the amount applied per hectare as low as possible while maintaining the dosage to the fly high enough to kill the heterozygous carriers of _R_ genes.

Insecticides such as dichlorvos or naled may be formulated with a synthetic chemical attractant (a lure), a number of which have been utilised for different insects. The best known for fruit flies is methyl eugenol, a male attractant used for controlling the oriental fruit fly _Dacus dorsalis_. In the Japanese Ryukyu islands, the species has been eliminated using this method (Koyama et al 1984). In this control programme, the mixture was applied to small fibre blocks distributed either from the ground or the air. In projects elsewhere (Steiner et al 1970, Koyama 1982) different cellulose-based materials (rolled cotton, boards, ropes etc.) have been used as a carrier for the mixture or it has been sprayed directly onto surfaces such as telegraph poles (Cunningham, personal communication). The method is known as male annihilation because it is the males which are mainly affected by it. Very small amounts of insecticide are required compared with the quantities needed for bait spraying (1/60 - 1/100 according to Cunningham, personal communication), where naled is used in the lures and malathion in the bait sprays. Success with methyl eugenol has encouraged the use of other powerful attractants: cuelure for the melon fly *D. orientalis*, trimedlure and capilure for the medfly.

Any mechanism extending the time of contact between insecticide and fly is likely to increase the dose taken up. The use of attractants is therefore expected to bring benefits not only in more efficient fly control but also in more effective resistance management.

2.3.3 Insecticide persistance and decay

The dosage of insecticide received by an insect from a residual deposit is dependent on the rate of decay in activity of the insecticide and the frequency with which it is applied. With an insecticide declining in activity and infrequently renewed, the dosage received by the insect may well drop to a level at which heterozygous carriers of *R* genes can survive. The problem has been modelled by Taylor and Georghiou (1982) and Mani and Wood (1984) and the conditions for maintaining a high dosage have been established. The safest type of deposit to use is one which decays to zero activity in less than a single generation of the target species. Otherwise the insecticide must be resprayed frequently, the rate being determined by various conditions: the extent of escape, the rate of decay of insecticide activity and the shape of the insecticide decay curve, the initial frequency of the *R* gene and whether mating occurs before or after exposure. An example for mosquitoes (*Anopheles culicifacies*) is given in Fig. 4. It shows the impact of three dosages of HCH applied to the inner walls of houses. The time to resistance, starting from an inital R gene frequency of 0.00001, is measured in generations, resistance being defined as the point at which *R* gene frequency reaches 0.14, a frequency at which

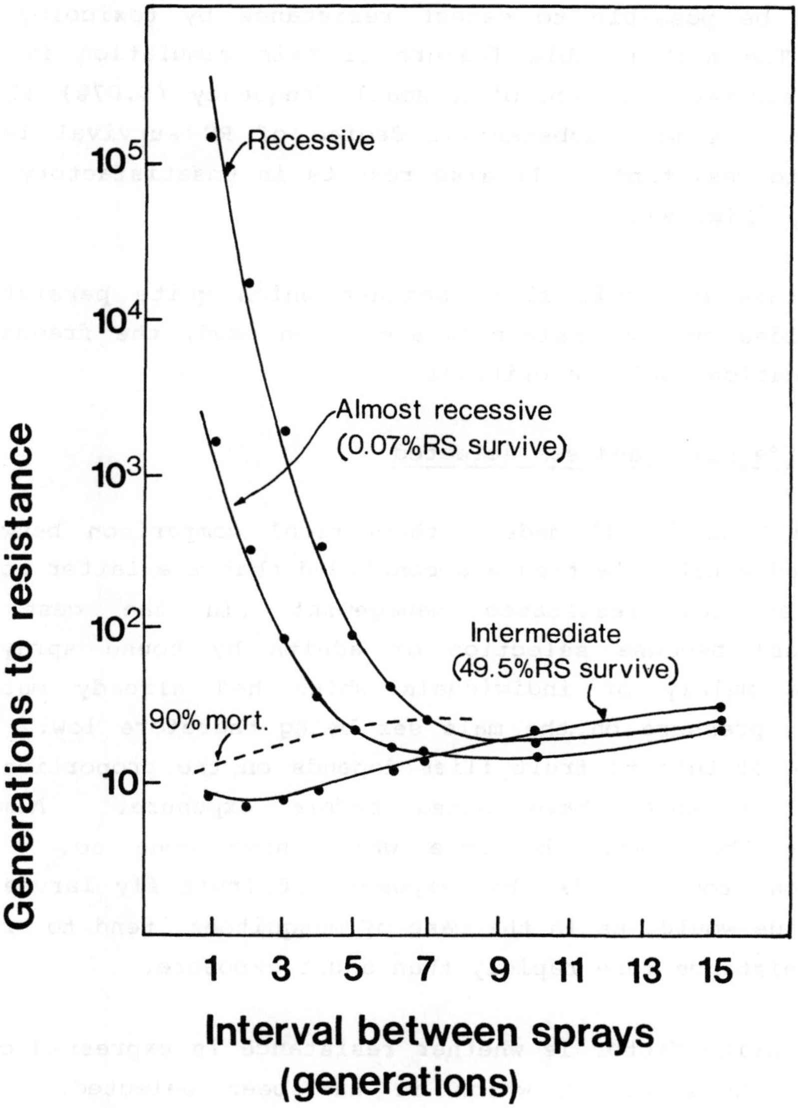

Fig. 4 Relationship between the number of generations required for R gene frequency in mosquitoes (Anopheles culicifacies) to increase from 0.00001 to 0.14 and the interval between sprayings of HCH to the inner walls of houses at dosages giving various degrees of RS survival, where the insecticide declines in potency to one tenth of its initial value within five generations, where mating occurs before selection and where 10% of the population escapes exposure. The broken line indicates the boundary of the region (below the line) in which mortality would be less than 90%. (Modified after Mani and Wood 1984).

it would be possible to detect resistance by toxicological tests. The most notable feature of this simulation is the very considerable impact of a small frequency (0.07%) of RS survivors. A more substantial degree of RS survival leads quickly to resistance. It also results in unsatisfactory low mortality (Fig. 4).

In the case of fruit flies against which quite persistant insecticides such as malathion are often used, the frequency of application could be critical.

2.3.4 Life stage and sex selected

Mani and Wood (1984) made a theoretical comparison between larval and adult selection and concluded that the latter would be better for resistance management (in the case of mosquitoes) because selection of adults by house spraying would be mainly of individuals which had already mated, selection pressure on the male sex being therefore low. The relevance of this to fruit flies depends on the proportion of adult flies which have mated before exposure. Almost certainly there will be some which have done so. The conclusion from this is that exposure of fruit fly larvae to insecticide would, as in the case of mosquitoes, tend to bring about resistance more rapidly than adult exposure.

A complicating factor is whether resistance is expressed only in the life stage at which it has been selected. The programme of laboratory selection of medfly larvae with dieldrin to produce the DiR strain, reported by Busch-Petersen and Wood (1983), led to a wide spectrum of cross resistance in larvae but no resistance at all in adults (Saaid and Wood unpublished). The effect of adult selection on the resistance of larvae is not known but adult resistance in the absence of larval resistance has been observed in other insects eg. mosquitoes (Wood 1983).

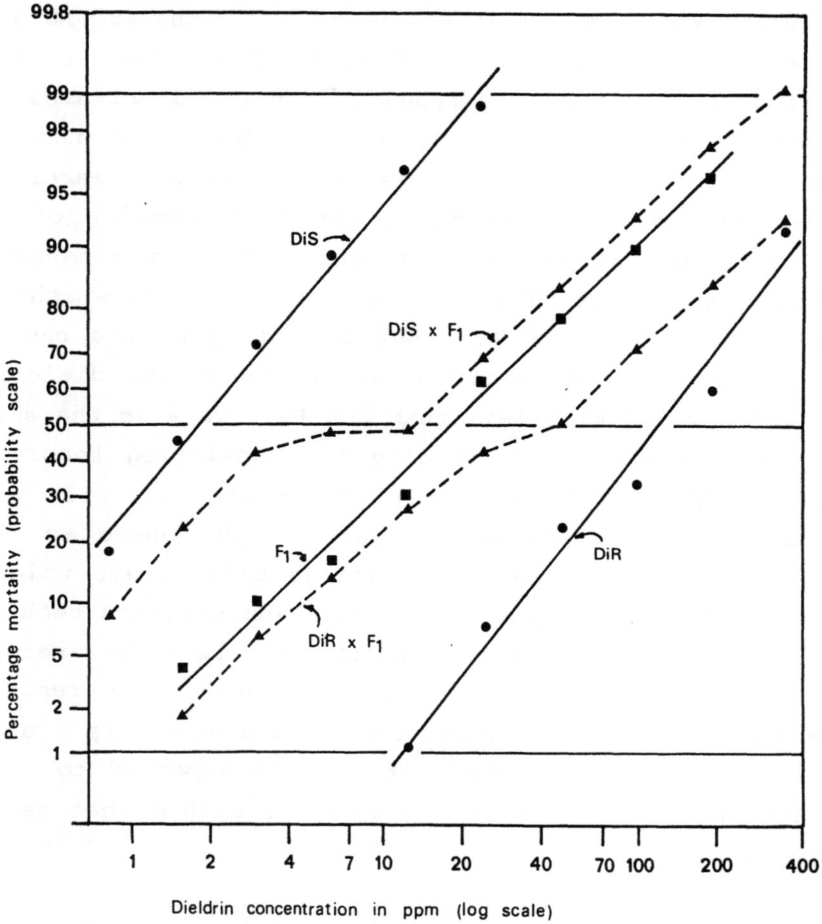

Fig. 5 Pattern of inheritance of the dieldrin resistance gene in larvae of Ceratitis captitata showing overlap between SS, SR and RR genotypes (in DiS, F_1 and DiR respectively) and a change in expression of these genotypes in the two backcrosses, where segregation, indicated by inflections on ld-p lines, is clearer than expected. (Modified after Busch-Petersen and Wood 1983).

The choice of whether to attack one or other life stage will also depend on economics and on how easy it is to make the relevant gene effectively recessive with an ecologically acceptable level of insecticide in the environment. Laboratory studies on larval resistance to dieldrin in the medfly by Busch-Petersen and Wood (1983) indicated uncertainty about what dosage would actually be needed to make the R gene effectively recessive. This was because (i) the SS, RS and RR genotypes overlapped in resistance (Fig. 5), (ii) the expression of the three genotypes in the parent strains (DiS, DiR) and in the F_1 was observed to change in the backcrosses, the pattern of segregation being clearer than expected (Fig. 5). According to the F_1 expression of RS, a dosage in excess of 400 ppm dieldrin would be necessary to make the R gene effectively recessive, a dosage which would also kill more than 90% RR. But according to the backcross expression of RS, 100% would be killed by 50 ppm dieldrin, a dosage which would kill less than 30% RR. This is the sort of problem that makes for difficulty in translating theory into practice. With fruit flies in the field, it will almost certainly be more difficult to get a high dosage to larvae than to adults so that methods such as bait sprays which are directed primarily at adults, and male annihilation techniques which are directed exclusively at adults, are to be preferred. Male annihilation is particularly favourable to resistance management because insecticidal pressure is applied essentially to one sex only. It will be expected to lead to resistance more slowly than a comparable method that destroys both sexes.

2.3.5 Uses of mixtures and rotations of insecticides

The combining of two or more insecticides, either in a mixture or by rotating their use in space or time, aims at achieving control while keeping the selection pressure below that required for the development of resistance. Interest in the concept has increased recently with the publication of several theoretical studies (Knipling and Klassen 1984, Mani 1985, Curtis 1985a). The earlier literature has been reviewed by Georghiou (1983).

a) Insecticide mixtures

The choice of insecticides suitable for combining together is conditioned by an absolute requirement that such chemicals should be sufficiently different in chemical structure and mode of action that no mechanism of cross resistance is possible. Then, if the gene for resistance to each insecticide is rare, the chance of any individual carrying both genes will be extremely rare.

Theoretical studies reported by Mani (1985) come down in favour of mixtures under the above conditions. However, the extent of advantage conferred by a mixture has been shown to depend on (i) the effective dominance of the resistance genes ($\underline{R}_1$, $\underline{R}_2$), (ii) the initial $\underline{R}_1$ and $\underline{R}_2$ frequencies, (iii) the extent of recombination between $\underline{R}_1$ and $\underline{R}_2$ and (iv) the initial value of linkage disequilibrium between $\underline{R}_1$ and $\underline{R}_2$.

Dominance R_1	R_2	initial frequency of R_1,R_2	recombination fraction (%)	t_2/t_1
0.0	0.0	0.001	0	0.002
0.0	0.2	0.001	0	0.045
0.0	0.6	0.001	0	0.114
0.0	1.0	0.001	0	0.186
0.2	0.2	0.001	0	0.333
0.2	0.2	0.001	50	0.002
0.0	0.0	0.0001	0	0.00
0.2	0.2	0.0001	0	0.407
0.2	0.2	0.0001	50	0.002

Table 3. Values of t_2/t_1 for different levels of dominance of two resistance genes (0.0 = R gene recessive; 1.0 = R gene dominant), different initial frequencies and different levels of recombination (0.1 = fully linked; 0.5 = no linkage), where t_1 = number of generations for resistance to evolve to both of two insecticides applied as a mixture, and t_2 = number of generations for resistance to evolve to two insecticides applied in sequence, and where there is no linkage disequilibrium between R_1 and R_2 (modified after Mani 1985).

The number of generations (t_1) required for resistance to evolve to two insecticides within a mixture was compared by Mani (1985) with the number of generations (t_2) required for resistance to arise to the two insecticides applied in sequence. The relative advantage is expressed by the ratio t_2/t_1, the smaller the value the greater being the relative advantage. A range of values of t_2/t_1 are given in Table 3 for various conditions of dominance of the two R genes (where 0.0 is fully recessive and 1.0 fully dominant), for different initial frequencies of R genes, and for different degrees of recombination between R_1 and R_2. It is clear that the effect of increasing the effective dominance of one R gene is considerable but not as great as when both R genes show a lesser degree of dominance. Reducing recombination towards zero has a considerable effect also. It may be noted, however, that under all conditions shown, the use of mixtures led to resistance more slowly than the use of insecticides in sequence.

To obtain a long delay in the evolution of resistance, the dosages of insecticides used in combination should be sufficiently high to kill the majority of heterozygotes i.e. to make both genes as recessive as possible within the constraints of cost (Mani 1985). If the genes are linked, the requirement for effective recessivity becomes more stringent, progressively so the more tightly the genes are linked.

An interesting point about recombination between R genes demonstrated by the model of Mani (1985) is that a threshold exists above which the significance of linkage tails off (Fig. 6). In fact under most conditions, R genes need to be within 20 units of recombination for linkage to be significant.

Knipling and Klassen (1984) argue the value of a "half strenth" mixture i.e. one in which both components are used at half the standard concentration. The value of this strategy depends, however, on the degree of "overkill" exerted by the standard concentration on the resistant heterozygote. Mani

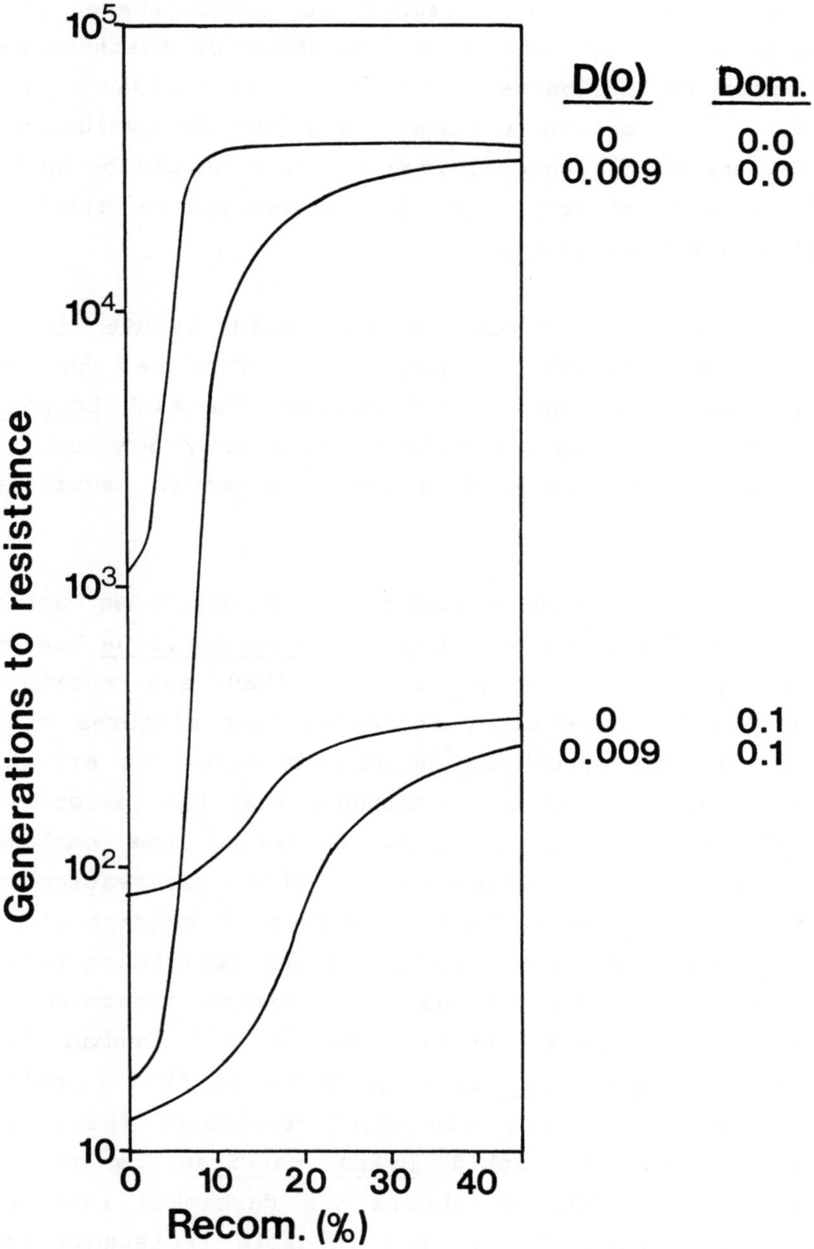

Fig. 6 The number of generations required for either of two resistance genes (R_1, R_2) to increase in frequency from 0.01 to 0.5, depending on the degree of recombination between the gene, the initial linkage disequilibrium (D(o)) and the dominance (Dom.) of the genes. (Data abstracted from Mani 1985).

(1985) has shown that at dosages at which there is no heterozygote mortality (effective dominance of resistance) a mixture would do no better than the insecticides applied sequentially. To obtain a large delay in the evolution of resistance, one of the insecticides at least should be applied at a dosage sufficiently high for a resistance allele to become effectively recessive.

A bonus in terms of cost saving would result if the insecticides were mutually synergistic. This was the case with ethion and a pyrethroid used against the tick Boophilus microplus (Nolan and Roulston 1979) when only 50% and 70%, respectively, of the normal dose was required for equivalent toxicity.

A laboratory study with mixtures of insecticides applied against larvae of the mosquito Culex quinquefasciatus has been carried out by A.T. Lagunes (unpublished 1980) and reported by Georghiou (1983). This study indicated that mixtures of the OP temephos and the pyrethroid permethrin were more effective in delaying resistance to both compounds that the insecticides used singly whereas combinations involving the carbamate propoxur hed no such advantage. This observation was interpreted in terms of different degrees of coadaptation of the three R genes and some negative cross resistance between organophosphates and pyrethroids. The latter phenomenon has been observed in spider mites (Chapman and Penman 1979) although in Musca domestica it is possible to find a positive correlation between OP and pyrethroid resistance (Sawicki et al 1984) and in the aphid Myzus persicae, a positive correlation between OP, pyrethroid and carbamate resistance (Devonshire and Moores 1982), the multiple resistance being associated in each case with a single detoxifying esterase.

Practical problems associated with the use of insecticides in mixtures have been discussed by Georghiou 1983. These include the need to ensure that both components of a mixture decay in activity at a similar rate. However the major constraint upon the use of mixtures is the danger of selecting an unexpected resistance mechanism giving a broad spectrum of resistance. This seems to make the practical use of mixtures a risky gamble without a lot of preliminary work on laboratory selection of wild-collected strains. Fortunately it seems unnecessary to consider this procedure at present in the case of fruit flies because of other strategies available to delay resistance. However it may be necessary to consider mixtures for future use and the basic laboratory studies could usefully be done. The aim is to be absolutely sure that insecticides being combined cannot be resisted by a common mechanism.

b) Insecticide rotations

The idea of alternating the use of two insecticides is theoretically less attractive than the use of mixtures as shown by the computer-based studies already discussed (Mani 1985). There is, however, one circumstance in which rotations could be of value. This would be when resistant individuals have a lower biotic fitness than the susceptibles so that their frequency declines during the intervals between applications. There are numerous instances of decreased fitness associated with resistance. The fitness disadvantage has been quantified under field conditions in a study on mosquitoes (Curtis et al 1978), the fitness differential between RR and SS being 0.44-0.97 depending on species and insecticide. In one well studied example, diazinon resistance in the sheep blow fly L. cuprina, the fitness of the homozygous susceptable genoptye was shown to increase with continued selection (McKenzie et al 1982). So that finally there was no difference in fitness between RR and SS when selection was relaxed (McKenzie and Whitten 1984).

The value of insecticide rotations in practice has been investigated in a field study of house flies by MacDonald et al (1983b). They alternated the use of permethrin and dichlorvos on a series of farms in Ontario, having obtained some encouraging results from earlier laboratory experiments

(MacDonald et al 1983a). They observed that resistance to permethrin developed more rapidly under continuous permethrin pressure than when permethrin and dichlorvos were alternated.

Georghiou et al (1983) investigated a four insecticide rotation (permethrin, temephos, propoxur, diflubenzuron) in laboratory population experiments with mosquito larvae (Culex quinquefasciatus). In long term rotation (several generations between alternations) temephos and permethrin manifested reciprocally suppressive effects as resistance to each other but propoxur caused longer retention of permethrin resistance. In short term rotations (insecticide changed every generation) permethrin resistance evolved to high levels in proportion to the frequency of its use. The long term alternations seemed to offer a better opportunity for the expression of reduced fitness, genetic instability of resistance and negative interaction between chemicals.

Rotation of BTI and chlorpyrifos were investigated by Mellon and Georghiou (personal communication) in a field experiment on C. quinquefasciatus but the results were difficult to interpret because of population mobility. Laboratory experiments have now been started to simulate the field experiment (Georghiou et al 1984).

The alternation of two or more insecticides in space (mosaic spraying) has been modelled by Curtis et al 1978. In theory there is less practical value in mosaics than mixtures (Curtis, quoted by Wood 1981b), Curtis 1985a) but this has not been tested by experiment.

2.4 Integration of insecticidal application with other control measures

Theoretically the integration of insecticide use with other methods of fly control should help to reduce selection pressure and the risk of resistance. In the medfly control programme in S.W. Mexico, aimed at total elimination of the pest, insecticides have been used (1) to reduce the population to a reasonable level before the SIT is initiated (2) to mop up pockets of infestation after SIT has been used (Hendrichs et al 1983). Lure-baited traps have also been used. This integrated programme did not lead to resistance and the pest has now apparently been cleared from Mexican territory. It cannot be doubted that combining different pest control techniques in this way has reduced selection pressure on R genes but whether in practice this was critical in preventing the evolution of resistance is not known. As already noted, there was no resistance either in California where, after an integrated ground-based control programme (incorporating SIT) which was unsuccessful, populations were finally eliminated by the aerial application of malathion bait sprays. Eradication programmes relying exclusively on insecticides have also eliminated this pest from Florida and Texas (Knipling 1979).

2.5 Eradication or population management?

It may be asked whether resistance is more likely to occur on programmes aimed at total eradiction of a pest or in those aimed at managing it at an acceptable level. Pros and cons can be argued for both approaches but management appears to offer less danger. The distinction between eradication and management is possibly less important than factors discussed earlier, particularly the way in which the insecticide is presented and whether insecticide usage is integrated with other methods of control. It has been argued already that the absence of resistance in fruit flies has probably been largely influenced by the combination of insecticides with baits or lures.

RESISTANCE TO THE STERILE INSECT TECHNIQUE

The possibility of resistance evolving to the sterile insect technique (SIT) is another matter to consider. Released sterile males seriously reduce the fitness of any female mating with them and must therefore act as a powerful selective force on a natural population. As Huettel (1977) has pointed out, an SIT programme is a large natural selection experiment. It is an exercise in post-mating reproductive isolation and, as such, it will produce intense selection for pre-reproductive isolating mechanisms. The basis of resistance could be any genetic variant favouring an increase in positive assortative mating between wild females and wild males.

Wild males can reasonably be expected to be different from factory-bred males from the outset. They will have been subject to laboratory selection pressures in crowded cages and have suffered the trauma of irradiation, mechanical handling and distribution. It comes as no surprise therefore that field tests of mating competitiveness have usually shown sterile males of several species to be less effective than wild males in fertilizing wild females (Curtis 1985b). In D. cucurbitae the competitiveness apparently declined from 80% to 20% during 13 generations of laboratory rearing (Itô and Koyama 1982). A study on the mosquito Culex tarsalis by Reisen et al (1982) showed competitiveness of 29% with wild females but 353% with laboratory-bred females (wild males = 100 in each case). The latter example indicates a high degree of assortative mating between the laboratory-bred flies as well as between the field flies. Recent studies on C. capitata support this finding. Investigations in field cages on the island of Procida by Robinson, Cirio, Hooper and Capparella (unpublished) showed that assortative mating occurred between released sterile males and sterile females, and that when sterile males were released alone, their ability to sterilise wild females was much improved.

It is clear, therefore, that assortative mating between wild males and females is already encouraged by (a) the lower fitness of factory-bred males, (b) the presence of factory-bred females. The question we are addressing here is whether the degree of assortative mating is likely to be increased further by genetic changes occurring in the wild flies in response to the presence of the sterile flies.

Although essentially passive, it is the female fruit fly which is discriminating in courtship. The male by contrast is "indiscriminately active", as in Drosophila spp. (Bateman 1948). The males of C. capitata take up a courting position ("lek") on the lower surfaces of leaves. They probably defend the lek like certain Drosophila spp. which have been studied (Ehrman, 1972), the fittest flies gaining the best positions to attract females. Having taken position, the male calls the females by pheromone emission and wing vibration, and when she alights on the leaf, he approaches her, undertakes a courtship display and finally mounts her (see Harris, this volume).

Male interactions precede courtship also in the olive fly D. oliae. These have been described by Cirio and Cervellati (1985) who observed a hierarchy of male dominance and noted aggressive interactions between males, and a preference by females to mate with the dominant male.

Speculation about the evolution of resistance to SIT is often in terms of "sexual selection" (for various definitions of this term see Ehrman 1972). The wild female is visualised as selecting between potential mates and choosing the fertile ones, or the wild male is visualised evolving an increased capacity to defend its lek. The possibility of female choice would clearly depend on whether a distinction could be made between fertile and sterile males on the basis of some aspect of physiology or behaviour. An increase in the capacity of wild males to defend their leks would depend on the heritability of this trait. The arguments assume that lekking and courtship remain significant after sterile male release.

In fact there is evidence (Cirio, personal communication) that the medfly lek breaks down at high densities. Howse (personal communication) believes that at high densities, the sterile males may confuse the females by greatly increasing the amount of pheromone in the environment. Prokopy and Hendrichs (1979) observed "rape" as an alternative mating strategy in field cage experiments. The breakdown of normal courtship patterns would be expected to reduce the chance of resistance by female choice. The impact of sterile males might then be more in terms of numbers than "quality", and carriers of potential resistance genes would be restricted to those females which could not be confused or which resisted rape. As to the male, defence of the lek would become irrelevant.

An alternative route to resistance might be via changes in response to ambient conditions. Zapien et al (1983) have reported that wild medflies in Guatemala have a greater tolerance to low light intensity and suboptimal temperature than factory-bred flies, in consequence of which they begin courtship earlier in the day. By contrast Cirio (personal communication) found that wild medflies on the island of Procida (Italy) mated over a longer period of the day than the released flies. Late in the afternoon there were two hours during which only the wild flies were mating. He attributes this variation to a different response to temperature. Given sufficiently intense selection pressure it is not difficult to imagine wild flies taking advantage of a capacity to mate under conditions in which factory flies are inactive. Whether and how quickly this might be translated into resistance would depend on (a) the heritability of the behaviour patterns involved and (b) the selection pressure.

The extent to which differences in behaviour between the Mexican and Italian populations are heritable is clearly important because of the possibility that natural mating barriers may exist in species subject to sterile insect releases. Studies on different natural populations of the screw worm fly Cochliomyia hominivorax have pointed to considerable heterogeneity in gross chromosomal morphology associated with variation in mating patterns. A difference of opinion exists concerning whether this variation represents a complex polymorphism with positive assortative homotypic mating (McInnis 1981, La Chance et al 1982, McInnis et al 1983) or whether the species is composed of a series of sympatric non-interbreeding "demes" (Richardson et al 1982a, 1982b). All agree, however, on the need to understand more about the variation involved so that the released males can be genetically matched to the target population (Richardson et al 1982a, McInnis et al 1983). In a polymorphic or polytypic population, it might be difficult to avoid selecting morphs or demes which were incompatible with the released males.

Whatever the pathway to resistance, the selection pressure towards resistance is exerted by the factory-bred males. The greater their fitness and the more numerous they are, the greater the selection pressure. Moreover, because of assortative mating between factory bred males and females (see above), the selection pressure is likely to be less if females are also released. Conversely selection pressure is expected to increase if the factory-bred females are removed by a sexing technique. The selection pressure is roughly defined by the proportion of wild females mated by sterile males. But this definition ignores the fact that there will be wild females that never meet a sterile male. The selection pressure is more strictly defined, therefore, as the proportion of wild females which when courted by a sterile male, mate with it.

The likelihood of resistance would be diminished the greater the degree of 'escape' from the factory-bred males i.e. the greater the proportion of wild females which never come into contact with sterile males and are therefore never exposed to selection. There is a parallel here with selection towards insecticide resistance. The development of resistance to SIT would also be delayed, as with resistance to insecticides, if the exposed population was open to ingress by unexposed immigrants.

How is resistance to SIT to be forstalled? The encouragement of escape and immigration would inhibit it but would also seriously diminish the effectiveness of the technique. The only practical way seems to be to make the factory male as near like the wild one as possible in those aspects of physiology and behaviour central to mating success with wild females. Identification of these aspects will require much greater knowledge than is currently available of courtship and mating by fruit flies in the wild and the underlying genetic and chromosomal variation. Genetic selection experiments can increase mating activity as shown by Dr. Harris working with the medfly (see this volume) but the genetic changes needed to improve fitness in the field will probably require more subtle selection programmes than those so far designed.

An alternative strategy could be to release sterile males in large enough numbers to produce such high ratios of sterile to wild males that the structure of the lek is broken down and mating proceeds more by rape than courtship, if this is possible. It is is interesting to note that Richardson et al (1982a) reports that certain local populations of the screw worm in southern Mexico which remained undiminished by release of sterile males at conventional densities, were brought under control by greatly increasing the density of releases.

CONCLUSIONS

1 Insecticide resistance

Escape, immigration and the integration of insecticide usage with other methods of control can all be expected to inhibit the development of resistance in fruit flies. However it is clear from considering actual cases that resistance has failed to evolve even in populations where insecticide pressure has been very high, where little or no escape or immigration has been possible and where there has been no attempt at integrated pest management. It is reasonable to suppose that in some of these cases the gene pool has been restricted to the extent that major _R_ genes are absent or associated with low fitness. However it seems unlikely that a genetic explanation can always be given in such cases. The common factor probably contributing most to the absence of resistance in fruitflies is the practice of formulating the insecticides with baits and chemical lures which greatly increase the dose to the target insect. The high dose probably renders any R genes which are present at low frequency, effectively recessive and therefore difficult to select. With further knowledge of resistance genes in fruit flies it should be possible to test this hypothesis by experiment.

2 Resistance to SIT

The more intensively and extensively SIT is used, the greater the chance of resistance arising to it by an increase in assortative mating between wild males and females. The way to avoid this seems to be to ensure that the released sterile males are at least as effective as the wild males and that in their courtship and mating activity they span the complete range of wild male behaviour. Failing this it may be necessary to increase greatly the density of released sterile males. The quality control of the released males, applying various genetic procedures including selection, is likely to become increasingly significant as knowledge of the wild insect increases.

Acknowledgement

I am indebted to Dr E. Busch-Petersen for a literature search on the historical use of different insecticides against the medfly and for permission to quote personal communications from Dr U. Cirio, Dr R. Cunningham, Dr G. Haniotakis, Dr P. Howse, Dr A. Robinson and Dr Y. Rössler. The typescript was patiently prepared by Mrs S.E. Hardman and the diagrams by Mr G.S. Proudlove.

REFERENCES

Awalladah, A. (1974) A laboratory test on the effect of chlorinated insecticides added to sand on the pupae of the medfly Ceratitis capitata (Wied.). Agr. Res. Rev. 52, 51-53.

Azab, A.K., Ali, A.M. and El-Hakim, A.M. (1974) On the tolerance of the Mediterranean fruit fly Ceratitis capitata (Wied.) to currently used insecticides. Agr. Res. Rev. 52, 73-80.

Bateman, A.J. (1948) Intra-sexual selection in Drosophila. Heredity 2, 349-368.

Busch-Petersen, E. and Wood, R.J. (1983) Insecticide resistance as a prospective candidate for the genetic sexing of the Mediterranean fruit fly Ceratitis capitata (Wied.) In: Fruitflies of Economic Importance (Ed. R. Cavalloro) Balkema, Rotterdam 1983 pp 12-189.

Chapman, R.B. and Penman, D.R. (1979). Negatively correlated cross-resistance to a synthetic pyrethroid in organophosphorus resistant Tetranychus urticae. Nature 281, 298-299.

Cirio, V. and Gervellati, M.C. (1985) Scelta sessuale in Dacus oliae Gmel. Atti XIV Congr. naz. ital. Ent., Palermo, Erice, Bagharia, 1985, 363-370.

Comins, H.N. (1977) The development of insecticide resistance in the presence of imigration. J. theor. Biol. 64, 177-197.

Curtis, C.F. (1985a) Theoretical models of the use of insecticide mixtures for the management of resistance. Bull. ent. Res. 75, 259-265.

Curtis, C.F. (1985b) Genetic control of insect pests: growth industry or lead balloon? Biol. J. Linn. Soc. (Lond.) (in press).

Curtis, C.F., Cook, L.M. and Wood, R.J. (1978) Selection for and against insecticide resistance, and possible methods for inhibiting the evolution of resistance in mosquitoes. Ecol. Ent. 3, 273-287.

Devonshire, A.L. and Moores, G.D. (1982) A carboxylesterase with broad substrate specificity causes organophosphorus, carbamate and pyrethroid resistance in peach-potato aphids (Myzus persicae). Pestic. Biochem. Physiol. 18, 235-246.

Ehrman, L. (1972) Genetics and Sexual Selection. In: Sexual Selection and the Descent of Man (Ed. B. Campbell) Heinemann, London 1972 pp. 1015-135

El-Sawy, S.E., Maher Ali, A., Faris, F.M. and Ezzat, M.A. (1965) Role of soil insecticides in reducing fruit-fly infestation. Bull. Soc. ent. Egypte 48, 159-162.

Ehler, L.E., Endicott, P.C., Hertlein, M.B. and Alvarado-Rodriguez, B. (1984) Medfly eradication in California: impact of malathion-bait sprays on an endemic gall midge and its parasitoids. Entomol. exp. appl. 36, 201-208.

Falconer, D.S. (1981) Introduction to Quantitative Genetics 2nd edition Longman, London, pp. VIII + 340.

Georghiou, G.P. (1981) The Occurrence of Resistance to Pesticides in Arthropods. FAO, Rome (1981) pp. XXV + 172.

Georghiou, G.P. (1983) Management of Resistance in Arthropod In: Pest Resistance to Pesticides (Eds. G.P. Georghiou and Tetsuo Saito) Plenum, New York, 1983, pp. 769-792

Georghiou, G.P. and Taylor, C.E. (1977) Genetic and biological influences in the evolution of insecticide resistance. J. Econ. Entomol. 70, 319-323.

Georghiou, G.P., Lagunes, A. and Baker, J.D. (1983) Effect of insecticide rotations on evolution of resistance. In: IUPAC Pesticide Chemistry Human Welfare and Environment (Eds. J. Miyamoto et al) Pergamon, Oxford 1983 pp. 13-189.

Georghiou, G.P., Worth, P. Mellon, R. and authors (1984) Insecticide resistance in mosquitoes: research on new chemicals and techniques for management. In: Mosquito Control Research Annual Report 1984, Entomology Extension, University of California, Davis pp. 96-99.

Haltebourg, M. (1966) Essais de lutte contre les larves de ceratite dans les abricots avant recolte. Al Awamia 19, 17-25.

Hendrichs, J., Ortiz, G., Liedo, P. and Schwartz, A. (1983) Six years of successful medfly program in Mexico and Guatemala. In: Fruitflies of Economic Importance (Ed. R. Cavalloro) Balkema, Rotterdam 1983 pp. 353-365.

Huettal, M.D. (1977) Measuring overall performance. In: Quality Control, An Idea Book for Fruit Fly Workers (Eds. E.F. Bollow and D.L. Chambers. International Organisation for Biological Control, 1977, pp. 14-16.

Itô, Y. and Koyama, J. (1982) Eradication of the melon fly: role of population ecology in the successful implementation of the sterile insect release method. Protection Ecology 4, 1-28.

Keiser, I. (1968) Residual effectiveness of foliar sprays against the oriental fruit fly, melon fly and Mediterranean fruit fly. J. Econ. Entomol. 61, 438-443.

Keiser, I., Schnieder, E.A. and Tomikawa, I. (1971) Species specificity among oriental fruit flies, melon flies and Mediterranean fruit flies in susceptibility to insecticides at several loci. J. Econ. Entomol. 64, 606-610.

Kikkawa, H. (1961) Genetical studies on the resistance to parathion in Drosophila melanogaster. 1. Gene analyses. Ann. Rept. Sci. Works, Faculty Sci., Osaka University 9, 1-20.

Kikkawa, H. (1964) The genetic study on the resistance to Sevin in Drosophila melanogaster. Botyu-kagaku 29, 42-46.

Knipling, E.F. (1979) The Basic Principles of Insect Population Suppression and Management. United States Dept. of Agriculture, Washington Agricultural Handbook No. 512 pp. ix + 659.

Knipling, E.F. and Klassen, W. (1984) Influence of insecticide use patterns on the development of resistance to insecticides - a theoretical study. The Southwestern Entomologist 9, 351-368.

Koren, B., Yawetz, A. and Perry, A.S. (1984) Biochemical properties characterising the development of tolerance to

malathion in Ceratitis capitata Weidemann (Diptera: Tephritidae). J. Econ. Entomol. 77, 864-867.
Koyama, J. (1982) The Japan and Taiwan projects on the control and/or eradication of fruit flies. In: Sterile Insect Technique and Radiation in Insect Control IAEA, Vienna 1982 pp. 39-51.
Koyama, J., Teruya, T., Tanaka, K. (1984) Eradication of the oriental fruit fly (Diptera: Tephrididae) from the Okinawa islands by the male annihilation method. J. Econ. Entomol. 77, 468-472.
La Chance, L.E., Bartlett, A.C., Bram, R.A. and others (1982) Mating types in screw worm populations? Science 218, 1142-1143.
Macdonald, R.S., Surgeoner, G.A., Solomon, K.R. and Harris, C.R. (1983). Laboratory studies on the effect of four spray regimes on the development of resistance to permethrin and dichlorvos in the house fly. J. Econ. Entomol. 76, 417-422.
Macdonald, R.S., Surgeoner, G.A. and Solomon, K.R. (1983) Development of resistance to permethrin and dichlorvos by the house fly (Diptera: Muscidae) following continuous and alternating insecticide use on four farms. Can. Ent. 115, 1555-1561.
McInnis, D.O. (1981) Cytogenetics of a local population of the screw worm Cochliomyia hominivorax, from North eastern Mexico. Ann. Entomol. Soc. Amer. 74, 582-589.
McInnis, D.O., Whitten, C.J., Mackley, J.W. and others (1983) Cytogenetic studies of the screw worm Cochliomyia hominivorax (Diptera: Calliphoridae), from Chiapas, Mexico. Ann. Entomol. Am. 76, 628-640.
McKenzie, J.A. and Whitten, M.J. (1984) Estimation of the relative viabilities of insecticide resistance genotypes of the Australian sheep blowfly, Lucilia cuprina. Aust. J. Biol. Sci 37, 45-52.
McKenzie, J.A., Whitten, M.J. and Adena, M.A. (1982) The effect of genetic background on the fitness of diazinon resistance genotypes of the Australian sheep blow fly, Lucilia cuprina. Heredity 49, 1-9.
Mani, G.S. (1985) Evolution of resistance in the presence of two insecticides. Genetics.
Mani, G.S. and Wood, R.J. (1984) The persistence and frequency of application of an insecticide in relation to the rate of evolution of resistance. Pestic. Sci. 15, 325-336.
Myburgh, A.C. (1961) Lebaycid as a cover spray for fruit fly control. S. Afr. J. agric. Sci. 4, 615-621.
Nolan, J. and Roulston, W.J. (1979) Acaricide resistance as a factor in the management of Acari of medical and veterinary importance. Recent Adv. Acarol. 11, 3.
Ogaki, M. and Tsukamoto, M. (1953) Genetical analysis of DDT resistance in some Japanese strains of Drosophila melenogaster. Botyu-kagaku 18, 100-104.
Orphanides, P.S., Kalmoukos, P., Betzios, B. and Kapetanakis, E. (1980) Development of resistance in Ceratitis capitata Wied. in laboratory under intermittent pressure of organophosphorus and chlorinated insecticides.

Annales de l'Institut Phytopathologique Benaki (New Series) 12, 198-207.
de Pietri-Tonelli, P., Barontini, A. and Santi, R. (1960) Experiments on the control of Ceratitis capitata Wied. Contr. 1st Ric. agr. Soc. Montecatini 3, 59-88.
Plapp, F.W. Jr. (1984) The genetic basis of insecticide resistance in the house fly: evidence that a single locus plays a major role in metabolic resistance to insecticides. Pestic. Biochem. Physiol. 22, 194-201.
Prokopy, R.J. and Hendrichs, J. (1979) Mating behaviour of Ceratitis capitata on a field-caged host tree. Ann. Entomol. Soc. Am. 72, 462-8.
Reisen, W., Milby, M.M., Asman, S.M. and others (1982) Attempted suppression of a semi-isolated population by the release of irradiated males: a second experiment using males from a recently colonized strain. Mosquito News 42, 565-575.
Richardson, R.H., Ellison, J.R. and Averhoff, W.W. (1982a) Autocidal control of screw worms in North America. Science, 215, 361-370.
Richardson, R.H., Ellison, J.R. and Averhoff, W.W. (1982b) Mating types in screw worm populations? Science 218, 1143-1145.
Sawicki, R.M. (1973) Resynthesis of multiple resistance to organophorphorus insecticides from strains with factors of resistance isolated from the SKA strain of house flies. Pestic. Sci. 4, 171-180.
Sawicki, R.M., Devonshire, A.L., Farnham, A.W. and others (1984). Factors affecting resistance to insecticides in house-flies, Musca domestica L. (Diptera: Muscidae). II Close linkage on autosome 2 between an esterase and resistance to trichlorphos and pyrethroids. Bull. ent. Res. 74, 197-206.
Steiner, L.F., Hart, W.G. Harris, E.J. and others (1970) Eradication of the oriental fruit fly from the Mariana Islands by the methods of male annihilation and sterile insect release. J. Econ. Entomol. 63, 131-135.
Taylor, C.E. and Georghiou, G.P. (1979) Suppression of insecticide resistance by alterations of gene dominance and migration. J. Econ. Entomol. 72, 105-109.
Taylor, C.E. and Georghiou, G.P. (1982) Influence of pesticide persistence in evolution of resistance. Environ. Entomol. 11, 746-750.
Wood, R.H. (1981a) Insecticide resistance: genes and mechanisms. In: Genetic Responses to Man Made Change (Eds. J.A. Bishop and L.M. Cook) Academic Press, London pp. 53-96.
Wood, R.J. (1981b) Strategies for conserving susceptibility to insecticides. Parasitology 82, 69-80.
Wood, R.J. and Busch-Petersen, E. (1982) Possible genetic sexing mechanisms for Ceratitis capitata Wied. In: Sterile Insect Technique and Radiation in Insect Control. IAEA, Vienna pp. 279-289.
Wood, R.J. and Cook, L.M. (1978) Estimating selection pressures on insecticide resistance genes (preliminary

note). WHO Working Document WHO/MAL 78.896, WHO/VBC 78.683 pp. 8.

Wood, R.J. and Harris, D.J. (1986) Artificial and natural selection in the Mediterranean fruit fly Ceratitis capitata. In: Fruitflies, their Biology, Natural Enemies and Control. (Eds. A.S. Robinson and G.H.S. Hooper) Elsevier, Amsterdam 1986.

Wood, R.J. and Mani, G.S. (1981) The effective dominance of resistance genes in relation to the evolution of resistance. Pestic. Sci. 12, 573-581.

Zapien, G., Hendrichs, J. Liedo, P. and Cisneros, A. (1983) Comparative mating behaviour of wild and mass-reared sterile medfly Ceratitis capitaa (Wied.) on a field cage host tree. II Female mate choice. In: Fruitflies of Economic Importance (Ed. R. Cavalloro) Balkema, Rotterdam pp. 397-409.

SOME ANALYTICAL MODELS FOR BIOTECHNICAL METHODS OF PEST CONTROL

P. van den Driessche
Department of Mathematics
University of Victoria
Victoria, B.C. V8W 2Y2 Canada

ABSTRACT

A review of some simple analytical models for pest control is presented. The sterile insect release method is considered in detail, and models are also discussed for the release of sex pheromones to attract and kill insects. It is shown how these models apply to the control of fruit flies and new models appropriate to this control are suggested.

1. INTRODUCTION

The search for effective and acceptable means of insect pest control can be aided by mathematical models which explain and predict behaviour under certain assumptions. Such models supplement biological field and laboratory experiments, and can be used to guide experimentalists. This review focusses on simple, analytical, deterministic models which predict qualitative behaviour. Both discrete and continuous time variation are considered; the former probably being more appropriate to fruit flies with discrete non-overlapping generations.

Insecticides have not been a complete answer to the problem of insect pest control; although they are indeed useful as a component of integrated pest management (IPM). In particular, the sterile insect technique (SIT) has been used practically in many fruit fly control programs, for example for suppression and eradication of the Mediterranean fruit fly, Ceratitis capitata and the melon fly, Dacus cucurbitae. Theoretical models for this technique are reviewed in section 2. Pheromone traps are frequently used to survey and monitor pest populations, and their use in IPM has been suggested, (Silverstein, 1981). Models in which

NATO ASI Series, Vol. G11
Pest Control: Operations and Systems Analysis
in Fruit Fly Management. Edited by M. Mangel et al.

sex pheromones are released to attract and trap insects are reviewed in section 3. Both these methods are included under biotechnical methods of insect pest control (Boller, 1983), and may play an important role in IPM (Batra, 1982).

2. STERILE INSECT TECHNIQUE (SIT)

Knipling was the first to suggest and model this technique in which artificially sterilized insects are introduced into the pest population (Knipling, 1955) and the SIT has since been used successfully against several fruit fly species. Knipling's model assumes the insect has discrete non-overlapping generations and density independent geometric population growth. A 1:1 sex ratio is assumed, and a constant number n of sterile males are introduced which are assumed to be equally competitive with wild males. The difference equation governing the population growth is therefore

$$f_{t+1} = \lambda f_t \left(\frac{f_t}{f_t + n} \right) \tag{2.1}$$

where f_t = number of fertile females (= males) at time t and λ (with $\lambda > 1$) is the reproductive potential. This model predicts a critical number of introduced steriles $n_c \equiv (\lambda-1) f_0$ for control. When $n < n_c$ the population increases and escapes control, but when $n > n_c$ the population goes extinct. The overflooding ratio defined as n/f_0 must therefore be greater than $(\lambda-1)$ for control to be effective.

In discussions at this workshop it was agreed that as the irradiation dose used to sterilize fruit flies is increased, so their competitiveness greatly decreases. To examine these interrelating factors, let k (with $0 \leq k \leq 1$) be the residual fertility of males, that is the proportion of males which have been artificially sterilized but remain fertile, and let c (with $0 \leq c \leq 1$) be the competitiveness of sterile males compared to that of the wild males (for which the competitiveness is taken as unity). The difference equation governing the number of fertile females then becomes

$$f_{t+1} = \lambda f_t \left(\frac{f_t + kcn}{f_t + cn} \right), \tag{2.2}$$

giving $$n_c = \frac{(\lambda-1) f_0}{(1-\lambda k) c} . \tag{2.3}$$

Thus the product of reproductive potential and residual fertility must be less than one for the SIT to be effective, and the number of steriles introduced must be large when their competitiveness is small. In a computer simulation (Klassen and Creech, 1971) incorporating these factors, it appears that, if the reproductive potential is large, then it is better to aim for a high level of sterility even if competitiveness decreases rather than a lower level of sterility without this decrease. The model above supports this qualitative result, but more detailed models are required to guide managers on this important practical question.

When density dependence is introduced into Knipling's model, the population is governed by the equation (Prout, 1978, Yamamura et al., 1978)

$$f_{t+1} = \lambda f_t \left(\frac{f_t}{f_t + n} \right) \Big/ \left[\frac{\lambda-1}{K} f_t \left(\frac{f_t}{f_t + n} \right) + 1 \right], \tag{2.4}$$

where K is the carrying capacity which is the equilibrium population density when no steriles are present. This model has $n_c = (\lambda-1)K/4$, with the population going extinct for $n > n_c$ and tending to a locally stable equilibrium in the interval $(K/2, K)$ for $n < n_c$. So, with a knowledge of the biological parameters λ and K, the number of steriles needed to cause extinction can be estimated. Note, that in contrast to the density independent model, this number is independent of the initial population.

In any practical SIT program there is the possibility of females which have already mated entering the region. This can be incorporated by considering the model (Prout, 1978)

$$f_{t+1} = \lambda\left(\frac{f_t^2}{f_t+n} + w\right) \Bigg/ \left[\frac{\lambda-1}{K}\left(\frac{f_t^2}{f_t+n} + w\right) + 1\right], \qquad (2.5)$$

where w is the number of immigrant fertilized females. For any $w > 0$ the model predicts that extinction is impossible, and for large w the effect of the SIT program is destroyed as steriles cannot substantially reduce the wild population.

The analogue of (2.4) for continuous population growth with density dependent deaths is (Barclay and Mackauer, 1980) the differential equation system:

$$\begin{aligned} \dot{f} &= a_1 f\left(\frac{f}{f+n}\right) - a_2 f - 2gf(f+n) \\ \dot{n} &= R - a_2 n - 2gn(f+n) \end{aligned} \qquad (2.6)$$

where $\cdot \equiv d/dt$, a_1, a_2 are density independent birth rate, death rate respectively, g is a crowding parameter $\propto 1/K$, R is the constant rate of release of sterile males ($\equiv$ sterile females), and the number of sterile females in the population is a variable. The biological assumptions of the model are discussed fully in the paper by Barclay and Mackauer and in a later paper by the first author (Barclay, 1984). This model predicts that, for small a_2/a_1, $R_c \simeq 2a_1^2/(27g)$. It shows that density dependence assists control in the SIT, and that the SIT is very effective for low population values, and thus can be an important part of an IPM program when the population has been reduced by other means, for example, insecticides. The inclusion of immigrant fertilized females in this model (with fixed n) can destroy the effect of steriles, as stated for the discrete case above. A generalization of this continuous model (Barclay, 1982) indicates again that male competitive ability is of primary importance in determining the outcome of the SIT with R_c being very large when this competitiveness is small.

For insects with several stages in their life cycle, the previous models must be extended. For a two-stage life cycle (Barclay, 1980) an extension of (2.6) shows that the SIT is favourable for species in which the larval stage is a problem, for example, fruit flies. Plant gives a theoretical study of the SIT to eradicate Ceratitis capitata taking age structure into account (Plant, 1986).

The sterile females of Ceratitis capitata also sting and damage fruit, and so it is vital to develop mechanisms for releasing sterile males only; genetic sexing is now feasible (Robinson, 1986). A continuous model for the release of sterile males only, which is a modification of (2.6), shows that the number of sterile males required to reach the critical level is about 20% higher than the number of males needed if both sexes are being released (Barclay and Mackauer, 1980). The release of sterile males only thus has great possibilities, and is a subject of current research.

Models discussed above have assumed no spatial variation, but in practical applications dispersion of insects may be important, for example, in the case of the olive fruit fly, Dacus oleae. As an attempt to incorporate spatial variation into the SIT model (Manoranjan and van den Driessche, 1985) consider a strip of territory surrounded by a hostile region. This is modelled by a reaction-diffusion equation and it is found that there is a critical strip width below which the population cannot sustain itself. Above this width the population is still unable to sustain itself unless the initial population is high enough. So the SIT can be evaluated as one component of an IPM program including, for example, crop cutting and insecticides.

3. PHEROMONES

For some insects, pheromones can be used to trap and kill one sex. For a species in which females emit pheromone which attracts males (for example, olive fruit fly, Dacus oleae) a model is constructed in which it is assumed that there are v_0 caged virgins (or pheromone equivalents) and that every male attracted to a cage is killed. This discrete, density independent model (Barclay and van den Driessche, 1983) is based on an earlier model (Knipling and McGuire, 1966). Assuming male polygamy and female monogamy, it has the form:

$$f_{t+1} = \frac{sm_t v_t}{v_0 + v_t} + sf_t$$

$$v_{t+1} = af_{t-T} + sv_t - \frac{sm_t v_t}{v_0 + v_t} \qquad (3.1)$$

$$m_{t+1} = af_{t-T} + \frac{sm_t v_t}{v_0 + v_t}$$

where f, v, m are the number of wild fertilized females, virgin females, males respectively, with $m \le v_0 + v$, a is the birth rate, s the survivorship and T the developmental time from eggs to adults. This model predicts a critical value $v_{0c}(a,s)$ so that when $v_0 > v_{0c}$ extinction occurs, with the method being most effective against species with low a and s.

As with the SIT, control is found to be very difficult with the immigration of even a moderate number of fertilized females (Barclay, 1984). Density dependence can also be included in the model, and is found (Fisher et al., 1985) to confer considerable reduction in v_{0c}, especially for high a and s; in fact this is even more pronounced than in the SIT models.

For insects in which male-produced pheromone attracts only females, it is found (Barclay, 1983) that the dynamics of pest control by releasing this pheromone and so killing females are similar to those of model (3.1) except when females remate regularly. At this workshop (Howse, 1986) there was a report of a new male pheromone for Ceratitis capitata which attracts both male and virgin females. This has the potential for use as a pest control method for killing both sexes, and a theoretical model for the use of this pheromone in traps is needed.

4. CONCLUSIONS

The two biotechnical methods of pest control considered, namely the SIT and release of pheromones to trap insects, look promising, especially as components of an IPM program. Immigration causes difficulty in both techniques, and, in addition, male competitiveness and residual fertility are problems in the SIT. Pheromones are used extensively in trapping insects for monitoring, and the models in section 3 consider their use in mass trapping for insect pest control. At high density, pheromones disrupt mate-finding (Silverstein, 1981) and cause confusion, this feature needs to be incorporated in the mathematical models. It is vital that models are developed alongside experimental work, so that modellers can be provided with accurate details of and realistic data for a particular insect pest. The theoretical predictions should then be field tested, and the simple models extended for individual situations incorporating as many as possible of the important biological and environmental variables. In this way analytic models prove to be a useful tool to guide experimental scientists and managers.

Acknowledgements: It is a pleasure to thank Hugh Barclay for discussions. This paper was started while the author was a visitor at the Mathematical Institute, University of Oxford and was partially supported by NSERC grants A-4645 and A-8965.

REFERENCES

Barclay, H.J. 1980. The sterile insect release method on species with two-stage life cycles. Res. Popul. Ecol. 21: 165-180.

Barclay, H.J. 1982. The sterile release method with unequal male competitive ability. Ecol. Modelling 15: 251-263.

Barclay, H.J. 1983. Models of pheromone release for pest control in which males attract females. In "Population Biology" ed. Freeman, H.I., Strobeck, C. Springer-Verlag Lecture Notes in Biomathematics 52: 389-393.

Barclay, H.J. 1984. Demographic consequences of monogamy and polygamy for a sterile release program, Protection Ecol. 6: 209-217.

Barclay, H.J., Mackauer, M. 1980. The sterile insect release method for pest control: a density-dependent model. Envir. Entom. 9: 810-817.

Barclay, H.J., van den Driessche, P. 1983. Pheromone trapping for insect pest control. Res. Pop. Ecol. 25: 105-115.

Batra, S.W.T. 1982. Biological control in agroecosystems. Science 215: 134-139.

Boller, E.F. 1983. Biotechnical methods for the management of fruit fly populations. In "Fruit Flies of Economic Importance" ed. Cavalloro, R., Balkema, page 342.

Fisher, M.E., van den Driessche, P., Barclay, H.J. 1985. A density dependent model of pheromone trapping. Theor. Pop. Biol. 27: 91-104.

Howse, P. 1986. Nature and function of medfly pheromones. This Volume.

Klassen, W., Creech, J.F. 1971. Suppression of pest population with sterile male insects. USDA Miscellaneous Publications 1182: 8 pp.

Knipling, E.F. 1955. Possibilities of insect control or eradication through the use of sexually sterile males, J. Econ. Entomol. 48: 459-462.

Knipling, E.F., McGuire, J.U. 1966. Population models to test theoretical effects of sex attractants used for insect control. Agric. Info. Bull. 308, USDA: 20 pp.

Manoranjan, V.S., van den Driessche, P. 1985. On a sterile insect release model. DM-393-IR.

Plant, R.E. 1986. Optimal sterile insect release. This Volume.

Prout, T. 1978. The joint effects of the release of sterile males and immigration of fertilized females on a density regulated population. Theor. Pop. Biol. 13: 40-71.

Robertson, A. 1986. Translocations and genetic sexing in Ceratitis capitata. This Volume.

Silverstein, R.M. 1981. Pheromones: background and potential for use in insect pest control. Science 213: 1326-1332.

Yamamura, N., Nakano, T., Teramoto, E. 1978. Catastrophic phenomena in ecological systems. International Symposium on Math. Topics in Biology, Japan: 16-23.

V. AFTERNOON WORKSHOPS

PROCEEDING OF THE AFTERNOON WORKSHOPS.

The afternoons were dedicated to various workshops and brainstorming sessions. The following are proceedings of these activities; notes were taken by numerous individuals and later transcribed by M. Mangel, who assumes all responsibilities for errors.

Contents:

1. REPORTS OF THE WORKING GROUPS

The working groups met separately on Monday, 5 August for most of the afternoon. The following is a summary of the deliberations of these groups.

1.a. Biology

This group considered the basic biology and ecology of fruit flies and discussed life history strategies (for example, survival, developmental rates and reproduction rates of species) with a view to separating between species. No obvious way appeared to the group. There was also a brief discussion of r and K selection; as well as considerable controversy about the need for introducing the concept of carrying capacity when discussing density dependence. Questions raised by this group include:

NATO ASI Series, Vol. G11
Pest Control: Operations and Systems Analysis in Fruit Fly Management. Edited by M. Mangel et al.

-- How does competition between fruit flies (e.g. medfly and Dacus) work and change population levels?

-- How does life history effect control, and what is the interaction between population biology and control.

-- What are the important genetical aspects for fruit fly control.

-- What details of host selection are important for control, especially switching to alternative hosts?

-- Should experimental emphasis be in the laboratory or in the field, and how can the two be correlated? For example, what is the field interpretation of life history tables?

-- What is the appropriate role of modelling, especially as regards hypothesis testing?

-- How is growth regulated and when is density dependence important? How does on characterize density dependence?

1.b. Control/Eradication and Policy

These two subgroups met together. First, J. Scribner presented a detailed description of the 1980-82 California medfly program. Second, R. Cunningham provided a review of control techniques. Third, Y. Rossler, T. Teruya and M. Aluja described control programs in Israel, Japan and Mexico respectively. This was followed by extensive discussion and summation of issues.

The main concerns in control programs fall within the areas of economics, policy, technology and biology:

ECONOMICS	POLICY
TECHNOLOGY	BIOLOGY

The following specifics appeared to be most important to the members of the working groups:

Economics

-- Cost/benefit analyses: who pays the costs and who benefits, how does one actually measure these costs and benefits?

-- Defining the economic nature of the pest: how does the pest affect the marketing of the fruit that it infests?

-- Multicrop-multipest analyses

Technology

-- How does one choose the intervention or control strategy (e.g. SIT, bait spray, male annihilation)?

-- What are the efficiencies of various intervention technologies and how are they assessed (e.g., aerial versus ground sprays or release of sterile insects)?

-- How does one develop "holistic" designs for integrated management (e.g., using orchard design as a control strategy)?

Policy

-- What are the effects of institutional infrastructure, constraints and pressures, on control and eradication programs?

-- Howe does one assess risks and other human factors in control and eradication programs? What are the real and perceived risks? How does one manipulate public perceptions?

-- What is the role of information and misinformation in determining public policy?

-- How do the politics of interregion and/or intercountry relationships affect control and eradication programs?

-- What are the legal aspects of control and eradication programs, including the legal basis for a control action, the effects of risks and damage, and regulatory constraints?

Biology

-- What is the effects of the evolution of resistance -- to both pesticides and sterile insect techniques -- on control and eradication programs?

-- What is the effect of natural reservoirs and alternate hosts?

-- What are the appropriate population dynamics and models?

-- How does one deal with multi-crop and pest interactions?

-- What is the ecological basis for spatial and/or temporal distribution of control or eradication policies?

-- How does one design good monitoring programs?

1.c. Trapping and Detection

Traps are used in three broad areas: 1) detection, 2) monitoring, and 3) control and eradication. They will be discussed separately.

Regarding detection, a summary of the current program in California was presented by R. Dowell. Traps are used to detect the presence of a fruit fly. The major advances include: a) the development of specific lures, such as methyl eugenol, cue lure, and sex phermones for olive fruit fly; b) improved trap designs, such as yellow traps for olive fruit fly or red spheres for apple maggot; and c) development of a medfly phermone.

Traps are used to monitor the level of fruit fly populations. Major advances include: a) the correlation of trap catch with population size and fruit damage in olive trees in Greece; b) improvements in the ability to estimate population size in relation to trap catch, especially for the olive fly.

Traps are used for control and eradication by reducing population levels; male annihilation is included here as well. Major recent advances include: 1) the continue successful use of methyl eugenol to eradicate the oriental fruit fly; 2) the use of red spheres to control apple maggot, at least on a small scale (less than 1.5 ha), 3) the use of plywood boards, impregnated with phermone, food lure, and insecticide, to control olive fly in a very large scale (2,2000,000 trees with one trap for every two trees) experiment in Greece.

This subgroup also addressed future research needs as follows:

Detection

-- There is a need for improved species specific lures.

-- What is the effect of environmental factors (wind, foliage density, host plant spacing, humidity, rainfall, temperature, trap site on host plant, season of the year, etc) on trap effectiveness?

-- How do changes in fruit fly behavior affect trap catches? Behavorial and ecological factors include: hunger and thirst, age, dispersal behavior, orchard structure, fly density.

-- What is the relationship between population size and trap density on the probability of detecting the population?

Monitoring

-- All of the above items for detection also pertain to monitoring.

-- What is the relationship between trap catch and damage?

-- What is the relationship between trap catch and population size?

Control and Eradication

-- There is considerable need to develop traps that are species specific.

-- There is a continued need for work on pesticide formulations for use with lures.

-- What is the relationship between pest kill and crop protection.

-- There is a need for the development of longer lasting lure/pesticide combinations.

-- When does one switch from traps to sprays as a method of control?

Finally, it was noted that human factors are a major cause in the need for good detection programs. In Sydney, Australia, a random check of incoming people gathered 1017 Kg of illegal citrus after 660 Kg were deposited in bins prior to the check. The experience in California is very similar.

2. FRUIT FLY "WAR GAMES"

In order to stimulate the group into considering the multifaceted and interdisciplinary aspects of any large area control or eradication program, one afternoon was dedicated to a simulated "war game" exercise. The entire group was presented with the following situation:

The sovereign state of Atlantis consists of three islands situated about 1000 km southwest of Gibraltar. The northern most island has an area of 5000 km^2 and has 90% of the 10 million population of Atlantis. The northern island is highly industrialized with a managanese nodule mining industry which utilizes the cheap electricity available from geothermal sources available on the volcanic island. The big island is less than 50 km to the south and is the largest of the three, with an area of 100,000 km and a population of 800,000. The northern portion of this island rises to a small mountain range about 150 meters in elevation. On this mountain range are orchards of peaches, apples, plums and other garden fruits and vegetables, mainly for consumption with Atlantis. The chief market is the densly populated northern island.

The southern plain of the big island provides an oceanic tropical climate and soil which is ideal for the growth of tropical fruits. A large commercial fruit growing operation exists in this area and specializes in tropical crops such as avacados, mangoes, guavas, sapote, etc. The chief market for these tropical fruits is the European Economic Community, of which Atlantis is a member. It exports fruit to countries with and without established medfly populations. At the time of development of the fruit industry, a deliberate decision was made not to compete in citrus with the already established citrus industries in the Mediterranean area. The export of these tropical fruits accounts for 30% of the GNP of the Atlantis; the per capita income of the country is about \$2000/year -- comparable to Portugal. This industry grosses about \$600 M per year.

The south island is 50 km away from the big island and has a land area of 20,000 km^2 with a population of 200,000 people. A national park and wildlife refuge occupies about one half of the island. The chief industry on the island is tourism. Tourists come to enjoy the salubrious climate, good beaches, beautiful scenery and the unique wildlife, endemic to Atlantis, which is protected in the National Park.

Atlantis has a constitutional government with Federal, Prefectural and Township levels. Thirty years previously, before the establishment of the modern tropical fruit industry, the country experienced a small fruit fly invasion. Draconian measures, such as orchard destruction, crop stripping, and DDT sprays were used to eradicate the pest at a very high cost to the government. As a result of this experience, a grid of traps has been maintained on the island for purposes of detection. This trapping program has been modernized through period technical consultation with international experts. Based on economic considerations, the government is able to fund only about 500 traps. Half of these traps are trimedlure baited Jackson traps and half are glass invaginated traps baited with a protein food attractant. On the Big island, traps are concentrated in the two towns around the airport and harbor. These traps are checked on a monthly schedule, rather than weekly, in order to save money.

The entire group at the ARW was presented with the situation just described and then the following information: a single female medfly with mature eggs was found in McPhail trap in the northeast corner of the tropical fruit growing area. Discussion brought out that the trap density which was being operated by Atlantis could discover an invasion in the parental generation only by incredible luck. The trapping grid scale was much too large to provide information on the location and size of the population of medflies. The consensus of the group was that the first action should be to increase the density of traps around the fly which was found and other areas most likely to contain other flies (such as the airport, harbor and towns). We did not, however, notify the EEC governments of the find, nor did we establish an internal quarantine on the first day. However, we all considered the biological consequences of undue delay in action, since the population modellers proposed that a new introduction of medfly into a favorable climate, in the absence of parasites and competitors, and at a favorable time of year could theoretically give rise to several hundred million progeny by the time growth would be slowed by the mild winter of December and January. The biological, economic, and social consequences of such a disaster would be large.

We decided that increased trapping would probably be the only response available to the Atlantis Department of Agriculture (ADA). The trapping program was run by the federal agency from the outset. It was decided that the most likely response would be to set out an additional 500 traps duringthe first day after the find.

The additional traps were placed in the field on the second day and checked on the third day. One the third day, 5 male flies were trapped in a single Jackson trap, baited with trimedlure, in the main city situated on the eastern, lee side that this level of catch in relation to the relatively low level of efficiency of the survey traps indicates that one is dealing with an established, breeding population of medflies at least in the F-1 generation. Only further intensification of trapping would tell the location and size of the infestation. Reluctantly, the agricultural attache of Atlantis notified counterparts at the EEC ministries to the presence of the medfly. At this point, susceptible fruit exports (valued at $50 M per month) were quarantined.

The entire group was now divided into three smaller working groups to consider what courses of action should be taken next. Before the working groups separated, discussion identified some of the main points to be considered in any large area program. Some of these are:

-- Information flow

- from scientists and technologists to politicians, policy makers, and administrators
- from program managers and policy people to the public
- from operational units with the program to program managers

-- Publicity and psychology of media information

-- Management: who runs the program and how

Economics: How does one measure the economic, social and political consequences of

- no action
- several different courses of action

-- Control or eradication technology:

- feasibility
- method or methods
- costs
- environmental consequences

Each of the individual working groups had a mixture of people from the various disciplines represented at the ARW. After about 1-1/2 hours of meeting in the subgroups, the three subgroups reconvened and a rappateour from each subgroup presented the results of that subgroup's work.

3. RESISTANCE TO STERILE INSECT METHODS

The third workshop was had a discussion of the possibility that resistance could arise to the sterile insect method. The entire group was presented with the following background material and then discussed the potential effects of such resistance.

BACKGROUND. The general mating strategy of tropical fruit flies involves lekking behavior. A lek is a site where males of a species aggregate and to where females of the species come for mating. The term "lek" was coined in Sweden in 1938 to describe the "dancing" grounds of a species of grouse. This mating strategy was observed in mammals, birds, fishes, and insects.

The understanding of this kind of mating behavior is extremely important for effective use of sterile insect techniques (SIT) as a means of autocidal control The typical lekking sequence is the following. The males arrive at the site (a leafy branch in the canopy) before the female. Usually one male arrives first and begins to display by fanning his wings and emitting phermone. Other males are attracted by the phermone. Intense competition by the males for the dominant location in the lek occurs; the largest male usually displaces smaller ones and assumes possession of the dominant site.

Females begin visiting lek sites as they near sexual maturation, but may not enter a lek site until fully sexually mature. The females will usually land near the lek and observe the male-male interaction.

The sex ratio of the mating arena is heavily skewed towards the males because they take part in the lekking activity throughout their reproductive life. Thus, there is an aggregation of males while females only visit the lek as sexually mature virgins or as older, egg laying females who have depleted their supply of sperm. Most females, because of high mortality rates in the field, probably only only mate one.

The skewed sex ratio allows the female to be very selective in her choice of a mate. She judges fitness of males by requiring an elaborate courtship of them. This courtship involves visual, chemical, and acoustic cues. Females control the mating.

In the laboratories in which sterile insects are reared, pupae of the same age are placed into egg collection cages. Flies emerge at about the same time and the sex ratio is about 50:50. All of the females probably mate in a short time, but all of the males probably don't have the opportunity to mate. It is likely that females exhibit discrimination in favor of more vigorous, fit males. Thus, mate selection may be made by the male and a fit male may control the mating.

In most mass rearing factories, ovipositing females are held for only a short time (about 10 days) and then all flies in the cage are discarded. A female who mates early and readily lays a large number of eggs during this short period and thus has more progeny in the next generation. Thus, this female has a larger number of her gene pool than a slower mating and/or poorer ovipositing female. Consequently, the mass rearing factory selects for fast mating flies.

As a result of more rapid mating, the courtship sequence may become abbreviated and certain cues required by wild females may be genetically lost by laboratory reared males.

DISCUSSION. In the discussion which followed this presentation, the following points were raised.

- It is of great advantage for the females to mate with fertile rather than unfertile individuals. Thus, it is important for us to know -- for purposes of SIT -- if mating depends upon female choice (over which we have no control) or male competitiveness (which can be affected by laboratory conditions). In general, one can view a SIT program as a large experiment in natural selection; perhaps this kind of view should drive the planning of SIT.

- The basis for resistance to SIT is any genetic mechanism which allows assortive mating between wild females and wild males. It is likely that wild males are more healthy and thus more competitive than laboratory reared males. In Japan, Ito has shown a reduction of 20% of competitiveness in the first generation of factory males. After 60 generations the colony in Mexico's large factory was completely changed. Asman at Berkeley has shown as assortive mating between laboratory and wild mosquitos. Variance in factory populations is important, but this is selected out in factory production. Too much efficiency in the factory leads to too homogeneous a population. In Drosophila, females prefer males with rare genotypes -- but such information is not known for true fruit flies. It is important to understand what competitiveness means in this context.

- In C. Capitata, assortive mating does occur in factory reared flies; techniques for male release only may help to solve this problem.

- Do genetic changes in wild flies increase assortive mating?

- At least in medfly, males defend their leks. Thus, male competitiveness becomes important for mating. In addition to defense of leks, males signal females with other kinds of courtship behavior. In addition, the time of day at which flies mate may be different for wild and laboratory flies. Thus, under this kind of intense selective pressure, wild flies should adapt towards some type of assortative mating including temporal selection. It may be possible that assortative lekking occurs as well. The two factors which should be considered are the heritability of such behavior and the actual selection pressure induced through the laboratory reared males. The number of genes involved in the selection for mating behavior is unknown, but selection for behavior has occurred in mosquitos (using chemicals). Presumably as the gap between the competitiveness of wild and laboratory flies increase, the selection pressure will increase. (An example of reduced competitiveness: sterilized males are less vigorous and thus use a reduced wing beat pattern in the lek.)

- A rough definition of selection pressure is the following one: it is the proportion of wild females which, when courted by a factory male, mate with it.

- Part of the discussion concerned why such resistance to SIT has not been observed and how one could avoid it. For example, in Israel where SIT is continually used, no resistance has been observed. In situations in which SIT is used for eradication, one would not expect resistance to develop because of extinction of the gene pool. Resistance to SIT may not be a problem because of immigration of wild flies. In situations in which 100% of the flies could be treated, one may want to consider allowing some refuges -- although this will increase the length of the program.

- One needs to know if SIT resistance is developing. Mating tests should be done every year. A baseline definition and measurement of competitiveness is needed. One must be able to assess if assortative mating is increasing. There are (political) difficulties with doing tests for assortative mating, which must occur in the wild, in a control or eradication area. In addition, funds for such testing are usually very limited. There are considerable difficulties with testing for such behavioral differences. For example, how does on distinguish between male deterioration and evolution of resistance? One must use field cages for the tests of resistance.

- The operational implications of SIT resistance can be considerable. For example, if laboratory flies are only 50% competitive, doubling the number of flies is not helpful if assortative mating is 100%. At high enough densities of male flies, female confusion may occur; in general one would expect that older females are less discriminatory. The operational implications of fewer high quality versus many low quality males need to be assessed. What should be the response of the operational community if resistance to SIT does develop.

SUMMARY OF THE DISCUSSION

1. Males compete, but female choice is most important for mating.

2. In female choice, there may be considerable differences between wild and factory males.

3. The development of resistance to SIT depends upon heritability and selection pressure.

4. Selection pressure can be defined as the frequency with which wild females courted by sterile males mate with the sterile males.

5. Short term resistance may be equivalent to reduced male competitiveness rather than any genetic behavioral changes. Such resistance can be tested for in the field.

6. It is not clear is resistance to SIT is a practical problem. In Israel it is not, presumably due to immigration of flies from refuges. The

problem of SIT resistance may arise when a quick eradication program in a small region fails.

4. ASSESSMENT OF THE CURRENT STATE OF RESEARCH AND RESEARCH RECOMMENDATIONS

The fourth workshop was an assessment of the current state of research and research recommendations. To develop these, the large group was broken into three smaller subgroups, containing a mix of disciplines. These subgroups met for about two hours and then the entire group reconvened. The leaders of each subgroup presented that subgroup's recommendations and discussion ensued. Finally, a small committee consisting of about five people prepared the final set of recommendations. These were presented to the entire group on the fifth workshop day, modified after discussion, and accepted by the entire group.

The research recommendations will be presented in a formal way, so that they can stand by themselves. One topic which received considerable discussion, but which was not adopted by the entire group was the development of a large computer based listing of people who have an expertise that could be applied to fruit fly problems. E. F. Boller, in Switzerland, keeps such a list of "World Wide List of Fruit Fly Experts", but Boller's list is essentially limited to entomologists. The idea of a second list would be to include modellers, economists, policy scientists, and social scientists who have some expertise that could be used in the event of a fruit fly problem. The majority of the participants at the ARW felt that difficulty with such a list is that it is too open-ended -- everyone can be included.

RESEARCH RECOMMENDATIONS BY SCIENTISTS AT THE NATO ADVANCED RESEARCH WORKSHOP "PEST CONTROL: OPERATIONS AND SYSTEMS ANALYSIS IN FRUIT FLY MANAGEMENT" (5-9 August 1985; Bad Windsheim, Germany)

Fruit flies are an international pest complex, causing widespread crop destruction in both developed and developing countries. Recent outbreaks of medfly in California costs aver $100 M to eradicate an ongoing fruit fly control programs in Mexico, Greece, Okinawa, Guatemala, Israel and Italy cost millions of dollars each year.

A group consisting of field laboratory biologists, mathematical modellers, economists and policy scientists, and operational program managers from around the world met in Bad Windsheim, Germany in August 1985 to discuss the fruit fly problem from the perspective of interdisciplinary research. This group identified three major problem areas which are impending further progress in fruit fly control and eradication. These are:

o Inadequate basic biological data on fruit fly behavior and ecology.

o Outmoded and/or inefficient control or eradication technologies and strategies.

o Lack of understanding of the political, social, and economic constrainings on effective control or eradication strategies.

<u>For these reasons, research is urgently needed in the following areas (not ranked according to priority)</u>:

o Development of methodologies for the assessment of potential losses, effectiveness and costs/benefits of different control and quarantine policies in specific situations.

o Understanding of the public decision making process in response to pest control emergencies.

- Means for monitoring the effectiveness and environmental impact of an ongoing control or eradication program.

- Determining the efficiency and predictive qualities of traps.

- Phermones for trapping, sexing, and control of fruit flies.

- Development of alternative insecticides, improved baits, and approved application strategies.

- Quality control and field effectiveness of flies produced for SIT.

- Genetics of fruit flies with respect particularly to the development of a sexing technique.

- Relative effectiveness of male only releases in sterile insect programs.

- Ecological constraints upon the colonization of a new environment.

- Population biology of fruit flies in the field.

- Evaluation of the potential of cultural and biological control.

- Species interactions between fruit flies and the taxonomy of species complexes.

- Field testing of mathematical models.

- Nutritional studies on fruit flies.

- Integration of fruit fly control into orchard management.

The following specific proposals are made:

THAT SUCH INTERDISCIPLINARY MEETINGS AS THIS ONE BE HELD ON A REGULAR BASIS. The present meeting has provided a unique opportunity for mathematicians, biologists, economists, psychologists, demographers, operational pest control experts, and decision makers to meet together to analyze problems in fruit fly pest control. The opportunity for an exceptionally wide exchange of view on pest control has clearly indicated areas for interdisciplinary action in the future. Such action needs to be monitored and reviewed.

THAT A GENETIC STOCK CENTER FOR FRUIT FLIES BE ESTABLISHED. Genetic studies are inhibited at the present time by quarantine restrictions. A stock center is required and would best be located in a temperate country at a center already active in fruit fly research. Such a center will coordinate genetic and cytogenetic studies of behavior and will facilitate the development of a genetic sexing system.

NATO ASI Series G

Vol. 1: **Numerical Taxonomy.** Edited by J. Felsenstein. 644 pages. 1983.

Vol. 2: **Immunotoxicology.** Edited by P.W. Mullen. 161 pages. 1984.

Vol. 3: **In Vitro Effects of Mineral Dusts.** Edited by E.G. Beck and J. Bignon. 548 pages. 1985.

Vol. 4: **Environmental Impact Assessment, Technology Assessment, and Risk Analysis.** Edited by V.T. Covello, J.L. Mumpower, P.J.M. Stallen, and V.R.R. Uppuluri. 1068 pages. 1985.

Vol. 5: **Genetic Differentiation and Dispersal in Plants.** Edited by P. Jacquard, G. Heim, and J. Antonovics. 452 pages. 1985.

Vol. 6: **Chemistry of Multiphase Atmospheric Systems.** Edited by W. Jaeschke. 773 pages. 1986.

Vol. 7: **The Role of Freshwater Outflow in Coastal Marine Ecosystems.** Edited by S. Skreslet. 453 pages. 1986.

Vol. 8: **Stratospheric Ozone Reduction, Solar Ultraviolet Radiation and Plant Life.** Edited by R.C. Worrest and M.M. Caldwell. 374 pages. 1986.

Vol. 9: **Strategies and Advanced Techniques for Marine Pollution Studies: Mediterranean Sea.** Edited by C.S. Giam and H.J.-M. Dou. 475 pages. 1986.

Vol. 10: **Urban Runoff Pollution.** Edited by H.C. Torno, J. Marsalek, and M. Desbordes. 893 pages. 1986.

Vol. 11: **Pest Control: Operations and Systems Analysis in Fruit Fly Management.** Edited by M. Mangel, J.R. Carey, and R.E. Plant. 465 pages. 1986.

MIX
Papier aus verantwortungsvollen Quellen
Paper from responsible sources
FSC® C105338

If you have any concerns about our products,
you can contact us on
ProductSafety@springernature.com

In case Publisher is established outside the EU,
the EU authorized representative is:
Springer Nature Customer Service Center GmbH
Europaplatz 3, 69115 Heidelberg, Germany

Printed by Libri Plureos GmbH
in Hamburg, Germany